Monographs in

Volume 9

Modular and Automorphic Forms & Beyond

Monographs in Number Theory

ISSN 1793-8341

Series Editors: Bruce C. Berndt
(*University of Illinois at Urbana-Champaign, USA*)

Wadim Zudilin
(*Radboud University, The Netherlands*)

Editorial Board Members:

Heng Huat Chan (*National University of Singapore, Singapore*)
William Duke (*University of California at Los Angeles, USA*)
Wen-Ching Winnie Li (*The Pennsylvania State University, USA*)
Kannan Soundararajan (*Stanford University, USA*)

Published

Vol. 9 *Modular and Automorphic Forms & Beyond*
by Hossein Movasati

Vol. 8 *Topics and Methods in q-Series*
by James Mc Laughlin

Vol. 7 *The Theory of Multiple Zeta Values with Applications in Combinatorics*
by Minking Eie

Vol. 6 *Development of Elliptic Functions According to Ramanujan*
by K. Venkatachaliengar, edited and revised by Shaun Cooper

Vol. 5 *Hecke's Theory of Modular Forms and Dirichlet Series (2nd Printing with Revisions)*
by Bruce C. Berndt & Marvin I. Knopp

Vol. 4 *Number Theory: An Elementary Introduction Through Diophantine Problems*
by Daniel Duverney

Vol. 3 *Analytic Number Theory for Undergraduates*
by Heng Huat Chan

Vol. 2 *Topics in Number Theory*
by Minking Eie

Vol. 1 *Analytic Number Theory: An Introductory Course*
by Paul T. Bateman & Harold G. Diamond

Monographs in Number Theory

Volume 9

Modular and Automorphic Forms & Beyond

Hossein Movasati

IMPA, Brazil

NEW JERSEY • LONDON • SINGAPORE • BEIJING • SHANGHAI • HONG KONG • TAIPEI • CHENNAI

Published by

World Scientific Publishing Co. Pte. Ltd.
5 Toh Tuck Link, Singapore 596224
USA office: 27 Warren Street, Suite 401-402, Hackensack, NJ 07601
UK office: 57 Shelton Street, Covent Garden, London WC2H 9HE

Library of Congress Control Number: 2021047906

British Library Cataloguing-in-Publication Data
A catalogue record for this book is available from the British Library.

Monographs in Number Theory — Vol. 9
MODULAR AND AUTOMORPHIC FORMS & BEYOND

Copyright © 2022 by World Scientific Publishing Co. Pte. Ltd.

All rights reserved. This book, or parts thereof, may not be reproduced in any form or by any means, electronic or mechanical, including photocopying, recording or any information storage and retrieval system now known or to be invented, without written permission from the publisher.

For photocopying of material in this volume, please pay a copying fee through the Copyright Clearance Center, Inc., 222 Rosewood Drive, Danvers, MA 01923, USA. In this case permission to photocopy is not required from the publisher.

ISBN 978-981-123-867-3 (hardcover)
ISBN 978-981-123-868-0 (ebook for institutions)
ISBN 978-981-123-869-7 (ebook for individuals)

For any available supplementary material, please visit
https://www.worldscienti ic.com/worldscibooks/10.1142/12325#t=suppl

Desk Editors: Vishnu Mohan/Tan Rok Ting

Typeset by Stallion Press
Email: enquiries@stallionpress.com

Printed in Singapore

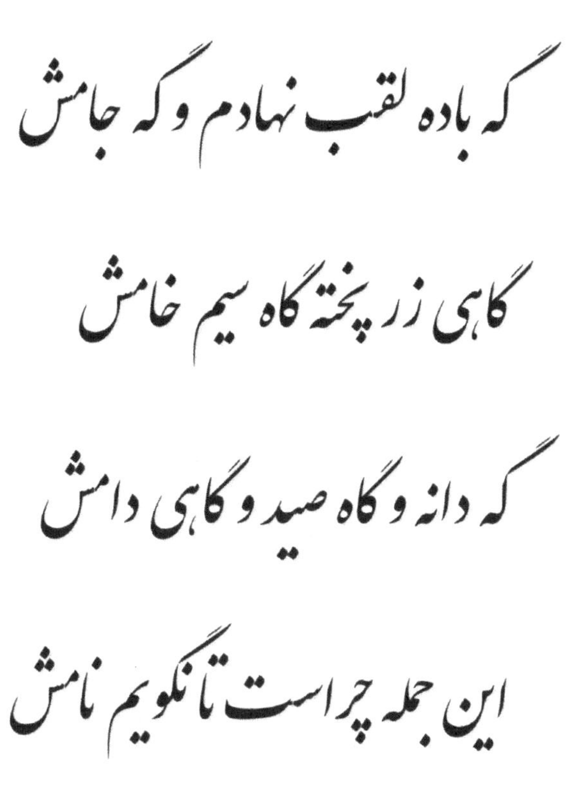

A poem by Jalal al-Din Muhammad Balkhi. Calligraphy: `nastaliqonline.ir`.

The names I gave him-wine, sometimes the cup.
At times he was raw silver, gold, refined.
A tiny seed, at times my prey, my trap-
All this because I could not say his name.

(Translation by Zara Houshmand in MOON & SUN: Rumi's rubaiyat).

Sometimes I called it wine and sometimes its glass,
I called it gold and then just usual brass.
Sometimes I called it seed, some other time prey or even trap,
why this and that? So that I do not tell you its name.

(Author's translation).

To my wife Sara Ochoa and my mother
Rogayeh Mollayipour

Preface

The guiding principle in this monograph is to develop a new theory of modular forms which encompasses most of the available theory of modular forms in the literature, such as those for congruence groups, Siegel and Hilbert modular forms, many types of automorphic forms on Hermitian symmetric domains, Calabi–Yau modular forms, with its examples such as Yukawa couplings and topological string partition functions, and even go beyond all these cases. Its main ingredient is the so-called *Gauss–Manin connection in disguise.* In the previous works, the emphasis was on examples and their applications, ranging from classical modular form theory to String Theory in Physics. The starting articles [Mov08b, Mov12b] were dedicated to the case of elliptic curves and modular forms. The case of Calabi–Yau threefolds is the topic of the joint article [AMSY16] and more details of this in the case of mirror quintic Calabi–Yau threefolds has been extensively elaborated in the monograph [Mov17b]. The article [Mov17c] mainly emphasizes the case of hypersurfaces with geometric questions in mind, such as Noether–Lefschetz and Hodge loci. Now, I feel that one has to develop the general theory of modular and automorphic forms, for the sake of its beauty and for the sake of completeness, even though in the moment it does not claim to solve any established conjecture in mathematics, and it is not clear how such a general theory might be related to some deep applications of modular forms, such as in the proof of Fermat's last theorem, or how it might enter into Langlands program. The text is written with the feeling that such a job must be done to pave the road for future investigations and applications. For this I would express myself with the following quotation from A. Weil's book [Wei99], "Where the road will lead remains in large part to be seen, but indications are not lacking that fertile country lies ahead." The present text is independent of these previous works, however, many of our motivations are spread among them. We have in mind an audience with a basic knowledge of Algebraic Geometry and Hodge Theory. Some basic knowledge in Complex Analysis, Algebraic Topology and Differential Equations would be useful for a smooth reading of the text.

Acknowledgments

The text is a culmination of many ideas throughout the last ten years of author's mathematical career. Here, I would like to thank my teachers in holomorphic foliations, César Camacho, Paulo Sad, Alcides Lins-Neto and my colleague Jorge Vitório Pereira who always kept me informed about the last developments in the subject. Special thanks go to Pierre Deligne who wrote many illuminating comments on many of my articles related to the present text. This includes the article [Mov12b] which is the starting point of the story presented here and [Mov18] which is written during the preparation of Chapters 9 and 10. My heartfelt thanks go to Shing-Tung Yau for his interest, support and the opportunity to present parts of this work at YMSC, Tsinghua university and CMSA, Harvard university. I would like to thank Tiago J. Fonseca who wrote an appendix to this book. Finally, thanks go to Younes Nikdelan, Roberto Vilaflor, Murad Alim and Jin Cao for many comments and corrections which improved the text.

Labyrinth of This Book

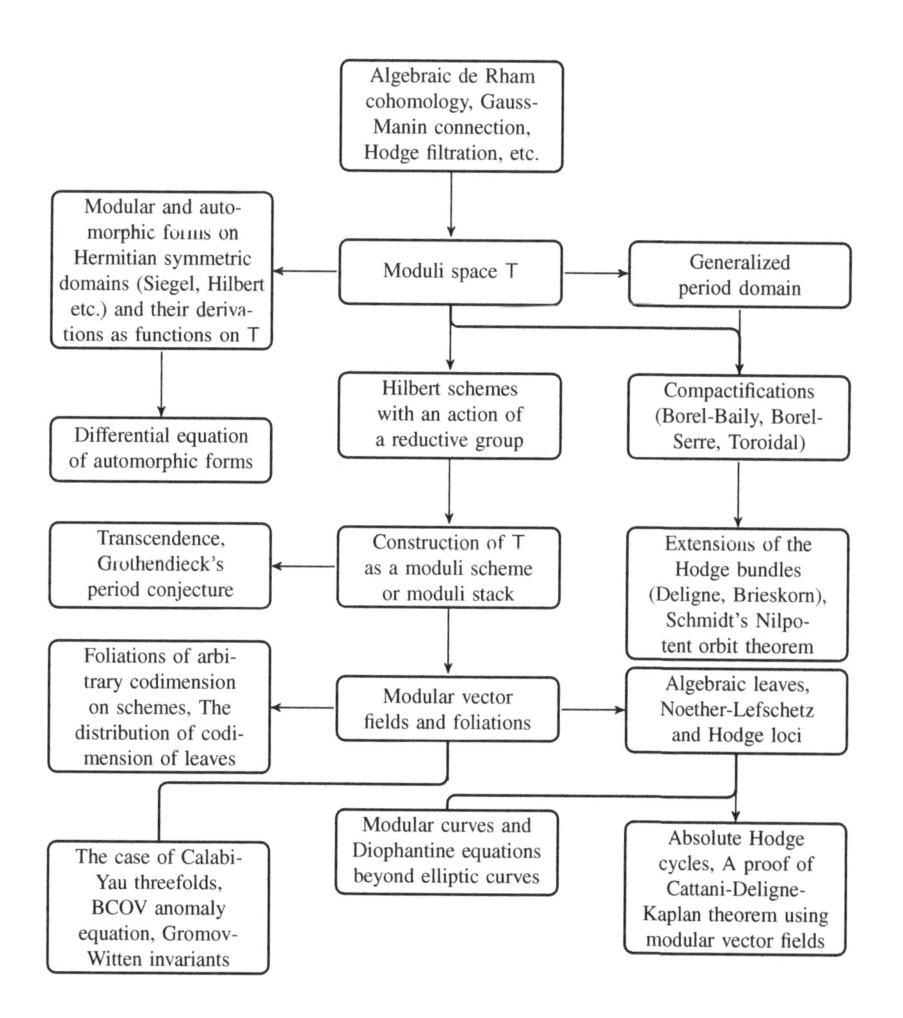

Contents

Chapter 1

Introduction

"When kings are building", says the German poet, "carters have work to do". Kronecker quoted this, in his letter to Cantor of September 1891, only to add, thinking of himself no doubt, that each mathematician has to be king and carter at the same time (A. Weil in his book [Wei99]).

The present book is the tale of a vast generalization of the differential equation

$$\begin{cases} q\frac{\partial E_2}{\partial q} = \frac{1}{12}(E_2^2 - E_4), \\ q\frac{\partial E_4}{\partial q} = \frac{1}{3}(E_2E_4 - E_6), \\ q\frac{\partial E_6}{\partial q} = \frac{1}{2}(E_2E_6 - E_4^2), \end{cases} \quad (1.1)$$

where E_i's are the Eisenstein series:

$$E_{2i}(q) := 1 + b_i \sum_{n=1}^{\infty} \left(\sum_{d|n} d^{2i-1} \right) q^n, \quad i = 1, 2, 3 \quad (1.2)$$

and $(b_1, b_2, b_3) = (-24, 240, -504)$. It was discovered by Ramanujan in [Ram16] and it is mainly known as Ramanujan's relations between Eisenstein series. He was a master of formal power series and had a very limited access to the modern mathematics of his time. In particular, he and many people in number theory didn't know that the differential equation (9.11) had been already studied by Halphen in his book [Hal86, p. 331], thirty years before S. Ramanujan and another equivalent version of the differential equation was derived by Darboux in [Dar78b]. Since then this differential equation has been rediscovered over and over, and it seems that the tale of its rediscovery will not end, and mathematicians of the future will have the joy of rediscovering it again. For a collections of such rediscoveries, see [CMN+18] and for one of the most celebrated applications of (1.1) in transcendental number theory, see Appendix A.

1.1 Gauss–Manin connection in disguise

My rediscovery of the Ramanujan's differential equation was through the computation of the Gauss–Manin connection for families of elliptic curves. In 2003, I started to think about Griffiths project [Gri70] in which he wanted to build a satisfactory theory of automorphic forms for the moduli of polarized Hodge structures, also called Griffiths period domain, see also [Mov08a]. However, as we read in [GS69, p. 254] "for some domains arising quite naturally in algebraic geometry, there is no theory of automorphic forms." It was not clear how the Griffiths' replacement for this under the name automorphic cohomology group, would play the role of classical automorphic forms. I started to think about its reformulation in a similar style as in K. Saito's article [Sai01], that is, inverting functions formed by integrals and obtaining automorphic forms. The role of multiple integrals with an arbitrary integrand was mainly missing in the literature, and I realized that even in the elliptic curve case, inserting elliptic integrals of the second kind in period maps, one gets new functions similar to modular forms. In [Mov08b, Mov08c], I computed an ordinary differential equation in three dimensions and I knew that a solution of this ODE involves the Eisenstein series E_4 and E_6. But what was the missing function? After few weeks of searching the literature I understood that I have rediscovered the Eisenstein series E_2 together with the Ramanujan's differential equation. This rediscovery was worth as I started to introduce a new moduli space T and a new period domain, and interpret the Ramanujan's differential equation as a vector field on both spaces.

The relation between modular forms in one hand and elliptic integrals and Picard–Fuchs equations on the other hand is as old as the modern mathematics, however, the first appearance of the relation between modular forms and the geometric incarnation of Picard–Fuchs equations under the name Gauss–Manin connection is in N. Katz's article [Kat73]. The

q variable of modular forms under the name canonical coordinate has produced the theory of ordinary crystals with fruitful applications in the arithmetic of elliptic curves, see Deligne's article [Del81] and the references therein, see also this article for a generalization in the case of $K3$ surfaces and Abelian varieties and [Yu13] for an attempt to use this in the case of Calabi–Yau varieties. What is missing in all these heavily arithmetic oriented topics, is the moduli space T. Soon I realized Darboux and Halphen's work on the topic, and in particular, Halphen's generalization using Gauss Hypergeometric function. Before going beyond elliptic curves, in [Mov12a] I reproduced Halphen's differential equation in a geometric context. What I called differential modular forms in the mentioned articles, were called quasi-modular forms in [KZ95]. The lecture notes [Mov12b] were supposed to be a complete exposition of the topic. P. Deligne in a personal communication (December 5, 2008) and in response to one of the early drafts of [Mov12b], called the Ramanujan's differential equation between Eisenstein series, the Gauss–Manin connection in disguise. This became the title of a series of articles mainly in the case of Calabi–Yau varieties [Mov17b, AMSY16, Mov17c, MN18, HMY17].

After my works in the case of elliptic curves and quasi-modular forms I directly started to think about modular form theories for non-rigid Calabi–Yau varieties, skipping the case of K3 surfaces. The main reason was that the corresponding period domains are not Hermitian symmetric, and so, the Griffiths project was not working in these cases. However, for the surprise of many mathematicians, physicists were producing many q-expansions attached to Calabi–Yau varieties and with rich enumerative properties. D. Zagier in few personal communications expressed his long wish to relate such q-expansions to modular forms, but it was not clear how this can be done. In 2014 I did a sabbatical year at Harvard and this gave origin to my collaborations [AMSY16, HMY17] with Shing-Tung Yau, Murad Alim, Emanuel Scheidegger and Babak Haghighat with whom I learnt a lot and this gave me the guideline in order to write the present book. I understood that the special geometry used in Topological String Theory, see for instance [CdlO91, CDLOGP91a, Str90, CDF$^+$97, Ali13a] is exactly the way physicist work with moduli spaces and generalizations of Ramanujan's vector field. Once again it was clear that when physicists do not find what they need in mathematics, they produce their own, even though it might not be in a polished format from a mathematician's point of view. All these together with my earlier works gave me the sufficient ingredients to formulate the new theory of modular forms attached to a very particular class of Calabi–Yau threefolds, see [Mov15a, Mov17b]. The present book has the ambition of even going further and construct modular form theories attached to arbitrary families of projective varieties.

1.2 Prerequisites

For definitions and preliminaries in Hodge theory needed for the purpose of the present text, the reader is referred to the first sections of Deligne's lecture notes in [DMOS82], Voisin's books [Voi02, Voi03] or the author's book [Mov19]. The reader who wants to follow this book with examples is recommended to have a look in the case of elliptic curves developed in [Mov12b] and the case of mirror quintic Calabi–Yau threefolds in [Mov17b].

The present text is not written linearly, and so, it is not recommended to read it linearly. The reader is recommended to start reading a chapter of his interest and then use a back and forth strategy to decipher objects and find the missing definitions.

1.3 A tale of love and madness

The present text is the tale of a kind of mathematics that I loved so much to write it, and in the same time I had a feeling of madness: why should I invest so much time on a topic whose fruitful applications, most probably, I will not see in my own lifetime, a kind of mathematics that apparently only for me revealed itself beautiful, a mathematics which looks like abstract nonsense. It seemed like a new land hidden behind the ocean, I will have only the joy of its discovery and not its hidden golds. An editor of one of the author's articles related to the present book, which was previously rejected by many other editors, had to confess that after many and many attempts to find a referee, he was not able to find one. As usual something which I enjoyed so much seemed to be not worthy of publication for others. It might be a tale of love for useless beauty and so, it might sound a tale of madness. It is the failure of my Grothendieck's moment in which the desire for writing in the most abstract and general context was high, but I felt the lack of capacity, time and motivation to do so. On the one hand, I believed that the generalization of automorphic forms, just for the sake of generalization, is alone enough to justify this book. On the other hand, I was completely stuck with the huge literature on automorphic forms and I was not able to pick a specific topic or conjecture and justify my text upon it. After studying the history of the theory of modular forms, which is more than hundred and fifty years of research, I came to the conclusion that it might not be possible to see the arithmetic applications of the generalized modular forms in my lifetime. Instead, I tried to push this text for applications in geometric objects like Noether–Lefschetz and Hodge loci, which will be hopefully useful for a better understanding of the Hodge conjecture. With all these contradictory feelings I took the job of writing this text, elaborating only few boxes in the main goals of the present book sketched in p. xiii, and I hope that at some point in the future it will be useful. At least I hope that some other people will have the same joy of exploring as I had during writing the present text.

During the preparation of the present text it happened that I made trivial and obvious mistakes, and they continued for years in the first drafts. This made me feel that I am not the right person to undertake the job of writing it. For instance, at some point I realized that for two functions $A \xrightarrow{f} B \xrightarrow{g} C$, I have used $f \circ g$ because of its pictorial similarity between two notations, and this has not produced a single contradiction in my thinking as the applications of this has been done with correct order of f and g. At some point I thought I am doing mathematical nonsense, but then I thought it does not matter, the joy of a mad man is more important for him than the outside world.

In 2019 when the text was more than 200 pages, I asked the opinion of Prof. P. Deligne about it, and he pointed out that I do not have a convincing proof for the fundamental proposition (Proposition 2.4) of the present text. For a moment I was perplexed, I felt that the whole theory might be an abstract nonsense, however, I went through it, and even though the proof of the fundamental proposition lacked rigor, something beautiful was there, and I was not supposed to judge the whole theory based on this. P. Deligne offered a proof based on some apparently well-known facts in SGA3 Vol. II, however, understanding the details was beyond my mathematical training.

1.4 Tupi words

There are many new mathematical objects in the present book which have not appeared in the literature, except for few of them in the author's previous works, and so they might deserve a proper name. For this I started to collect names in Tupi. This is an extinct language which was spoken mainly by natives of south and southeast Brazil. At some point I gave up this linguistic project, as it needed time, effort and an expert's advice. Below, the reader finds a suggestion of tupi words for few mathematical objects in this book that I was able to gather. They are not used in the main body, however, they might be used in the future development of the mathematics of the present work.

Aupaba	[Homeland, origin], the domain V of a family $Y \to V$ of smooth projective varieties, §3.6
Ibicatu	[Good land], the parameter space T of an enhanced family X/T in §3.4
Ibiporanga	[Pretty land], the moduli space T in §3.11
Ibicuí	[Sand], the moduli space S in §3.11
Ibipiranga	[Red land], moduli of Hodge decompositions, §3.12
Ybacoby	[Blue sky], generalized period domain U and $\mathsf{\Pi}$, §8.3
Itaquerejuguá	[Precious stone], the marked variety, §3.2
Itacuatiara	[Written or painted stone], enhanced smooth projective variety/scheme, §3.4
Itapiranga	[Red stone], a variety with Hodge decomposition defined over a field, §3.10
Itaoby	[Blue stone], varieties introduced in §2.17 using infinitesimal variation of Hodge structures. These are also called R-varieties and are natural generalizations of Calabi–Yau varieties
Nhauumboca, Ocuera	[House made of clay, House in the past], a family $Y \to V$ of smooth projective varieties, §3.6
Itaoca	[House made of stone], enhanced family $\mathsf{X} \to \mathsf{T}$, §3.4
Amanoca	[House of rain], full enhanced family $\mathsf{X} \to \mathsf{T}$, §3.4
Amanocuera	[House of rain in the past], weakly enhanced varieties, §3.5
Jatapy	[To make fire], enhanced family with an action of a reductive group, §3.7
Ocatu	[Good house], enhanced family with constant Gauss–Manin connection, §6.13
Ocuama	[Future house], the universal family $\mathsf{X} \to \mathsf{T}$, §3.11
Atauúba	[Fire arrow], modular vector fields in the moduli space T, §6.11
Amana	[Rain], the algebraic group G, §3.3
Amandy	[Rain's water], the orbits of the algebraic group G, §3.3
Atá	[Fire], reductive group $\mathbf{G}$, §2.11

I wished to find better names for the following objects and I failed. Scheme theoretic leaf in §5.4. Smooth and reduced leaf in §5.5. N-Smooth leaf in §5.8. Modular foliations $\mathscr{F}(\mathsf{C})$ in §6.3. Trivial modular foliation in §6.4. Modular foliation with trivial character in §6.5. Space of leaves $\mathscr{L}(\mathsf{C})$ in §6.6. Foliation $\mathscr{F}(2)$ in §§6.9 and 6.10. Constant foliation in §6.12. Gauss–Manin connection matrix in §3.8. Generalizations of Yukawa couplings Y in §6.11. Constant vector fields in §6.12. Monodromy group $\Gamma_{\mathbb{Z}}$ in §4.3. Monodromy covering $\mathbb{H}$ in §4.3. Generalized period map P in §4.4. τ-map in §8.5. t-map in §8.5. Constant period vector C in §6.3.

Chapter 2

Preliminaries in Algebraic Geometry

Modern algebraic geometry has deservedly been considered for a long time as an exceedingly complex part of mathematics, drawing practically on every other part to build up its concepts and methods and increasingly becoming an indispensable tool in many seemingly remote theories (J. Dieudonné in [Die72, p. 827]).

2.1 Introduction

Research in modern algebraic geometry requires a hard training in its grandiose foundation, gathered in the texts EGA, SGA and FGA, by Dieudonné, Grothendieck and Serre among many other contributors. And once this is done, the new mathematics to be created is hardly free of references to it. There are few practical reasons to undergo this training. This includes the unification of Diophantine equations (schemes over $\mathrm{Spec}(\mathbb{Z})$) with complex analytic spaces, the need for spaces with nilpotent functions, and to be able to talk about varieties and families of varieties with a single notation. In the present text all these reasons are relevant, however, we would like to avoid the unnecessary introduction of general machineries and objects, and so, we highlight the basic concepts related to schemes needed for the purpose of the present text. The strategy is to formulate definitions and propositions in a scheme theoretic style, and once this becomes cumbersome, the language of varieties over algebraically closed fields, and then of complex analytic varieties is adopted and used.

2.2 The base ring and field

We consider a commutative ring R with multiplicative identity element $1 \neq 0$. We also consider projective varieties defined over R. This uses a finite number of elements of R, therefore, in practice we can assume that the ring R is finitely generated, and so

$$\mathsf{R} := \frac{\mathbb{Z}[t_1, t_2, \ldots, t_r]}{I}, \text{ where } I \subset \mathbb{Z}[t_1, t_2, \ldots, t_r], \text{ is an ideal.}$$

We assume that R is an integral domain. In other words, it is without zero divisors, i.e. if for some $a, b \in \mathsf{R}$, $ab = 0$ then $a = 0$ or $b = 0$. In particular, it is reduced, that is, it has not nilpotent elements. The characteristic of R is the smallest $p \in \mathbb{N}$ such that the sum of $1 \in \mathsf{R}$, p-times, is zero. It is either a prime number or zero. In the first case, we can write $\mathsf{R} = \frac{\mathbb{F}_p[t_1, t_2, \ldots, t_r]}{I}$, where I is now an ideal in $\mathbb{F}_p[t_1, t_2, \ldots, t_r]$. In the second case, we have $\mathbb{Z} \subset \mathsf{R}$, and we define N to be the greatest positive integer which is invertible in R and we have $\mathsf{R} = \frac{\mathbb{Z}[\frac{1}{N}, t_1, t_2, \ldots, t_r]}{I}$. The primes dividing the number N are called the bad primes of the ring R. By our assumptions R is Noetherian, that is, every ideal of R is finitely generated. We denote by k the field obtained by the localization of R over $\mathsf{R} \setminus \{0\}$ and by $\bar{\mathsf{k}}$ the algebraic closure of k.

Proposition 2.1. *Let R be a finitely generated, commutative ring with multiplicative identity element $1 \neq 0$, of characteristic zero and without zero divisors. There exists an embedding $\mathsf{R} \hookrightarrow \mathbb{C}$ which makes R a subring of $\mathbb{C}$.*

Proof. We take the quotient field k of R and construct an embedding of fields $\mathsf{k} \hookrightarrow \mathbb{C}$. Since we have a canonical injective morphism $\mathsf{R} \hookrightarrow \mathsf{k}$ of rings this would prove the proposition. Let k be generated by $a_1, a_2, \ldots, a_r$ over $\mathbb{Q}$. The proof is by induction on the number r. The case $r = 0$ is trivial as we have a unique embedding $\mathbb{Q} \hookrightarrow \mathbb{C}$. Let us assume the proposition for $r - 1$. We define $\tilde{\mathsf{k}}$ to be the subfield of k generated by $a_1, a_2, \ldots, a_{r-1}$. By the hypothesis of induction we have an embedding $\tilde{\mathsf{k}} \hookrightarrow \mathbb{C}$. Let

$$A := \{P \in \tilde{\mathsf{k}}[x] \mid P(a_r) = 0\}.$$

If $A = \{0\}$ then we choose a transcendental number b over $\tilde{\mathsf{k}}$ and we construct $\mathsf{k} \hookrightarrow \mathbb{C}$ by sending a_r to b. If not then A is an ideal of $\tilde{\mathsf{k}}[x]$. This ideal is generated by a unique monic polynomial $P(x)$. We take a number $b \in \mathbb{C}$ such that the minimal polynomial of b over $\tilde{\mathsf{k}}$ is $P(x)$ and send a_r to b. □

When we choose a transcendental number a then we have an infinite number of options, whereas when we choose an algebraic number over $\tilde{\mathsf{k}}$ we have a finite number. For this reason, the number of embeddings $\mathsf{R} \hookrightarrow \mathbb{C}$ is infinite if the transcendental degree of the quotient field k of R is strictly bigger than zero. The main motivation behind Proposition 2.1 is the following version of Lefschetz principal. For a property P talking about schemes (and foliations on them) defined over a ring R, we may assume that $\mathsf{R} \subset \mathbb{C}$ and it has finite transcendence degree over rational numbers. Therefore, in order to prove P, we can use all the possible transcendental methods.

Later in §3.2 we will need to fix a subring $\mathfrak{R} \subset \mathsf{R}$. In most of the cases this is going to be

$$\mathfrak{R} = \mathbb{Z}\left[\frac{1}{N}\right], \quad \text{or} \quad \mathfrak{R} := \mathbb{F}_p.$$

The quotient field of $\mathfrak{R}$ is denoted by $\mathfrak{k}$. The most famous rings in the present text are as follows:

$$\mathsf{R} := \mathbb{Z}\left[t_1, t_2, t_3, \frac{1}{6(27t_3^2 - t_2^3)}\right], \quad \mathfrak{R} := \mathbb{Z}\left[\frac{1}{6}\right]. \tag{2.1}$$

This appears in the study of elliptic curves in Chapter 9.

2.3 Schemes over integers

A basic knowledge of algebraic geometry of schemes would be enough for our purposes, see for instance the first chapters of Hartshorne's book [Har77] or Eisenbud and Harris's book [EH00].

We will need schemes T over $\mathfrak{R}$ which are covered with a finite number of affine schemes of type $\mathrm{Spec}(\mathsf{R}_i)$, $i = 1, 2, \ldots$, where R_i is the ring in §2.2. In most of the cases $\mathsf{T} := \mathrm{Spec}(\mathsf{R})$. For our purpose, we make the following definition.

Definition 2.1. A parameter scheme T over $\mathfrak{R}$ satisfies the following properties:

(1) The morphism of schemes $\mathsf{T} \to \mathrm{Spec}(\mathfrak{R})$ is of finite type. This means that there is a covering of T by open affine subsets $\mathsf{T}_i := \mathrm{Spec}(\mathsf{R}_i)$ such that R_i is a finitely generated $\mathfrak{R}$-algebra.
(2) T is irreducible, that is, the underlying topological space does not contain two proper disjoint nonempty open sets. In particular, T is

connected, that is, it cannot be written as a disjoint union of two nonempty open sets.

(3) T is reduced, that is, T is covered by open sets $\mathsf{T}_i := \mathrm{Spec}(\mathsf{R}_i)$ such that R_i is reduced.

Note that by definition an integral scheme is reduced and irreducible, and so, a parameter scheme T is integral. The ring (resp. field) of regular (resp. rational) functions on T is denoted by $\mathfrak{k}[\mathsf{T}]$ (resp. $\mathfrak{k}(\mathsf{T})$). The most general example of a parameter scheme of the present text will be introduced in §3.6.

Definition 2.2. Let T be a parameter scheme over $\mathfrak{R}$. A point $t \in \mathsf{T}$ is a prime ideal $\mathfrak{p}$ of R_i, where $\mathrm{Spec}(\mathsf{R}_i)$ is some open subset of T. It is called a closed point if $\mathfrak{p}$ is maximal, and hence, the quotient $\mathsf{R}_i/\mathfrak{p}$ is a field. This is called the residue field. An $\mathfrak{R}$-valued point comes further with an isomorphism of rings $\mathsf{R}_i/\mathfrak{p} \cong \mathfrak{R}$, or equivalently, a surjective morphism of rings $a : \mathsf{R}_i \to \mathfrak{R}$, for which the kernel of a is the prime ideal used in the earlier definition. The set of $\mathfrak{R}$-values points of T is denoted by $\mathsf{T}(\mathfrak{R})$.

2.4 Differential forms on schemes

We will need the sheaf of differential 1-forms in T. It is enough to define it for the case of affine schemes $\mathsf{T} := \mathrm{Spec}(\mathsf{R})$. In this case for the definition of $\Omega_{\mathsf{R}/\mathfrak{R}} = \Omega^1_{\mathsf{T}}$ and also Ω^i_{T} see §10.3 of [Mov19].

Definition 2.3. The dimension of the parameter scheme T is the smallest number a such that $\Omega^{a+1}_{\mathsf{T}}$ is the torsion sheaf and Ω^a_{T} is not. The sheaf Ω^a_{T} is called the canonical sheaf of T.

Recall that a sheaf on T is called a torsion sheaf if any section s of this sheaf is annihilated by some non-zero section of $\mathscr{O}_{\mathsf{T}}$ in the same open set as of s. In practice, we take smooth parameter schemes of dimension a, and hence, $\Omega^{a+1}_{\mathsf{T}}$ is the zero sheaf. For singular T, $\Omega^{a+1}_{\mathsf{T}}$ might be non-zero and it is well-known as Milnor or Tjurina module in singularity theory, see for instance [Mov19, Chapter 10].

The canonical sheaf of T is an invertible sheaf, that is, in local charts it is free of rank 1. Therefore, it comes from a Cartier divisor in T, see Hartshorne's book [Har77, Proposition 6.13, p. 144]. When $\mathfrak{R} = \mathfrak{k}$ is an algebraically closed field, then the dimension of T is the maximum dimension among irreducible components of T. The following definition will be mainly used in Chapter 5.

Definition 2.4. Let T be a parameter scheme of dimension n and let Ω be a submodule of the $\mathscr{O}_\mathsf{T}$-module Ω^m_T for some $m \geq 1$. We have

$$\Omega \wedge \Omega^{n-m}_\mathsf{T} \subset \Omega^n_\mathsf{T}. \tag{2.2}$$

and so we have an ideal sheaf $\mathrm{ZI}(\Omega) \subset \mathscr{O}_\mathsf{T}$, which we call it the zero ideal, such that the left-hand side of (2.2) is equal to $\mathrm{ZI}(\Omega) \cdot \Omega^n_\mathsf{T}$. The zero scheme of Ω is defined to be

$$\mathrm{ZS}(\Omega) := \mathrm{Spec}(\mathscr{O}_\mathsf{T}/\mathrm{ZI}(\Omega)).$$

2.5 Vector fields

Definition 2.5. Let T be a parameter scheme. The sheaf of vector fields is

$$\Theta_\mathsf{T} := (\Omega^1_\mathsf{T})^\vee$$

An element in $\Theta_\mathsf{T}(U)$ for some open set $U \subset \mathsf{T}$, is by definition an $\mathscr{O}_\mathsf{T}(U)$-linear map $\Omega^1_\mathsf{T}(U) \to \mathscr{O}_\mathsf{T}(U)$ and it is called a vector field in U.

If T is a smooth variety over an algebraically closed field $\mathfrak{k}$, a vector field can be also interpreted as a section of the tangent bundle of T. The $\mathscr{O}_\mathsf{T}$-module of vector fields Θ_T is isomorphic to the sheaf of derivations.

Definition 2.6. A map $\mathsf{v} : \mathscr{O}_\mathsf{T} \to \mathscr{O}_\mathsf{T}$ is called a derivation if it is $\mathfrak{R}$-linear and it satisfies the Leibniz rule

$$\mathsf{v}(fg) = f\mathsf{v}(g) + \mathsf{v}(f)g, \quad f, g \in \mathscr{O}_\mathsf{T}.$$

We denote by $\mathrm{Der}(\mathscr{O}_\mathsf{T})$ the $\mathscr{O}_\mathsf{T}$-module of derivations.

Proposition 2.2. *We have an isomorphism of $\mathscr{O}_\mathsf{T}$-modules*

$$\Theta_\mathsf{T} \cong \mathrm{Der}(\mathscr{O}_\mathsf{T}),$$
$$\mathsf{v} \mapsto (f \mapsto \mathsf{v}(df)).$$

Proof. This isomorphism maps the vector field v to the corresponding derivation $\check{\mathsf{v}}$ obtained by

$$\check{\mathsf{v}}(f) = \mathsf{v}(df). \tag{2.3}$$

Since in local charts Ω^1_{T} is generated as $\mathscr{O}_{\mathsf{T}}$-module by $d\mathscr{O}_{\mathsf{T}}$, the equality (2.3) also defines its inverse $\check{\mathsf{v}} \mapsto \mathsf{v}$. □

In $\mathrm{Der}(\mathscr{O}_{\mathsf{T}})$ we have the Lie bracket $[\mathsf{v}_2, \mathsf{v}_2]$, $\mathsf{v}_1, \mathsf{v}_2 \in \mathrm{Der}(\mathscr{O}_{\mathsf{T}})$ defined by

$$[\mathsf{v}_1, \mathsf{v}_2] := \mathsf{v}_1 \circ \mathsf{v}_2 - \mathsf{v}_2 \circ \mathsf{v}_1.$$

We have to show that $[\mathsf{v}_1, \mathsf{v}_2]$ is a derivation. It is $\mathfrak{R}$-linear because v_1 and v_2 are. The Leibniz rule

$$[\mathsf{v}_1, \mathsf{v}_2](fg) = f[\mathsf{v}_1, \mathsf{v}_2](g) + g[\mathsf{v}_1, \mathsf{v}_2](f)$$

is left as an exercise to the reader.

Definition 2.7. Let $f : \mathsf{T}_1 \to \mathsf{T}_2$ be morphism of $\mathfrak{R}$-schemes and v_i, $i = 1, 2$ be vector fields on T_i, $i = 1, 2$. We say that f maps v_1 to v_2 if the following diagram commutes:

$$\begin{array}{ccc} \Omega^1_{\mathsf{T}_2} & \to & \Omega^1_{\mathsf{T}_1} \\ \downarrow & & \downarrow \\ \mathscr{O}_{\mathsf{T}_2} & \to & \mathscr{O}_{\mathsf{T}_1} \end{array} \tag{2.4}$$

where the down arrows are respectively v_2 and v_1.

Note that for f as above we have the induced map in the sheaf of differential 1-forms, however, we do not have a morphism $\Theta_{\mathsf{T}_1} \to \Theta_{\mathsf{T}_2}$.

Definition 2.8. Let $\mathsf{T}_1, \mathsf{T}_2$ be two $\mathfrak{R}$-schemes and v be a vector field in T_1. A parallel extension of v in $\mathsf{T}_1 \times \mathsf{T}_2$ is a vector field $\check{\mathsf{v}}$ in $\mathsf{T}_1 \times \mathsf{T}_2$ such that under the first projection $\mathsf{T}_1 \times \mathsf{T}_2 \to \mathsf{T}_1$ it maps to v and under the second projection $\mathsf{T}_1 \times \mathsf{T}_2 \to \mathsf{T}_2$ it maps to the zero vector field, see Fig. 2.1.

Proposition 2.3. *The parallel extension of a vector field exists and it is unique.*

Proof. By definition $\mathscr{O}_{\mathsf{T}_1 \times \mathsf{T}_2} = \mathscr{O}_{\mathsf{T}_1} \otimes_{\mathfrak{R}} \mathscr{O}_{\mathsf{T}_2}$ and hence

$$\Omega^1_{\mathsf{T}_1 \times \mathsf{T}_2} = \Omega_{\mathsf{T}_1} \otimes_{\mathfrak{R}} \mathscr{O}_{\mathsf{T}_2} + \mathscr{O}_{\mathsf{T}_1} \otimes_{\mathfrak{R}} \Omega^1_{\mathsf{T}_2}. \tag{2.5}$$

For a $\mathscr{O}_{\mathsf{T}_1}$-linear map $\mathsf{v} : \Omega^1_{\mathsf{T}_1} \to \mathscr{O}_{\mathsf{T}_1}$, its parallel extension $\Omega^1_{\mathsf{T}_1 \times \mathsf{T}_2} \to \mathscr{O}_{\mathsf{T}_1 \times \mathsf{T}_2}$ evaluated at the first piece (resp. second piece) in (2.5) is $\mathsf{v} \otimes \mathrm{Id}$ (resp. zero). □

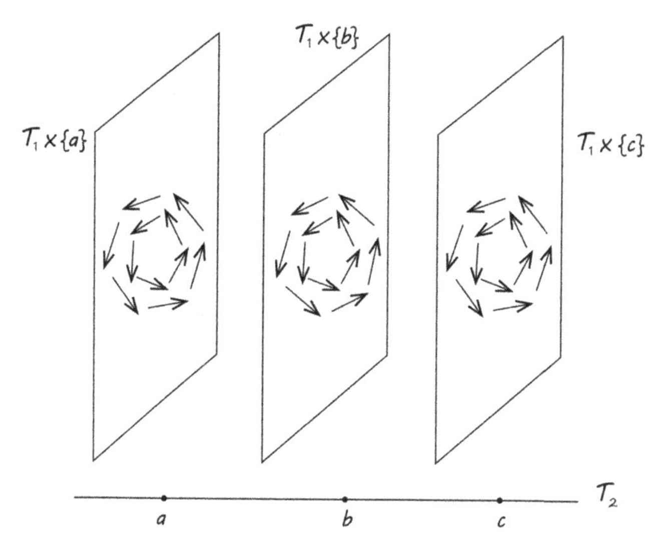

Fig. 2.1. Parallel extension.

There are some other differential geometric objects that will be useful later. The definition of Lie derivative is taken from Cartan's formula

$$\mathscr{L}_{\mathsf{v}} : \Omega^i_{\mathsf{T}} \to \Omega^i_{\mathsf{T}}, \quad \mathscr{L}_{\mathsf{v}} := d \circ i_{\mathsf{v}} + i_{\mathsf{v}} \circ d \tag{2.6}$$

where i_{v} is the contraction of differential forms with the vector field v. We have the identities

$$\mathscr{L}_{[\mathsf{v}_1,\mathsf{v}_2]} = [\mathscr{L}_{\mathsf{v}_1}, \mathscr{L}_{\mathsf{v}_2}], \tag{2.7}$$

$$i_{[\mathsf{v}_1,\mathsf{v}_2]} = [\mathscr{L}_{\mathsf{v}_1}, i_{\mathsf{v}_2}] = [i_{\mathsf{v}_1}, \mathscr{L}_{\mathsf{v}_2}]. \tag{2.8}$$

2.6 Projective schemes over a ring

In the present text we will need another class of schemes which are usually projective. For simplicity, we first explain this for schemes over the ring R defined in §2.2. We work with finite schemes X over R and we write X/R or say that X is an R-scheme. This means that X is covered with a finite number of open sets $U_i,\ i = 1, 2, \ldots, s$ such that $U_i := \mathrm{Spec}(A_i)$ and each A_i is a finitely generated R-algebra. In practice, we will only encounter projective schemes

$$X \subset \mathbb{P}^N_{\mathsf{R}} := \mathbb{P}^N_{\mathbb{Z}} \times_{\mathbb{Z}} \mathrm{Spec}(\mathsf{R})$$

which is automatically closed. This can be thought as

$$X := \mathrm{Proj}(\mathsf{R}[x_0, x_1, \ldots, x_N]/I),$$

where I is a homogeneous ideal of the graded ring $\mathsf{R}[x_0, x_1, \ldots, x_N]$ with usual weights $\deg(x_i) := 1$. Later, we will also consider arbitrary weights. We denote by

$$X_{\mathsf{k}} := X \times_{\mathsf{R}} \mathsf{k}$$

the variety over the quotient field k of R. This is also called the generic fiber of X. In a similar way

$$X_{\bar{\mathsf{k}}} := X \times_{\mathsf{R}} \bar{\mathsf{k}}$$

is the variety over the algebraic closure $\bar{\mathsf{k}}$ of k. For a prime ideal $\mathfrak{p}$ of R, we denote by

$$X_{\mathfrak{p}} := X \times_{\mathsf{R}} \mathsf{R}/\mathfrak{p}$$

the fiber of X over $\mathfrak{p} \in \mathrm{Spec}(\mathsf{R})$. This is an scheme over the residue ring $\mathsf{R}/\mathfrak{p}$. It is also called X modulo the prime ideal $\mathfrak{p}$. For any other ring $\mathsf{R} \subset \tilde{\mathsf{R}}$ we also define $X_{\tilde{\mathsf{R}}} := X \times_{\mathsf{R}} \tilde{\mathsf{R}}$ which is an scheme over $\tilde{\mathsf{R}}$. An R-scheme X is called smooth if $X_{\bar{\mathsf{k}}}$ is smooth. We denote by $X(\mathsf{R})$ and $X(\mathsf{k}) := X_{\mathsf{k}}(\mathsf{k})$ the set of R- and k-rational points of X. For two R-schemes X_1, X_2 we will consider morphisms, isomorphism and birational maps of R-schemes $X_1 \to X_2$.

We introduce the above notation for families of projective schemes. A scheme over R gives us a morphism of schemes $X \to \mathsf{T}$, where $\mathsf{T} := \mathrm{Spec}(\mathsf{R})$. In general, we will take T a parameter scheme described in §2.3. A point $t \in \mathsf{T}$ is a prime ideal $\mathfrak{p}$ of R_i, where $\mathrm{Spec}(\mathsf{R}_i)$ is some open subset of T. This gives us the projective scheme X_t over $\mathsf{R}/\mathfrak{p}$. In this way, we can talk about the family of projective schemes with fibers $X_t, t \in \mathsf{T}$. We will also use the following equivalent notations

$$\{X_t\}_{t \in \mathsf{T}}, \qquad X/\mathsf{T}, \qquad X \subset \mathbb{P}^N_{\mathsf{T}}, \qquad X \to \mathsf{T}.$$

Note that if $\mathfrak{p}$ is a maximal prime ideal then $\mathsf{R}/\mathfrak{p}$ is the residue field. If R is a finitely generated $\mathfrak{k}$-algebra and $\mathfrak{k}$ is algebraically closed then all the residue fields are isomorphic to $\mathfrak{k}$ and X can be identified with the set $X(\mathfrak{k})$ of its closed points.

Definition 2.9. Let X be a projective scheme over T and let $\pi : X \to \mathsf{T}$ be the corresponding morphism of algebraic varieties over the field $\mathfrak{k}$. A point $t_0 \in \mathsf{T}(\bar{\mathfrak{k}})$ is called a regular value of π if the derivative of π at any point $x_0 \in X(\bar{\mathfrak{k}})$ with $\pi(x_0) = t_0$ is surjective. It is called critical otherwise. The corresponding fiber X_{t_0} is called respectively a regular and critical fiber. The morphism π is called smooth if it has only regular values.

For a more general definition of a smooth morphism of schemes see [Har77, Chapter 3, §10].

2.7 Algebraic de Rham cohomology and Hodge filtration

Let X be a smooth projective variety over the ring R. We are going to use the algebraic de Rham cohomology $H^*_{\mathrm{dR}}(X)$ of X and its Hodge filtration, both defined over R. The main references on this topic are the original article of Grothendieck [Gro66], Hartshorne's work [Har75] and Deligne's lecture notes in [DMOS82]. In all these articles, it is assumed that R is a field of characteristic zero, however, for our purpose we need to relax this condition and work over a ring. This forces us to assume that the ring R is the function ring of a smooth affine variety $\mathsf{T} := \mathrm{Spec}(\mathsf{R})$ over a field $\mathfrak{k}$ of characteristic zero. Moreover, the morphism $X \to \mathsf{T}$ is smooth, see Definition 2.9. For families of affine varieties one actually needs that the field $\mathfrak{k}$ to be of characteristic zero, see [Mov12b, §2.4] for the discussion of de Rham cohomology of compact and punctured elliptic curves. In a personal communication with P. Deligne (December 13, 2010) he writes: "For affine varieties, dR cohomology is indeed bad in char. p: usually infinite dimensional (H^0 contains all pth powers of functions). The projective case is better, and for an abelian variety of dimension g, one gets the expected exterior algebra over H^1, which is of dimension $2g$, extension of H^1 of $\mathscr{O}$ by H^0 of Ω^1, both of dimension g." Using Brieskorn modules for tame polynomials which is a finer version of algebraic de Rham cohomology one can also work with a general Cohen–Macaulay ring R and the family $X \to \mathsf{T}$ might have isolated singular fibers, see [Mov19]. We believe that we can relax the conditions over $\mathfrak{k}$ in a great amount and even we could work over a ring $\mathfrak{R}$ instead of $\mathfrak{k}$. However, for the lack of references in the literature we will limit ourselves to a field $\mathfrak{k}$ of characteristic 0. Further discussions of this type are being collected in the book [Mov20a].

Definition 2.10. For a smooth projective scheme X of dimension n over the ring R the algebraic de Rham cohomologies

$$H^m_{\mathrm{dR}}(X) := \mathrm{Im}(\mathbb{H}^m(X, \Omega^*_{X/\mathsf{R}}) \to \mathbb{H}^m(X, \Omega^*_{X/\mathsf{k}})),\ m = 0, 1, 2, \ldots, 2n,$$

are R-modules in a natural way, where k is the quotient field of R.

For other m's by definition we have $H^m_{\mathrm{dR}}(X) = \{0\}$. Further, we have the cup product which is a R-bilinear map

$$H^{m_1}_{\mathrm{dR}}(X) \times H^{m_2}_{\mathrm{dR}}(X) \to H^{m_1+m_2}_{\mathrm{dR}}(X), \quad (\alpha, \beta) \mapsto \alpha \cup \beta. \tag{2.9}$$

In each $H^m_{\mathrm{dR}}(X)$ we have the Hodge filtration

$$0 = F^{m+1} \subset F^m \subset \cdots \subset F^1 \subset F^0 = H^m_{\mathrm{dR}}(X)$$

which is defined by

$$F^q = F^q H^m_{\mathrm{dR}}(X) = \mathrm{Im}(\mathbb{H}^m(X, \Omega^{\bullet \geq q}_{X/\mathsf{R}}) \to \mathbb{H}^m(X, \Omega^\bullet_{X/\mathsf{k}})).$$

Each F^q is also an R-module. The cup product and the Hodge filtration satisfy the relations

$$F^i H^{m_1}_{\mathrm{dR}}(X) \cup F^j H^{m_2}_{\mathrm{dR}}(X) \subset F^{i+j} H^{m_1+m_2}_{\mathrm{dR}}(X). \tag{2.10}$$

The embedding $X \hookrightarrow \mathbb{P}^N_{\mathsf{R}}$ gives us an element

$$\theta \in F^1 H^2_{\mathrm{dR}}(X) \tag{2.11}$$

which we call it the polarization. For m even number we write

$$\theta^{\frac{m}{2}} := \underbrace{\theta \cup \theta \cup \cdots \cup \theta}_{\frac{m}{2}\text{-times}} \in H^m_{\mathrm{dR}}(X).$$

In particular, for $m = 2n$ we get the element θ^n in the top cohomology $H^{2n}_{\mathrm{dR}}(X)$. The trace map

$$\mathrm{Tr} : H^{2n}_{\mathrm{dR}}(X) \to \mathsf{R}$$

is an R-linear map. Let

$$L^i : H^*_{\mathrm{dR}}(X) \to H^*_{\mathrm{dR}}(X), \quad \alpha \mapsto \alpha \cup \theta^i.$$

The mth primitive cohomology is defined to be

$$H^m_{\mathrm{dR}}(X)_0 := \ker(L^{n-m+1} : H^m_{\mathrm{dR}}(X) \to H^{2n-m+2}_{\mathrm{dR}}(X)).$$

The cup product composed with the trace map gives us the bilinear maps

$$\langle \cdot, \cdot \rangle : H^m_{\mathrm{dR}}(X) \times H^{2n-m}_{\mathrm{dR}}(X) \to \mathsf{R}, \quad (\alpha, \beta) \mapsto \mathrm{Tr}(\alpha \cup \beta) \tag{2.12}$$

and

$$\langle \cdot, \cdot \rangle : H^m_{\mathrm{dR}}(X) \times H^m_{\mathrm{dR}}(X) \to \mathsf{R}, \quad (\alpha, \beta) \mapsto \mathrm{Tr}(\alpha \cup \beta \cup \theta^{n-m}). \tag{2.13}$$

Note that the second bilinear map depends on the polarization.

Definition 2.11. By abuse of classical notations we define

$$H^{j,m-j}(X) := F^j H^m_{\rm dR}(X)/F^{j+1}H^m_{\rm dR}(X) \cong H^{m-j}(X, \Omega^j_X). \qquad (2.14)$$

We say that $\alpha \in H^m_{\rm dR}(X)$ is of type $(j, m-j)$ if $\alpha \in F^j H^m_{\rm dR}(X)$ and $\alpha \notin F^{j+1}H^m_{\rm dR}(X)$. We write

$$\mathrm{Type}(\alpha) := (j, m-j).$$

The Hodge numbers are

$$\mathsf{h}^{i,j} := \dim H^i(X, \Omega^j_X), \quad 0 \le i+j \le 2n.$$

The Betti numbers are

$$\mathsf{b}_m := \mathsf{h}^{m,0} + \mathsf{h}^{m-1,1} + \cdots + \mathsf{h}^{0,m}. \qquad (2.15)$$

The dimensions of the Hodge filtration of the de Rham cohomologies of X are

$$\mathsf{h}^i_m := \mathsf{h}^{m,0} + \mathsf{h}^{m-1,1} + \cdots + \mathsf{h}^{i,m-i}. \qquad (2.16)$$

We use subscript m to denote objects attached to the mth cohomology. When we do not use other cohomologies we usually drop this subscript. For instance, we write $\mathsf{b} = \mathsf{b}_m$ to denote the mth Betti number. The following statements are well-known.

(1) The number $\mathrm{Tr}(\theta^n)$ is the degree of the projective scheme $X \subset \mathbb{P}^N_{\mathsf{R}}$.
(2) The top cohomology is an R-module of rank one and the trace map is an isomorphism of R-modules.
(3) The bilinear maps 2.12 and (2.13) are non-degenerate.
(4) Hard Lefschetz theorem: the map $L^{n-m}: H^m_{\rm dR}(X) \to H^{2n-m}_{\rm dR}(X)$ is an isomorphism of R-modules.
(5) Lefschetz decomposition: we have

$$\bigoplus_q H^{m-2q}_{\rm dR}(X)_0 \cong H^m_{\rm dR}(X)$$

which is given by $\bigoplus_q L^q$.

In the complex context, $\mathsf{R} = \mathbb{C}$ the trace map is just

$$\mathrm{Tr}(\alpha) := \frac{1}{(2\pi i)^n} \int_{X(\mathbb{C})} \alpha. \tag{2.17}$$

The appearance of powers of $2\pi i$-factors in period manipulations is also formulated in terms of the so-called Tate twist, see Deligne's Lecture notes [DMOS82]. For the purpose of the present text, it is more convenient to write these factors explicitly and not to use the Tate twist notation. For a projective variety X over a field $\mathfrak{k}$ of characteristic zero, there is no canonical inclusion of $H^{i,m-i}(X)$ inside $H^m_{\mathrm{dR}}(X)$. Therefore, there is no canonical isomorphism between $H^m_{\mathrm{dR}}(X)$ and $\bigoplus_{i=0}^m H^{i,m-i}(X)$. For $\mathfrak{k} = \mathbb{C}$, such a canonical inclusion and isomorphism exist and are given by harmonic forms, see for instance [Voi02].

Proposition 2.4. *Let $X_t, t \in \mathsf{T}$ be a family of smooth projective varieties and let X, X_0 be two regular fibers of this family. We have an isomorphism*

$$(H^*_{\mathrm{dR}}(X), F^*, \cup, \theta) \overset{\alpha}{\simeq} (H^*_{\mathrm{dR}}(X_0), F^*_0, \cup, \theta_0). \tag{2.18}$$

Proof. It suffices to consider families over the field of complex numbers. Further, it is enough to prove the isomorphism (2.18) for $X = X_t$ with t in a small neighborhood $U = (\mathsf{T}, 0)$ of $0 \in \mathsf{T}$ for which we have used the usual/analytic topology of T. We can take sections $\alpha_{m,i}$ of the cohomology bundle $H^m_{\mathrm{dR}}(X/\mathsf{T})$ in U such that θ^i's are included in this basis, and moreover, it is compatible with the Hodge filtration. However, we need that the cup product to be constant (independent of t) in this basis which is not clear why this must be the case. The following proof is due to P. Deligne (personal communication May 12, 2019).

Let G be the linear algebraic group of automorphisms of $(H^*_{\mathrm{dR}}(X_0), \cup, \theta)$. By Ehresmann's fibration theorem we have C^∞ isomorphisms $X_t \cong X_0$, $t \in U$ which gives us unique isomorphisms $i : (H^*_{\mathrm{dR}}(X_t), \cup, \theta) \cong (H^*_{\mathrm{dR}}(X_0), \cup, \theta)$. Note that the uniqueness follows from the fact that U is simply connected. The Hodge decomposition in $H^*_{\mathrm{dR}}(X_t)$ is given by the action of the multiplicative group $\mathbb{G}_m = (\mathbb{C} - \{0\}, \cdot)$: multiplication by z^{p-q} on $H^{p,q}$. This composed with i gives us a holomorphic family of algebraic group morphisms $i_t : \mathbb{G}_m \to G$ for all $t \in U$. We need to prove that i_t's are conjugate, that is, there is a holomorphic map $g : U \to G$ such that $i_t(z) = g(t)^{-1} i_0(z) g(t)$. In order to prove this we use SGA3 Vol. II, see [DG70]. "IX 3 uses cohomology to obtain infinitesimal statements. XI 4 proves representability of the functor

M of subgroupschemes of multiplicative type. XI 5 puts it all together to prove that for an affine smooth groupscheme G/S, and M the scheme parametrizing subgroupscheme of multiplicative type, M is smooth over S and the action by conjugation of G on M gives a smooth morphism (action, Id_M) : $G \times M \to M \times M$" (P. Deligne, personal communication, August 14, 2019). □

A very important fact is that there is no canonical way to choose the isomorphism (2.18). This is the driving philosophy behind the present text and it is the main starting observation for the creation of new theories of modular forms.

2.8 Block matrix notations

In many occasions we fix a number $0 \leq m \leq 2n$ and work only with the mth cohomology of varieties. In this case, we usually omit the subscript m, for instance we write $\mathsf{b}_m = \mathsf{b}$. We will usually use $\mathsf{b} \times \mathsf{b}$ matrices. For a $\mathsf{b} \times \mathsf{b}$ matrix M we denote by M^{ij}, $i, j = 0, 1, 2, \ldots, m$ the $\mathsf{h}^{m-i,i} \times \mathsf{h}^{m-j,j}$ submatrix of M corresponding to the decomposition (2.15):

$$M = [M^{ij}] = \begin{pmatrix} M^{00} & M^{01} & M^{02} & \cdots & M^{0m} \\ M^{10} & M^{11} & M^{12} & \cdots & M^{1m} \\ M^{20} & M^{21} & M^{22} & \cdots & M^{2m} \\ \vdots & \vdots & \vdots & \ddots & \vdots \\ M^{m0} & M^{m1} & M^{m2} & \cdots & M^{mm} \end{pmatrix}.$$

We call M^{ij}, $i, j = 0, 1, 2 \ldots, m$ the (i, j)th Hodge block of M. For a $\mathsf{b} \times 1$ matrix M we denote by M^i, $i = 0, 1, 2, \ldots, m$ the $\mathsf{h}^{m-i,i} \times 1$ submatrix of M corresponding to the decomposition (2.15):

$$M = \begin{pmatrix} M^0 \\ M^1 \\ M^2 \\ \vdots \\ M^m \end{pmatrix}.$$

For any property "P" of matrices we say that the property "block P" or "Hodge block P" is valid if the property P is valid with respect to the Hodge blocks. For instance, we say that a matrix M is block upper triangular if $M^{ij} = 0$, $i > j$. In many occasions, writing the general matrix might be confusing and so we reproduce the matrices for $m = 3$, 4 or 5. The general format of the matrix can be easily guessed and reproduced from this case.

2.9 Moduli space

Let $k = \bar{\mathfrak{k}}$ be an algebraically closed field and let $X \subset \mathbb{P}^N$ be a smooth projective variety over k. We want to talk about all possible deformations of X, and the moduli space of X. However, the construction of moduli spaces in Algebraic Geometry is usually a hard task, and it needs mastering of many techniques such as Geometric Invariant Theory, see [MFK94]. In this section, we briefly describe what we need to know about moduli spaces in a more intuitional language. Further discussion of this topic will be done in §3.11.

Definition 2.12. By an algebraic deformation of a projective variety $X \subset \mathbb{P}^N_k$ over k we mean any fiber of a smooth proper family $\{X_t\}_{t \in \mathsf{T}}$, $X_t \subset \mathbb{P}^N_k$ of projective varieties over k with T smooth and connected. This is obtained by taking a closed subvariety of $\mathbb{P}^N_k \times \mathsf{T}$ and projecting it to T. If necessary, we might replace T with an open subset of T.

Let M be a moduli of projective varieties X over k. By this we mean a set of algebraic varieties such that

(1) For any two variety $X_1, X_2 \in \mathsf{M}$, there is a proper family $\{X_t\}_{t \in \mathsf{T}}$, $X_t \subset \mathbb{P}^N$ of algebraic varieties over k such that X_1 and X_2 are two regular fibers of the family, that is, for some $t_1, t_2 \in \mathsf{T}(k)$ we have $X_1 = X_{t_1}$, $X_2 = X_{t_2}$.
(2) Any algebraic deformation of $X \in \mathsf{M}$ in $\mathbb{P}^N$ and over k is in M.

For $k \subset \mathbb{C}$ it follows from Ehresmann's theorem (and its generalization to singular varieties) that all the varieties in M are diffeomorphic and so from the topological point of view there is no difference between them. The algebraic structure distinguishes the elements of M. Hodge numbers do not depend on the particular choice of $X \in \mathsf{M}$ because they are topological invariants. "To say that "Hodge numbers are topological" is abusive. In a family, $h^{p,q}$ is (locally) constant because it is upper semi-continuous (as the dimension of a coherent cohomology group), while their sum for $p + q = n$ is constant, being topological by degenerescence of Hodge to de Rham" (P. Deligne, personal communication, May 12, 2019).

2.10 Flatness condition

Let X be a projective scheme over a parameter scheme T. It seems that in our way we will need to assume that X is flat. The geometric notation $X \to \mathsf{T}$ is more convenient for this purpose.

Definition 2.13. Let $\pi : X \to \mathsf{T}$ be a morphism of schemes. It is called flat if for all $x \in X$ with $t := \pi(x)$, $\mathscr{O}_{X,x}$ is a flat $\mathscr{O}_{\mathsf{T},t}$-module, that is, for any finitely generated ideal $I \subset \mathscr{O}_{\mathsf{T},t}$ the map $I \otimes \mathscr{O}_{X,x} \to \mathscr{O}_{X,x}$ is injective.

For more details on flat morphisms see Hartshorne's book [Har77, p. 253]. The first important property of flat morphisms is that the dimension of fibers does not change, see [Har77, Proposition 9.5, p. 256]. In our study of foliations in Chapter 5, we will consider fibrations that might have fibers of different dimensions and hence they do not enjoy flatness. The following theorem is the main reason for us to assume the flatness condition throughout the present text. We will discuss Hilbert schemes in §2.11.

Theorem 2.5. *A projective scheme $X \subset \mathbb{P}^N_{\mathsf{T}}$ over a reduced and connected scheme T is flat if and only if all the fibers of $X \to \mathsf{T}$ have the same Hilbert polynomial.*

See for instance [EH00, Proposition III-56], and the references therein.

2.11 Hilbert schemes

In this section, we gather some well-known facts concerning Hilbert schemes. Our main references are Eisenbud and Harris's book [EH00] and Mumford, Fogarty and Kirwan's book [MFK94]. Let k be a field and $X \subset \mathbb{P}^N_{\mathsf{k}}$ be a projective scheme. By definition X is given by a homogeneous ideal

$$I \subset \mathsf{k}[x] := \mathsf{k}[x_0, x_1, \ldots, x_N].$$

Definition 2.14. The Hilbert function of X is

$$\mathrm{HF}(\cdot) = \mathrm{HF}(X, \cdot) : \mathbb{N} \to \mathbb{N}, \quad \mathrm{HF}(X, \nu) := \dim_{\mathsf{k}}(\mathsf{k}[x]/I)_\nu.$$

The Hilbert series is

$$\mathrm{HS}(t) = \mathrm{HS}(X, t) := \sum_{\nu} \mathrm{HF}(X, \nu) t^\nu.$$

Theorem 2.6 (Hilbert). *There is a unique polynomial $\mathrm{HP}(X, \nu)$ in ν such that $\mathrm{HF}(X, \nu) = \mathrm{HP}(X, \nu)$ for all sufficiently large ν.*

The polynomial in the above theorem is called Hilbert polynomial. For a proof see [EH00, Theorem III-35, p. 125]. The following examples are useful to carry in mind:

(1) For the projective space $X = \mathbb{P}^N_{\mathsf{k}}$ we have

$$\mathrm{HS}(t) := \frac{1}{(1-t)^{N+1}}, \quad \mathrm{HF}(\nu) = \mathrm{HP}(\nu) = \binom{\nu+N}{N}.$$

(2) For a complete intersection X of type $(d_1, d_2, \ldots, d_s)$ we have

$$\mathrm{HS}(t) := \frac{(1-t^{d_1})(1-t^{d_2})\cdots(1-t^{d_s})}{(1-t)^{N+1}}.$$

(3) For a hypersurface X of degree d in $\mathbb{P}^N$:

$$\mathrm{HS}(t) := \frac{1-t^d}{(1-t)^{N+1}}, \quad \mathrm{HF}(\nu) = \mathrm{HP}(\nu) = \binom{\nu+N}{N} - \binom{\nu-d+N}{N}. \tag{2.19}$$

We consider the set of all projective varieties $X \subset \mathbb{P}^N_{\mathsf{k}}$ with a given Hilbert polynomial P. In fact, we have the Hilbert scheme $\mathrm{Hilb}_P(\mathbb{P}^N)$ whose k-rational closed points will be a good substitute for M introduced in §2.9.

Let P be a fixed Hilbert polynomial. Let us consider the following functor:

$$h : \{\text{ schemes}\} \to \{\text{sets}\}$$

where for a scheme T, $h(\mathsf{T})$ is the set of schemes $X \subset \mathbb{P}^N_{\mathsf{T}}$ flat over T whose fibers over points of T have the Hilbert polynomial P. The functor h is representable and a more precise result is given below.

Theorem 2.7. *There is a projective scheme* $\mathrm{Hilb}_P(\mathbb{P}^N_{\mathbb{Z}}) \subset \mathbb{P}^M_{\mathbb{Z}}$ *over* $\mathbb{Z}$ *and a closed subscheme*

$$W \subset \mathbb{P}^N_{\mathbb{Z}} \times \mathrm{Hilb}_P(\mathbb{P}^N_{\mathbb{Z}})$$

which is universal, that is, for any closed subscheme $X \subset \mathbb{P}^N_{\mathbb{Z}} \times \mathsf{T}$ *flat over* T *whose fibers over points of* T *have the Hilbert polynomial* P*, there is a unique morphism* $f : \mathsf{T} \to \mathrm{Hilb}_P(\mathbb{P}^N_{\mathbb{Z}})$ *such that* X *is the pull-back of* W *under the map*

$$\mathrm{Id} \times f : \mathbb{P}^N_{\mathbb{Z}} \times \mathsf{T} \to \mathbb{P}^N_{\mathbb{Z}} \times \mathrm{Hilb}_P(\mathbb{P}^N_{\mathbb{Z}}).$$

For more details and further references on Hilbert schemes see the book [MFK94, Chapter 0, §5], see also [EH00, p. 263]. An important feature of this theorem is that $\mathrm{Hilb}_P(\mathbb{P}^N_{\mathbb{Z}})$ is projective and so there is a homogeneous ideal $I \subset \mathbb{Z}[t] = \mathbb{Z}[t_0, t_1, \ldots, t_M]$ such that

$$\mathrm{Hilb}_P(\mathbb{P}^N_{\mathbb{Z}}) = \mathrm{Proj}(\mathbb{Z}[t]/I).$$

The reductive group

$$\mathbf{G} := \mathrm{GL}(N+1) = \mathrm{Aut}(\mathbb{P}^N_{\mathbb{Z}})$$

acts from the left on $\mathbb{P}^N_{\mathbb{Z}}$ and hence it induces an action on the Hilbert scheme:

$$\mathbf{G} \times_{\mathbb{Z}} \mathrm{Hilb}_P(\mathbb{P}^N_{\mathbb{Z}}) \to \mathrm{Hilb}_P(\mathbb{P}^N_{\mathbb{Z}}), \quad (\mathbf{g}, t) \mapsto \mathbf{g} \cdot t.$$

This follows from the universal property of the Hilbert scheme.

2.12 Group schemes and their action

In this section, we recall some basic definitions related to group schemes and their actions. As usual all schemes are over $\mathfrak{R}$. For missing definitions see [MFK94].

Definition 2.15. A group scheme $\mathbf{G}$ acts from the left on $\pi : \mathsf{X} \to \mathsf{T}$ (or simply on X/T) if it acts from the left on both X and T and its action commutes with the morphism $\mathsf{X} \to \mathsf{T}$, that is,

$$\begin{array}{ccc} \mathbf{G} \times \mathsf{X} & \to & \mathsf{X} \\ \downarrow & & \downarrow \\ \mathbf{G} \times \mathsf{T} & \to & \mathsf{T} \end{array} \tag{2.20}$$

commutes, where the first down arrow is $\mathrm{Id} \times \pi$. In geometric terms, this means that

$$\pi(\mathbf{g} \cdot x) = \mathbf{g} \cdot \pi(x), \quad x \in \mathsf{X}, \quad \mathbf{g} \in \mathbf{G}.$$

It follows that the action of $\mathbf{g} \in \mathbf{G}$ on X induces an isomorphism $\mathsf{X}_t \to \mathsf{X}_{\mathbf{g} \cdot t}$, $x \mapsto \mathbf{g} \cdot x$ for any $t \in \mathsf{T}$.

In a similar way we define a right action. In Chapter 3, we will consider the action of two groups in the same time.

Definition 2.16. Let $\mathbf{G}$ and G be two group schemes acting on an scheme X from the left and right, respectively. We say that the actions of $\mathbf{G}$ and

G are independent from each other if the canonical compositions

$$(\mathbf{G} \times \mathsf{X}) \times \mathsf{G} \to \mathsf{X} \times \mathsf{G} \to \mathsf{X},$$
$$\mathbf{G} \times (\mathsf{X} \times \mathsf{G}) \to \mathbf{G} \times \mathsf{X} \to \mathsf{X},$$

are the same. In geometric words

$$\mathrm{g}\cdot(t \bullet \mathsf{g}) = (\mathrm{g}\cdot t) \bullet \mathsf{g}, \quad \mathrm{g} \in \mathbf{G},\ t \in \mathsf{X},\ \mathsf{g} \in \mathsf{G}. \tag{2.21}$$

2.13 Stable points

For the preparation of this section we have used Newstead's book [New78], see also his lecture notes in Guanajuato. Let $X \subset \mathbb{P}^N_{\mathfrak{R}}$ be a projective scheme over $\mathfrak{R}$. We consider a group scheme $\mathbf{G}$ which acts linearly from the left on X, that is, we have a representation $\mathbf{G} \to \mathrm{GL}(N+1)$, $\mathbf{G}$ acts on $\mathbb{P}^N_{\mathfrak{R}}$ through this representation, and X is invariant under this action. It turns out that $\mathbf{G}$ acts on the space of polynomials $\mathfrak{R}[X_0, X_1, \ldots, X_N]$. For a homogeneous $\mathbf{G}$-invariant polynomial f, let

$$X_f := \{x \in X | f(x) \neq 0\} \tag{2.22}$$

which is a $\mathbf{G}$-invariant affine open subset of X.

Definition 2.17. A point $x \in X$ is called semistable for the action of $\mathbf{G}$ if there exists a $\mathbf{G}$-invariant polynomial f such that $x \in X_f$. It is called stable if it has finite stabilizer (or equivalently $\dim(\mathbf{G} \cdot x) = \dim(\mathbf{G})$) and there is an f as above such that $\mathbf{G}$ acts on X_f and all the orbits of $\mathbf{G}$ in X_f are closed.

Definition 2.18. Let X and $\mathbf{G}$ as before. A morphism of $\mathfrak{R}$-schemes $\phi : X \to Y$ is called a good quotient of X by $\mathbf{G}$ if

(1) ϕ is an affine and surjective morphism. Recall that ϕ is affine if the inverse image of every affine Zariski open set in Y is affine.
(2) ϕ is constant on orbits.
(3) For U a Zariski open subset of Y, the induced homomorphism $\phi^* : \mathscr{O}_Y(U) \to \mathscr{O}_X(\phi^{-1}(U))^{\mathbf{G}}$ is an isomorphism. Here, $\mathscr{O}_X(\phi^{-1}(U))^{\mathbf{G}}$ is the ring of $\mathbf{G}$-invariant functions.
(4) If W is a Zariski-closed $\mathbf{G}$-invariant subset of X, then $\phi(W)$ is also closed in Y.
(5) If W_1 and W_2 are Zariski closed $\mathbf{G}$-invariant subsets of X and $W_1 \cap W_2 = \emptyset$, then $\phi(W_1) \cap \phi(W_2) = \emptyset$.

In this case we write $Y = \mathbf{G}\backslash\backslash X$. It is called a geometric quotient if it is a good quotient, and an orbit space, that is, we have a bijection between the set of orbits $\{\mathbf{G} \cdot x,\ x \in X\}$ and Y which maps $\mathbf{G} \cdot x$ to $\phi(\mathbf{G} \cdot x)$. In this case we write $Y = \mathbf{G}\backslash X$.

Let X^{ss} and X^s be the set of semistable and stable points of X, respectively.

Theorem 2.8. *Let* $\mathbf{G}$ *be a reductive group acting linearly on a projective variety* X. *Then*

(1) *There exists a good quotient* $\phi : X^{ss} \to Y$ *and* Y *is projective.*
(2) *The image* Y^s *of the morphism* ϕ *restricted to* X^s *is a Zariski open subset of* Y *and* $Y^s = \mathbf{G}\backslash X^s$ *is a geometric quotient of* X^s.
(3) *For* $x_1, x_2 \in X^{ss}, \phi(x_1) \neq \phi(x_2)$ *if and only if*

$$\mathbf{G} \cdot x_1 \cap \mathbf{G} \cdot x_2 \cap X^{ss} = \emptyset.$$

(4) *For* $x \in X^{ss}$, x *is stable if and only if* x *has finite stabiliser and* $\mathbf{G} \cdot x$ *is closed in* X^{ss}.

The fact that Y is projective implies that we have a natural compactification of the orbit space X^s. As a corollary of Theorem 2.8 we have the following proposition.

Proposition 2.9. *Let* $f : X_1 \to X_2$ *be a morphism of* $\mathfrak{R}$*-schemes, and consider an action of the reductive group* $\mathbf{G}$ *from the left on* f *(see Definition 2.15) and two points* $x_1 \in X_1$, $x_2 \in X_2$. *Assume that* f *is a finite morphism, that is, there is a covering of* X_2 *by open affine subsets* $U_i = \mathrm{Spec}(\mathsf{R}_i)$ *such that* $f^{-1}(U_i) = \mathrm{Spec}(\check{\mathsf{R}}_i)$ *is affine and* $\check{\mathsf{R}}_i$ *is a finitely generated* R_i*-algebra. If* x_2 *is stable (resp. semistable) for the action of* $\mathbf{G}$ *then* x_1 *is also stable (resp. semistable) for the action of* $\mathbf{G}$.

This, for instance, will be used in Theorem 3.5. Note that families of projective varieties over a Hilbert scheme is not finite, and so this observation cannot be applied in this case. For this reason many times, coarse moduli spaces exist, however the universal families do not.

2.14 Group actions and constant vector fields

Let G be an algebraic group scheme over $\mathfrak{R}$ and let v be a vector field in G. We are going to consider the parallel extension of v in $\mathsf{G} \times \mathsf{G}$. For this we have to consider v a vector field in the first or second factor of the product $\mathsf{G} \times \mathsf{G}$. This will not be important for the definition below:

Definition 2.19. The Lie algebra of G is

$$\mathrm{Lie}(\mathsf{G}) := \{\mathsf{v} \in H^0(\mathsf{G}, \Theta_{\mathsf{G}}) |\ \text{the parallel transport of } \mathsf{v}$$
$$\text{in } \mathsf{G} \times \mathsf{G} \text{ is mapped to } \mathsf{v} \text{ under the multiplication morphism } \mathsf{G} \times \mathsf{G} \to \mathsf{G}\}.$$

In the rest of this section we will work with an algebraic group G over an algebraically closed field $\mathfrak{k}$. The reader might try to formulate and prove the scheme theoretic version of what follows.

Proposition 2.10. *If G is an algebraic group over an algebraically closed field $\mathfrak{k}$ and 1 is its identity element then the evaluation at $1 \in \mathsf{G}$ map*

$$\mathrm{Lie}(\mathsf{G}) \mapsto \mathbf{T}_1\mathsf{G}, \tag{2.23}$$

induces an isomorphism of $\mathfrak{k}$-vector fields, where $\mathbf{T}_1\mathsf{G}$ is the tangent space of G at 1.

Proof. The inverse of the map (2.23) is obtained in the following way. Any element $\mathsf{g} \in \mathsf{G}$ induces an isomorphism $i_{\mathsf{g}} : \mathsf{G} \to \mathsf{G},\ x \mapsto x \cdot \mathsf{g}$. For a vector $\mathsf{v}_1 \in \mathbf{T}_1\mathsf{G}$ we can consider an element $\mathsf{v} \in \mathrm{Lie}(\mathsf{G})$ such that $\mathsf{v}_{\mathsf{g}} \in \mathbf{T}_{\mathsf{g}}\mathsf{G}$ at the point g is the push-forward of v_1 under the isomorphism

$$D_1 i_{\mathsf{g}} : \mathbf{T}_1\mathsf{G} \to \mathbf{T}_{\mathsf{g}}\mathsf{G}.$$

The vector field v is characterized by the fact that it is invariant under i_{g} for all g. The last part of the proof has inspired the definition of $\mathrm{Lie}(\mathsf{G})$. □

Let us consider $\mathrm{Lie}(\mathsf{G})$ as a constant sheaf with values in $\mathrm{Lie}(\mathsf{G})$ defined in (2.19). For simplicity, we do not produce new notations. We have a canonical inclusion of sheaves

$$\mathrm{Lie}(\mathsf{G}) \subset \Theta_{\mathsf{G}}.$$

In a similar way, we have also a canonical inclusion of sheaves

$$\mathrm{Lie}(\mathsf{G})^{\vee} \subset \Omega^1_{\mathsf{G}}.$$

Proposition 2.11. *We have*

$$\Theta_{\mathsf{G}} = \mathrm{Lie}(\mathsf{G}) \otimes_{\mathfrak{k}} \mathscr{O}_{\mathsf{G}},$$
$$\Omega^1_{\mathsf{G}} = \mathrm{Lie}(\mathsf{G})^{\vee} \otimes_{\mathfrak{k}} \mathscr{O}_{\mathsf{G}}.$$

Proof. This follows from the fact that at each point g of G the elements of Lie(G) evaluated at g form a basis of $\mathbf{T}_{\mathsf{g}}\mathsf{G}$. □

Definition 2.20. An element of the Lie algebra Lie(G) is denoted by $\mathfrak{g}$. Later, we will also use another (reductive) group $\mathbf{G}$ and an element of Lie($\mathbf{G}$) is also denoted by $\mathfrak{g}$.

Proposition 2.12. *Let* G *be an algebraic group acting from the right on a variety* T *non-trivially, all defined over an algebraically closed field* $\mathfrak{k}$. *There is a canonical homomorphism of Lie algebras*

$$i : \mathrm{Lie}(\mathsf{G}) \to H^0(\mathsf{T}, \Theta_{\mathsf{T}}), \quad \mathfrak{g} \mapsto \mathsf{v}_{\mathfrak{g}} \tag{2.24}$$

which is uniquely characterized by the following property: for $\mathfrak{g} \in \mathrm{Lie}(\mathsf{G})$ *viewed as a vector field in* G, *its parallel transport in* $\mathsf{T} \times \mathsf{G}$ *is mapped to* $\mathsf{v}_{\mathfrak{g}}$ *under the morphism of group action* $\mathsf{T} \times \mathsf{G} \to \mathsf{T}$.

Note that the proposition can be also stated for group schemes.

Proof. Let $\mathfrak{g} \in \mathrm{Lie}(\mathsf{G})$ and consider its parallel transport $\check{\mathfrak{g}}$ in $\mathsf{T} \times \mathsf{G}$. The vector field $\mathsf{v}_{\mathfrak{g}}$ as an element of $(\Omega^1_{\mathsf{T}})^\vee$ is the following. It sends the differential form $\alpha \in \Omega^1_{\mathsf{T}}$ to the pull-back of α under the action morphism $\mathsf{T} \times \mathsf{G} \to \mathsf{T}$ and then evaluated at $\check{\mathfrak{g}}$. After this evaluation one gets a regular function in $\mathsf{T} \times \mathsf{G}$ and one has to check that it is a pull-back of some regular function in T by the projection map $\mathsf{T} \times \mathsf{G} \to \mathsf{T}$ in the first coordinate. □

One might be interested in cases, where the map i is an injection. For this one might impose conditions on the action of the algebraic group G, for instance, the action is not trivial in the sense that the action morphism $\mathsf{G} \times \mathsf{T} \to \mathsf{T}$ is not constant (its image is not a point). It is well-known that if the action is free then i is injective, see [Ham17, Proposition 3.4.3].

Definition 2.21. Following [LM87, Chapter IV], see also [Ham17, §3.4] we call $\mathsf{v}_{\mathfrak{g}}$ the fundamental vector field corresponding to $\mathfrak{g} \in \mathrm{Lie}(\mathsf{G})$. We also call i in (2.24) the fundamental vector field map.

One can also describe $\mathsf{v}_{\mathfrak{g}}$ in a more geometric fashion. For $t \in \mathsf{T}$ let

$$j : \mathsf{G} \to \mathsf{T}, \quad j(\mathsf{g}) = t \bullet \mathsf{g}. \tag{2.25}$$

Let us identify Lie(G) with the tangent space of G at $1 \in \mathsf{G}$. For $\mathfrak{g} \in \mathrm{Lie}(\mathsf{G})$ we have to define $\mathsf{v}_{\mathfrak{g},t}$, $t \in \mathsf{T}$ which must be a vector in the tangent space of T at t. The vector $\mathsf{v}_{\mathfrak{g},t}$ is defined to be the image of $\mathfrak{g}$ under the derivation of j. Note that under the derivation of the action morphism $\mathsf{T} \times \mathsf{G} \to \mathsf{T}$ over the point $(t, 1)$, the vector $(w, \mathfrak{g})$ maps to $w + \mathsf{v}_{\mathfrak{g},t}$.

Proposition 2.13. *For a regular function $f \in \mathscr{O}_\mathsf{T}$ we have*

$$d(f \circ j)(\mathfrak{g}) = df(\mathsf{v}_\mathfrak{g}), \quad \forall \mathfrak{g} \in \mathrm{Lie}(\mathsf{G}),$$

where the first d refers to the differential operator in G *and the second d refers to the differential operator in* T.

Proof. This follows from the fact that under j the vector field $\mathfrak{g} \in \Theta_\mathsf{G}$ is mapped to the vector field $\mathsf{v}_\mathfrak{g} \in \Theta_\mathsf{T}$. □

Later in §5.10 we will use the foliation $\mathscr{F}(\mathsf{G})$ induced by the image of the map (2.24). This is one of our main examples of foliations with leaves of different codimensions.

2.15 Gauss–Manin connection

The Gauss–Manin connection from a topological point of view is simple to describe, however, it becomes computationally complicated from an algebraic point of view. The topological description is as follows.

Let $\mathsf{X} \to \mathsf{T}$ be a family of smooth projective varieties over $\mathbb{C}$. By Ehresmann's fibration theorem, this is a locally trivial C^∞ bundle over T, and hence, it gives us the cohomology bundle

$$H := \bigcup_{t \in \mathsf{T}} H^m(\mathsf{X}_t, \mathbb{C})$$

over T whose fiber at $t \in \mathsf{T}$ is the mth cohomology of the fiber X_t, for more details see [Mov19, Chapter 6]. This bundle has special holomorphic sections s such that for all $t \in \mathsf{T}$ we have $s(t) \in H^m(\mathsf{X}_t, \mathbb{Q})$. These are called flat sections. In a small neighborhood U of $t \in \mathsf{T}$ we can find flat sections $s_1, s_2, \ldots, s_\mathsf{b}$ such that any other holomorphic section in U can be written as $s = \sum_{i=1}^{\mathsf{b}} f_i s_i$, where f_i's are holomorphic functions in U. The Gauss–Manin connection on H is the unique connection on H with the prescribed flat sections:

$$\nabla : H \to \Omega_\mathsf{T} \otimes_{\mathscr{O}_\mathsf{T}} H, \quad \nabla\left(\sum_{i=1}^{\mathsf{b}} f_i s_i\right) = \sum_{i=1}^{\mathsf{b}} df_i \otimes s_i,$$

see Fig. 2.2 for a pictorial description of Gauss–Manin connection. The algebraic description of the Gauss–Manin connection is done by N. Katz

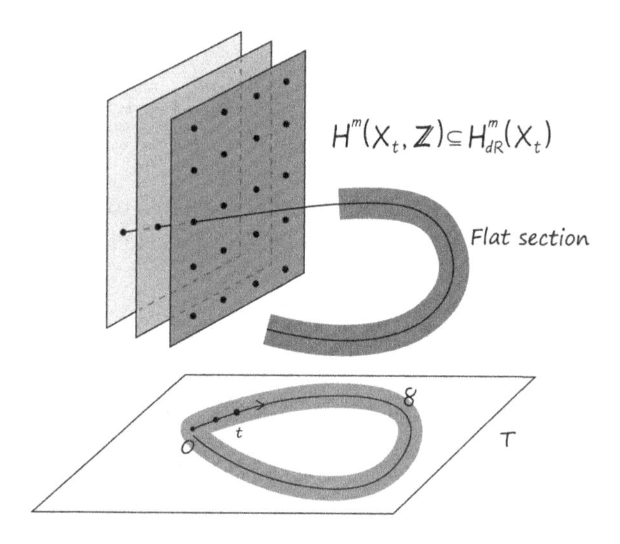

Fig. 2.2. Gauss–Manin connection.

and T. Oda in [KO68], see also Deligne's Bourbaki seminar [Del69]. Its computation usually produces huge polynomials and the available algorithms work only for families of lower dimensional varieties with few parameters. For more on this topic see [Mov11b, Mov12b]. In this text we will only need the following information about the algebraic Gauss–Manin connection.

Let $\mathsf{X} \to \mathsf{T}$ be a family of smooth projective varieties over a field $\mathfrak{k}$ of characteristic 0 as in §2.6. For a fixed $1 \leq m \leq 2n$, the mth de Rham cohomology bundle $H^m(\mathsf{X}/\mathsf{T})$, which one must look at it as a free sheaf on T, enjoys a canonical connection

$$\nabla : H^m_{\mathrm{dR}}(\mathsf{X}/\mathsf{T}) \to \Omega^1_{\mathsf{T}} \otimes_{\mathscr{O}_{\mathsf{T}}} H^m_{\mathrm{dR}}(\mathsf{X}/\mathsf{T}), \quad m = 0, 1, \ldots, 2n$$

which is called the Gauss–Manin connection of the family $\mathsf{X} \to \mathsf{T}$. It satisfies the following properties:

(1) The polarization $\theta \in H^2_{\mathrm{dR}}(\mathsf{X}/\mathsf{T})$ is a flat section, that is,

$$\nabla(\theta) = 0. \tag{2.26}$$

(2) For $\alpha \in H^{m_1}_{\mathrm{dR}}(\mathsf{X}/\mathsf{T})$ and $\beta \in H^{m_2}_{\mathrm{dR}}(\mathsf{X}/\mathsf{T})$ we have

$$\nabla(\alpha \cup \beta) = \nabla(\alpha) \cup \beta + (-1)^{m_1} \alpha \cup \nabla(\beta). \tag{2.27}$$

(3) We have

$$\nabla(F^i H^m_{\mathrm{dR}}(\mathsf{X}/\mathsf{T})) \subset F^{i-1} H^m_{\mathrm{dR}}(\mathsf{X}/\mathsf{T}), \quad i = 1, 2, \ldots, \tag{2.28}$$

which is called the Griffiths transversality.

(4) ∇ sends the primitive cohomology $H^m(\mathsf{X}/\mathsf{T})_0$ to itself, and hence, it respects the Lefschetz decomposition. This follows from item (1) and (2).

The Gauss–Manin connection induces maps

$$\begin{aligned} &\nabla_i : \Omega^i_{\mathsf{T}} \otimes_{\mathscr{O}_{\mathsf{T}}} H^m_{\mathrm{dR}}(\mathsf{X}/\mathsf{T}) \to \Omega^{i+1}_{\mathsf{T}} \otimes_{\mathscr{O}_{\mathsf{T}}} H^m_{\mathrm{dR}}(\mathsf{X}/\mathsf{T}), \\ &\nabla_i(\alpha \otimes \omega) = d\alpha \otimes \omega + (-1)^i \alpha \wedge \nabla\omega, \ \alpha \in \Omega^i_{\mathsf{T}}, \ \omega \in H^m_{\mathrm{dR}}(\mathsf{X}/\mathsf{T}), \end{aligned} \tag{2.29}$$

for $i = 0, 1, 2, \ldots$, and it is an integrable connection, that is,

$$\nabla_{i+1} \circ \nabla_i = 0, \quad i = 0, 1, 2, \ldots. \tag{2.30}$$

2.16 Infinitesimal variation of Hodge structures

The infinitesimal variation of Hodge structures, IVHS for short, is a partial data of the Gauss–Manin connection and cup product for families of projective varieties. It is also a computable part of it by means of closed formulas, at least for hypersurfaces. It was developed by Griffiths and his coauthors in a series of papers [CG80, CGGH83], and it produced many applications such as Torelli problem for hypersurfaces, see also Harris' expository article [Har85] on this topic. In this section, we introduce a slightly more general version of IVHS by considering the whole cohomology ring, whereas in the literature one defines it in a fixed cohomology. There is a close relation between our IVHS and the theory of modular vector fields developed in Chapter 6. This relation has been partially discussed in [Mov17c].

Let Y/V be a family of smooth projective varieties. By Griffiths transversality theorem, the Gauss–Manin connection of Y/V induces maps

$$\nabla_k : \bigcup_{t \in V} H^k(Y_t, \Omega^{m-k}_{Y_t}) \to \Omega^1_V \otimes_{\mathscr{O}_V} \bigcup_{t \in V} H^{k+1}(Y_t, \Omega^{m-k-1}_{Y_t}), \quad k = 0, 1, \ldots, m. \tag{2.31}$$

Here, we have used the canonical identification in (2.14). From now on we use the notation

$$H_t^{m-k,k} := H^k(Y_t, \Omega_{Y_t}^{m-k}).$$

One usually compose ∇_k with vector fields in V and arrives at the bilinear map in the first entry of

$$\delta = \delta_{m,k} = \delta_k \ : \ \mathbf{T}_t V \to \mathrm{Hom}(H_t^{m-k,k}, H_t^{m-k-1,k+1}), \tag{2.32}$$

$$H_t^{m-k,k} \times H_t^{m'-k',k'} \to H_t^{m+m'-k-k',k+k'}, \ (\alpha,\beta) \mapsto \alpha \cup \beta, \tag{2.33}$$

$$\theta \in H_t^{1,1}, \ \ \mathrm{Tr} : H_t^{n,n} \cong \mathsf{k}, \ \ \mathrm{Tr}(\alpha) := \frac{\alpha}{\theta^n}. \tag{2.34}$$

The second entry is induced from the cup product (2.13) in the de Rham cohomology of fibers of Y/V and for simplicity we have also denoted it by $\cup$. The element $\theta \in H_t^{1,1}$ is induced by polarization and Tr is induced by the trace map in de Rham cohomology. We will drop the sub indices of δ; being clear in the context where it acts. In the literature instead of (2.33) one mainly finds

$$\begin{aligned} Q = Q_{m,k} = Q_k : \ H_t^{m-k,k} \times H_t^{k,m-k} \to \mathsf{k}, \quad k = 0,1,\ldots,m, \\ Q(\alpha,\beta) := \mathrm{Tr}(\alpha \cup \beta \cup \theta^{n-m}). \end{aligned} \tag{2.35}$$

It is a non-degenerate bilinear map.

Proposition 2.14. *We have the following equalities*

$$\delta(w) \circ \delta(v) = \delta(v) \circ \delta(w), \qquad \forall v, w \in \mathbf{T}_t V, \tag{2.36}$$

$$\delta(v)(\alpha \cup \beta) = \delta(v)(\alpha) \cup \beta + \alpha \cup \delta(v)(\beta), \tag{2.37}$$

$$\forall \alpha \in H_t^{m'-k',k'}, \quad \beta \in H_t^{m'-k',k'}, \quad v \in \mathbf{T}_t V,$$

$$\delta(v)(\theta) = 0, \quad v \in \mathbf{T}_t V. \tag{2.38}$$

Note that for Q the equality (2.37) becomes:

$$Q_{k+1}(\delta_k(v)(\alpha), \beta) + Q_k(\alpha, \delta_{m-k-1}(v)(\beta)) = 0,$$

$$\forall \alpha \in H_t^{m-k,k}, \quad \beta \in H_t^{k+1,m-k-1}, \quad v \in \mathbf{T}_t V.$$

Proof. The proposition follows from similar equalities for the Gauss–Manin connection and cup product in cohomology, see (2.30), (2.27) and (2.26). □

The equalities (2.36) and (2.37) in our context of enhanced families is given in (6.30) and (3.22), respectively.

Definition 2.22. The collection of data (2.32), (2.33) and (2.34) with (2.36), (2.37) and (2.38) is called the infinitesimal variation of Hodge structures at the point $t \in V$.

For now we do not need the integral cohomology $H^m(Y_t, \mathbb{Z}) \subset H^m_{\mathrm{dR}}(Y_t)$ and so we have omitted it from the above definition. One may use a theorem of Griffiths which says that δ_k is the composition of the Kodaira–Spencer map

$$\mathbf{T}_t V \to H^1(Y_t, \Theta_{Y_t}) \tag{2.39}$$

with

$$\delta = \delta_{m,k} = \delta_k : H^1(Y_t, \Theta_{Y_t}) \to \mathrm{Hom}(H^k(Y_t, \Omega^{m-k}_{Y_t}),\ H^{k+1}(Y_t, \Omega^{m-k-1}_{Y_t})) \tag{2.40}$$

$$\delta_{m,k}(\mathsf{v})(\omega) = i_{\mathsf{v}}\omega,$$

where i_{v} is the contraction of differential forms along vector fields. For the definition of Kodaira–Spencer map see [Voi02, 9.12], or [Mov20a]. Sometimes in the literature, for the definition of IVHS, δ in (2.32) is replaced with δ in (2.40).

We will need the version of IVHS for primitive cohomologies. For this we will need to define the part of $H^1(Y_t, \Theta_{Y_t})$ responsible for deformations of Y_t inside a projective space. Note that we have

$$\delta_{2,1} : H^1(Y_t, \Theta_{Y_t}) \to \mathrm{Hom}(H^1(Y_t, \Omega^1_{Y_t}),\ H^2(Y_t, \mathscr{O}_{Y_t})) \tag{2.41}$$

and the polarization $\theta \in H^1(Y_t, \Omega^1_{Y_t})$.

Definition 2.23. The primitive part of the deformation space of Y_t is defined in the following way:

$$H^1(Y_t, \Theta_{Y_t})_0 := \{v \in H^1(Y_t, \Theta_{Y_t}) | \delta_{2,1}(v)(\theta) = 0\}. \tag{2.42}$$

This is a natural definition because the polarization as a global section of $H^2_{\mathrm{dR}}(Y/V)$ is flat for the Gauss–Manin connection. Note that the image of the Kodaira–Spencer map is in $H^1(Y_t, \Theta_{Y_t})_0$, and so, we can also consider the Kodaira–Spencer map as

$$\mathbf{T}_t V \to H^1(Y_t, \Theta_{Y_t})_0. \tag{2.43}$$

We may also define the primitive cohomologies

$$H^k(Y, \Omega_{Y_t}^{m-k})_0 := \{\omega \in H^k(Y, \Omega_{Y_t}^{m-k}) | \omega \cup \theta^s = 0\}, \tag{2.44}$$

where $s = n - m + 1$ for $m \leq n$ and $s = 1$ for $m \geq n$, and it is easy to see that δ sends primitive pieces to each other.

2.17 R-varieties

This section is the continuation of §2.16. We define the notion of an R-variety which is the generalization of many classical varieties in the literature, such as abelian and Calabi–Yau varieties. For simplicity, we use the notation $X = Y_t$. Let us consider the direct sum

$$\mathrm{IVHS} := \bigoplus H^*(X, \Omega_X^*), \tag{2.45}$$

equipped with the cup product $\cup$ and polarization $\theta \in H^1(X, \Omega_X^1)$ and call it an IVHS ring.

Definition 2.24. A linear map $\check{\delta} : \mathrm{IVHS} \to \mathrm{IVHS}$ is called an IVHS map if it satisfies all the properties of δ as before, that is, it sends the graded pieces of IVHS as in (2.32) and it satisfies (2.37) and (2.38).

We denote by Δ the set of all such $\check{\delta}$'s. Note that Δ is not a group and composition of two $\check{\delta}$'s is no more an IVHS map. By definition, we have a linear map

$$H^1(X, \Theta_X)_0 \to \Delta, \quad v \mapsto \delta(v). \tag{2.46}$$

The property (2.36) means that the two IVHS maps $\delta(v)$ and $\delta(w)$ commute and it does not appear in the above definition. We have canonical projections

$$\Delta \to \mathrm{Hom}(H^k(X, \Omega_X^{m-k})_0, \ \ H^{k+1}(X, \Omega_X^{m-k-1})_0), \tag{2.47}$$

and denote its image by $\Delta_{m,k}$.

Definition 2.25. A smooth projective variety X of dimension n over $\mathfrak{k}$ is called an R-variety if there is $0 \leq m \leq 2n$ and $0 \leq k \leq m$ such that the composition $H^1(X, \Theta_X)_0 \to \Delta \to \Delta_{m,k}$ is an isomorphism of $\mathfrak{k}$-vector spaces. An R-family $Y \to V$ is a family whose fibers are all R-varieties.

We could also consider direct sum of many $\Delta_{m,k}$ in which case we get an even more general definition. For lack of examples, we content ourselves with Definition 2.25. In general, the map (2.47) is not surjective and so the following is a weaker version of the above definition.

Definition 2.26. A smooth projective variety X of dimension n over $\mathfrak{k}$ is called an R-variety if there exists $0 \leq m \leq 2n$ and $0 \leq k \leq m$ such that the following is an isomorphism of $\mathfrak{k}$-vector spaces:

$$H^1(X, \Theta_X)_0 \to \mathrm{Hom}(H^k(X, \Omega_X^{m-k})_0,\ H^{k+1}(X, \Omega_X^{m-k-1})_0). \tag{2.48}$$

The class of R-varieties includes Calabi–Yau varieties in §13.2, Abelian varieties in §11.3 and smooth hypersurfaces of degree d and dimension n with $d|(n+2)$ in §12.4.

2.18 Full Hilbert schemes

Let $V \subset \mathrm{Hilb}_P(\mathbb{P}_{\mathfrak{k}}^N)$ be a Zariski open subset of a Hilbert scheme parameterizing deformations of smooth projective varieties $Y_0 \subset \mathbb{P}_{\mathfrak{k}}^N$ and and let $Y \to V$ be the corresponding family of smooth projective varieties. Recall from §2.11 that the reductive group $\mathbf{G}$ acts from the left on V and its Lie algebra can be interpreted as a $\mathfrak{k}$-vector space of vector fields $\mathsf{v}_{\mathfrak{g}}$ in V. The following property seems to be valid in many interesting cases such as hypersurfaces.

Definition 2.27. A Hilbert scheme $\mathrm{Hilb}_P(\mathbb{P}_{\mathfrak{k}}^N)$ is called full if the Kodaira–Spencer map

$$\mathbf{T}_t V \to H^1(Y_t, \Theta_{Y_t})_0 \tag{2.49}$$

is surjective for all $t \in V$ and its kernel is given by vector fields $\mathsf{v}_{\mathfrak{g},t},\ \ \mathfrak{g} \in \mathrm{Lie}(\mathbf{G})$.

The surjectivity means that we capture all deformations of Y_t within the Hilbert scheme. The assertion about the kernel is also natural because it says that via the Kodaira–spencer map $H^1(Y_t, \Theta_{Y_t})_0$ is identified as a tangent space of the moduli space $\mathbf{G}\backslash V$ at t. Our main example for full Hilbert schemes are parameter spaces of hypersurfaces, see Chapter 12.

Chapter 3

Enhanced Schemes

In one of the seminar programs that we had with the physicists at IAS, my wish was not to have to rely on Ed Witten but instead to be able to make conjectures myself. I failed! I did not understand enough of their picture to be able to do that, so I still have to rely on Witten to tell me what should be interesting (P. Deligne in [RS14, p. 185]).

3.1 Introduction

In this chapter, we introduce projective varieties enhanced with elements in their algebraic de Rham cohomologies. Later, in §3.11 we will discuss the construction of the moduli of such objects. Before doing this we have to analyze enhanced varieties in families and this is one of the main reasons why in this chapter we elaborate the concept of an enhanced projective scheme. One can even do it in the context of stacks, however, for the lack of motivation we avoid this. For a history and the main references in the literature on this topic see the introduction of §3.11. Our main examples are the case of elliptic curves, which is originally treated in the author's lecture notes [Mov12b], and the case of mirror quintic treated in the book [Mov17b].

3.2 A marked projective variety

Recall our terminology of algebraic de Rham cohomology in §2.7. For the definition of enhanced schemes we need to fix a field $\mathfrak{k} \subset \mathsf{R}$ which in most of the cases is going to be $\mathbb{Q}$. For arithmetic purposes it would be essential to proceed with the ring $\mathbb{Z}[\frac{1}{N}]$ for some natural number N, and not the field of rational numbers. In general, we will consider parameter schemes T over $\mathfrak{k}$ for which $\mathsf{T} := \mathrm{Spec}(\mathsf{R})$ is going to be a particular case, see §2.3. We also fix a projective scheme X_0 over $\mathfrak{k}$. From a geometric point of view, see §2.9, we fix a point X_0 in the moduli space M. Frequently, we will need to fix a basis

$$\alpha_{m,i}, \quad m = 0, 1, 2, \ldots, 2n, \quad i = 1, 2, \ldots, \mathsf{b}_m$$

of the free $\mathfrak{k}$-module $H^m_{\mathrm{dR}}(X_0)$. Let us write

$$\alpha_m := \begin{pmatrix} \alpha_{m,1} \\ \alpha_{m,2} \\ \alpha_{m,3} \\ \vdots \\ \alpha_{m,\mathsf{b}_m} \end{pmatrix}.$$

This basis has the following properties:

(1) It is compatible with the Hodge filtration of $H^*_{\mathrm{dR}}(X_0)$.
(2) It is compatible with the Lefschetz decomposition of $H^*_{\mathrm{dR}}(X_0)$. In particular, $\alpha_{2n-m} = \alpha_m \cup \theta^{n-m}, \quad m \leq n$ and $\alpha_{2n} = \theta^n$.

For many examples such as elliptic curves, we will take $\alpha_{2n} = \frac{1}{d}\theta^n$, where d is the degree of X_0. In the complex context, this is equivalent to say that the integration of α_{2n} over X_0 is one.

We write the bilinear form (2.13) in this basis $\alpha_{m,i}$ and define

$$\Phi_m := [\langle \alpha_m, \alpha_m^{\mathsf{tr}} \rangle] = [\langle \alpha_{m,i}, \alpha_{m,j} \rangle], \quad m = 0, 1, \ldots, n. \tag{3.1}$$

In general, we define the $\mathsf{b}_{m_1} \times \mathsf{b}_{m_2}$ matrices $\Phi_{m_1,m_2,i}$ with entries in $\mathfrak{k}$ through the equality

$$[\alpha_{m_1} \cup \alpha_{m_2}^{\mathsf{tr}}] = \sum_{i=1}^{\mathsf{b}_{m_1+m_2}} \Phi_{m_1,m_2,i}\alpha_{m_1+m_2,i}. \tag{3.2}$$

Note that using Hodge blocks, the matrix $\Phi_{m_1,m_2,i}$ has many zero blocks. For instance, $\Phi_{m,2n-m,1}$ is block upper anti-triangular, that is, it has the format

$$\Phi_m = \begin{pmatrix} 0 & 0 & 0 & 0 & * \\ 0 & 0 & 0 & * & * \\ 0 & 0 & * & * & * \\ 0 & * & * & * & * \\ * & * & * & * & * \end{pmatrix}$$

(a sample for $m = n = 5$). It is quite reasonable to take a basis of the de Rham cohomology of the marked point X_0 such that the matrix Φ_m has the simplest form, that is, with many zero entries. This will be done case by case. Sometime our notations such as $H^m_{\mathrm{dR}}(X_0)_0$, carry a subindex 0 which refers to primitive cohomology. We sometimes omit this subindex, being clear in the context with which cohomology we are working with, primitive or usual de Rham cohomology.

3.3 An algebraic group

Recall that we have fixed an algebraic scheme X_0 over the field $\mathfrak{k}$. The algebraic group

$$\mathsf{G} := \mathrm{Aut}(H^*_{\mathrm{dR}}(X_0), F^*_0, \cup, \theta_0)$$

is a group scheme over $\mathfrak{k}$ and plays an important role throughout the present text. By definition, for $m = 0, 1, 2, \ldots, 2n$ we have a b_m-dimensional representation of G. Using Hard Lefschetz theorem and by our choice of the basis of $H^*_{\mathrm{dR}}(X_0)$, b_m and b_{2n-m} dimensional representations are the same. We sometimes fix a basis α_m of $H^m_{\mathrm{dR}}(X_0)$ and write the representation of g as a $\mathsf{b}_m \times \mathsf{b}_m$ block upper triangular matrix:

$$\mathsf{g}_m = \begin{pmatrix} * & * & * & * & * \\ 0 & * & * & * & * \\ 0 & 0 & * & * & * \\ 0 & 0 & 0 & * & * \\ 0 & 0 & 0 & 0 & * \end{pmatrix}$$

(a sample for $m = 4$). The matrix g_m is defined through the equality

$$\mathsf{g}(\alpha_m^{\mathsf{tr}}) = \alpha_m^{\mathsf{tr}} \cdot \mathsf{g}_m. \tag{3.3}$$

It satisfies the following equalities:

$$\mathsf{g}_m^{\mathsf{tr}} \Phi_m \mathsf{g}_m = \Phi_m. \tag{3.4}$$

We have the usual left action of G on $H^*_{\mathrm{dR}}(X_0)$, however, we transform it into the right action by taking dual of $\mathfrak{k}$-vector spaces:

$$H^*_{\mathrm{dR}}(X_0)^\vee \times \mathsf{G} \to H^*_{\mathrm{dR}}(X_0)^\vee, \ (\omega, \mathsf{g}) \mapsto \omega \bullet \mathsf{g}, \tag{3.5}$$

where $\omega \bullet \mathsf{g} \in H^*_{\mathrm{dR}}(X_0)^\vee$ maps $a \in H^*_{\mathrm{dR}}(X_0)$ to $\omega(\mathsf{g}(a))$. We can also see this right action in the equality (3.3). The following proposition will be useful later in the discussion of Hodge cycles. Recall the Hodge block notation of matrices in §2.8.

Proposition 3.1. *For m an even number, the map (2.13) induces a well-defined non-degenerate bilinear map*

$$H^{\frac{m}{2},\frac{m}{2}}(X_0) \times H^{\frac{m}{2},\frac{m}{2}}(X_0) \to \mathfrak{k}$$

and so

$$(\mathsf{g}_m^{\frac{m}{2},\frac{m}{2}})^{\mathsf{tr}} \Phi_m^{\frac{m}{2},\frac{m}{2}} \mathsf{g}_m^{\frac{m}{2},\frac{m}{2}} = \Phi_m^{\frac{m}{2},\frac{m}{2}}. \tag{3.6}$$

Proof. The proof follows from (2.10) and (2.11). It is non-degenerate because (2.13) is so. □

When m is fixed in the context, we sometimes omit the subscript m and, for instance, identify g with g_m. The Lie algebra of G is given by

$$\begin{aligned} \mathrm{Lie}(\mathsf{G}) = \{\mathfrak{g} \in \mathrm{End}(H^*_{\mathrm{dR}}(X_0), F_0^*, \theta_0) \,|\, \mathfrak{g}\alpha \cup \beta + \alpha \cup \mathfrak{g}\beta = 0, \\ \forall \alpha, \beta \in H^*_{\mathrm{dR}}(X_0)\}. \end{aligned} \tag{3.7}$$

For more information on algebraic groups the reader is referred to [Bor91, Bor01, Spr98]. See also [Mur05] for a fast overview of the main results for algebraic groups. In many interesting cases such as elliptic curves and mirror quintic Calabi–Yau threefolds, the algebraic group G is a Borel subgroup of $\mathrm{GL}(N)$.

3.4 Enhanced varieties

Recall our notations in §§2.3 and 2.6 of projective schemes over parameter scheme T, which in turn, is a scheme over the field $\mathfrak{k}$.

Definition 3.1. An enhanced scheme is a pair $(\mathsf{X}/\mathsf{T}, \alpha)$, where X is a smooth projective scheme over T and α is an isomorphism

$$(H^*_{\mathrm{dR}}(\mathsf{X}/\mathsf{T}), F^*, \cup, \theta) \overset{\alpha}{\cong} (H^*_{\mathrm{dR}}(X_0), F_0^*, \cup, \theta_0) \otimes_{\mathfrak{k}} \mathcal{O}_{\mathsf{T}}. \tag{3.8}$$

By (3.8) we mean the following. For each m there is an isomorphism of sheaves $\alpha : H^m_{\rm dR}(\mathsf{X}/\mathsf{T}) \to H^m_{\rm dR}(X_0) \otimes_{\mathfrak{k}} \mathcal{O}_{\mathsf{T}}$ such that

(1) It respects the Hodge filtration, that is, for all p

$$\alpha(F^p H^m_{\rm dR}(\mathsf{X}/\mathsf{T})) = F^p H^m_{\rm dR}(X_0) \otimes_{\mathfrak{k}} \mathcal{O}_{\mathsf{T}}.$$

(2) It respects the cup product, that is,

$$\alpha(\omega_1 \cup \omega_2) = \alpha(\omega_1) \cup \alpha(\omega_2)$$

for all $\omega_i \in H^{m_i}_{\rm dR}(\mathsf{X}/\mathsf{T}), \quad i = 1, 2.$

(3) It sends $\theta \in H^2_{\rm dR}(\mathsf{X}/\mathsf{T})$ to $\theta_0 \otimes 1 \in H^2_{\rm dR}(X_0) \otimes_{\mathfrak{k}} \mathcal{O}_{\mathsf{T}}$.

Note that (3.8) induces isomorphisms in the level of fibers:

$$(H^*_{\rm dR}(X_t),\ F^*_t,\ \cup,\ \theta_t) \overset{\alpha_t}{\cong} (H^*_{\rm dR}(X_0), F^*_0, \cup, \theta_0). \tag{3.9}$$

One usually take X_0 a fiber of $X \to \mathsf{T}$ over $0 \in \mathsf{T}$ and in this way the notation X_0 for a marked projective variety is justified.

One could generalize Definition 3.1 by adding more structure to X, such as torsion point structure in the case of elliptic curves, fixed algebraic cycles in X, morphisms from a fixed variety to X and so on. We will introduce such enhanced schemes case by case. For instance see Chapter 10 for the case in which X is a product of two elliptic curves.

Let us discuss Definition 3.1 in the geometric context, that is, to define enhanced families. The cohomology bundle $H^*_{\rm dR}(\mathsf{X}/\mathsf{T})$ and its Hodge filtration bundle becomes trivial via the map α. Further, we have a global section $\theta \in H^2_{\rm dR}(\mathsf{X}/\mathsf{T})$. From now on, when we talk about an enhanced scheme or family (X, α), we simply write

$$H^*_{\rm dR}(\mathsf{X}/\mathsf{T}) \overset{\alpha}{\cong} H^*_{\rm dR}(X_0) \otimes_{\mathfrak{k}} \mathcal{O}_{\mathsf{T}}$$

instead of (3.8), keeping in mind that α preserves the Hodge filtration, cup product and the polarization.

Definition 3.2. An enhanced projective scheme $(X/\mathsf{T}, \alpha)$ is full if we have an action of the algebraic group G from the right on both X and T such that it commutes with the morphism $\mathsf{X} \to \mathsf{T}$, and it is compatible with the isomorphism (3.8), that is, the induced left action of G on $H^*_{\rm dR}(\mathsf{X}/\mathsf{T})$

$$\mathsf{G}\times H^*_{\rm dR}(\mathsf{X}/\mathsf{T})\to H^*_{\rm dR}(\mathsf{X}/\mathsf{T}),$$
$$(\mathsf{g}\bullet s)(t):=s(t\bullet \mathsf{g})\bullet \mathsf{g}^{-1},\quad \mathsf{g}\in \mathsf{G},\ s\in H^*_{\rm dR}(\mathsf{X}/\mathsf{T}),\quad t\in \mathsf{T}$$

under the isomorphism α is the canonical left action of G on $H^*_{\rm dR}(X_0)\times \mathcal{O}_{\mathsf{T}}$:

$$\mathsf{G}\times \left(H^*_{\rm dR}(X_0)\times \mathcal{O}_{\mathsf{T}}\right)\to \left(H^*_{\rm dR}(X_0)\times \mathcal{O}_{\mathsf{T}}\right),\quad \mathsf{g}\bullet(\omega,f):=(\mathsf{g}(\omega),\mathsf{g}\bullet f).$$

Note that G acts from the right on T and from the left on the space of functions on T and also the sheaf of sections of the cohomology bundle $H^*_{\rm dR}(\mathsf{X}/\mathsf{T})$. In terms of fibers of $X\to\mathsf{T}$, Definition 3.2 says that we have an isomorphism

$$f_t: X_{t\bullet \mathsf{g}}\cong X_t,\quad x\mapsto x\bullet \mathsf{g}^{-1}, \tag{3.10}$$

and the following diagram commutes:

$$\begin{array}{ccc} H^*_{\rm dR}(X_t) & \stackrel{f_t^*}{\to} & H^*_{\rm dR}(X_{t\bullet \mathsf{g}}) \\ \alpha_t\downarrow & & \downarrow\alpha_{t\bullet\mathsf{g}} \\ H^*_{\rm dR}(X_0) & \stackrel{\mathsf{g}}{\to} & H^*_{\rm dR}(X_0). \end{array}$$

Therefore, we have an isomorphism of enhanced schemes over k

$$(X_{t\bullet\mathsf{g}},\alpha_{t\bullet\mathsf{g}})\cong (X_t,\mathsf{g}\circ\alpha_t),\ t\in\mathsf{T},\ \mathsf{g}\in\mathsf{G}, \tag{3.11}$$

where $\circ$ is the usual composition of functions.

Example 3.1. Our main example of full enhanced schemes is the following three parameter family of elliptic curves. This is the main ingredient of the theory of quasi-modular forms in [Mov12b]:

$$X\ :\ y^2-4(x-t_1)^3+t_2(x-t_1)+t_3=0,$$
$$\mathsf{T}:=\mathrm{Spec}\left(\mathsf{k}\left[t_1,t_2,t_3,\frac{1}{27t_3^2-t_2^3}\right]\right).$$

Note that X is written in the affine coordinate (x,y). The algebraic group G is

$$\mathsf{G}=\left\{\begin{bmatrix} k & k' \\ 0 & k^{-1}\end{bmatrix}\ \middle|\ k'\in\mathsf{k}, k\in\mathsf{k}-\{0\}\right\} \tag{3.12}$$

and its action on X is given by

$$(x,y,t_1,t_2,t_3)\bullet\mathsf{g}:=(k^2x-k'k,\ \ k^3y,\ \ t_1k^{-2}+k'k^{-1},\ \ t_2k^{-4},\ \ t_3k^{-6}),$$

for more details see [Mov12b, Proposition 6.1] and Chapter 9.

Definition 3.3. A morphism $(X_1/\mathsf{T}_1, \alpha) \to (X_2/\mathsf{T}_2, \beta)$ of two enhanced projective schemes is a commutative diagram

$$\begin{array}{ccc} X_1 & \to & X_2 \\ \downarrow & & \downarrow \\ \mathsf{T}_1 & \to & \mathsf{T}_2 \end{array}$$

such that

$$\begin{array}{ccc} H^*_{\mathrm{dR}}(X_2/\mathsf{T}_2) & \to & H^*_{\mathrm{dR}}(X_1/\mathsf{T}_1) \\ \downarrow & & \downarrow \\ H^*_{\mathrm{dR}}(X_0) \otimes_{\mathfrak{k}} \mathcal{O}_{\mathsf{T}_2} & \to & H^*_{\mathrm{dR}}(X_0) \otimes_{\mathfrak{k}} \mathcal{O}_{\mathsf{T}_1} \end{array}$$

is also commutative.

Remark 3.1. The most similar concept to our enhanced varieties is the notion of frame bundle used in topology, see [Ham17, §4.4]. The compatibility of $\alpha_{m,i}$'s with the Hodge filtration and the constancy of the cup product make our notion much finer than the notion of frame bundle. Moreover, note that due to the automorphisms of projective varieties the projection $\mathsf{T} \to \mathsf{T}/\mathsf{G}$ is not necessarily a fiber bundle.

3.5 Weakly enhanced varieties

In this section, we introduce the content of §§3.3 and 3.4 removing the cup product structure. The main reason for this is that many geometric problems related to Hodge loci do not need the cup product structure of de Rham cohomologies. However, for the introduction of geometric automorphic forms and topological string partition functions the cup product structure becomes an essential ingredient. The new notations reproduced in this section are obtained by putting tilde on the old notations.

Definition 3.4. We define the algebraic group

$$\tilde{\mathsf{G}} := \mathrm{Aut}(H^*_{\mathrm{dR}}(X_0), F^*_0, \theta_0)$$

and in a canonical way we have b_m-dimensional representations of $\tilde{\mathsf{G}}$ given by (3.3). A weakly enhanced scheme $(\tilde{X}/\tilde{\mathsf{T}}, \alpha)$ is given by

$$(H^*_{\mathrm{dR}}(\tilde{X}/\tilde{\mathsf{T}}), F^*, \theta) \overset{\alpha}{\simeq} (H^*_{\mathrm{dR}}(X_0), F^*_0, \theta_0) \otimes_{\mathfrak{k}} \mathcal{O}_{\tilde{\mathsf{T}}}. \tag{3.13}$$

In a similar way as in Definition 3.2 we can define full weakly enhanced schemes.

We have many functions on the space $\tilde{\mathsf{T}}$ that can be constructed as follows. Similar to the case of enhanced varieties, we fix a basis of $H^*_{\mathrm{dR}}(X_0)$ as in §3.2. The pull-back of this basis by the isomorphism α gives us global sections of $H^*_{\mathrm{dR}}(\tilde{X}/\tilde{\mathsf{T}})$. We denote it again by $\alpha_{m,i}$. The cup product in $H^*_{\mathrm{dR}}(\tilde{X}/\tilde{\mathsf{T}})$ is no more constant. We write the equalities (3.2) and we get matrices $\Phi_{m_1,m_2,i}$ whose entries are functions on the space $\tilde{\mathsf{T}}$. Let

$$f : \tilde{\mathsf{T}} \to \mathbb{A}^s_{\mathfrak{k}} \tag{3.14}$$

be the map given by all such functions. Let us assume that we have a point $p_0 \in \tilde{\mathsf{T}}$ such that the fiber of $\tilde{X}/\tilde{\mathsf{T}}$ over p_0 is X_0 and α in (3.13) induces the identity map in $(H^*_{\mathrm{dR}}(X_0), F^*_0, \theta_0)$.

Proposition 3.2. *Let* $\mathsf{T} := f^{-1}(f(p_0))$. *We have a canonical enhanced family* X/T *such that the following diagram commutes:*

$$\begin{array}{ccc} \mathsf{T} & \hookrightarrow & \tilde{\mathsf{T}} \\ \uparrow & & \uparrow \\ X & \hookrightarrow & \tilde{X} \end{array}$$

Moreover, if $\tilde{X}/\tilde{\mathsf{T}}$ *is full then* X/T *is also full.*

Proof. We define $\mathsf{X} := \pi^{-1}(\mathsf{T})$ and X/T is the desired enhanced family. □

The map (3.14) is a morphism of algebraic schemes over $\mathfrak{k}$ whose fibers are either empty or enhanced families of projective schemes (with possibly different marked varieties X_0). If there is no danger of confusion, we will drop tilde sign from our notations above; being clear which we mean: enhanced or weakly enhanced case. In particular, the algebraic group G will be either $\tilde{\mathsf{G}}$ (weakly enhanced case) or G in §3.3.

3.6 Constructing enhanced schemes

Let $\pi : Y \to V$ be a family of smooth projective varieties defined over $\mathfrak{k}$. In this section, we construct a family $\tilde{\mathsf{X}} \to \tilde{\mathsf{T}}$ of (weakly) enhanced projective schemes using π. This is done by adding additional parameters apart from those in V. Our main example for $\pi : Y \to V$ comes from an irreducible component of a Hilbert scheme and the corresponding family of projective varieties. We have to remove singular fibers in order to get $Y \to V$. We would like to construct the total space $\tilde{\mathsf{T}}$ of all the basis of the de Rham cohomology bundles $H^*_{\mathrm{dR}}(Y_t)$, $t \in V$ compatible with the

Hodge filtration. Once the variety $\tilde{\mathsf{T}}$ over $\mathfrak{k}$ and the canonical projection $\tilde{\mathsf{T}} \to V$ is constructed, $\tilde{\mathsf{X}}$ is the fiber product of $Y \to V$ and $\tilde{\mathsf{T}} \to V$. We give explicit construction of affine charts for $\tilde{\mathsf{T}}$ and $\tilde{\mathsf{X}}$.

Around any point of V we can find a Zariski open neighborhood V^k and global sections $\omega_m^k = [\omega_{m,1}^k, \omega_{m,2}^k, \ldots, \omega_{m,\mathsf{b}_m}^k]^{\mathsf{tr}}$ of the relative de Rham cohomology sheaf of Y/V such that ω_m^k at each fiber $H_{\mathrm{dR}}^*(Y_t)$, $t \in V^k$ form a basis compatible with the Hodge filtration. Let $S_m^k = [S_{m,ij}^k]$ be a Hodge block lower triangular $\mathsf{b}_m \times \mathsf{b}_m$ matrix with unknown coefficients $S_{m,ij}^k$. We consider $S_{m,ij}^k$ as variables and define

$$U^k := \mathrm{Spec}\left(\mathfrak{k}\left[S_{m,ij}^k, \frac{1}{\det(S_m^k)}, \ m = 0, 1, 2, \ldots, 2n, \ i, j = 1, 2, \ldots, \mathsf{b}_m\right]\right).$$

The variety U^k is a Zariski open subset of $\mathbb{A}_{\mathfrak{k}}^N$, where

$$N = \sum_{m=0}^{2n} \frac{1}{2}(\mathsf{b}_m^2 + \sum_i (\mathsf{h}^{m-i,i})^2).$$

We consider the morphism of schemes $\tilde{\mathsf{X}}_k \to \tilde{\mathsf{T}}_k$ over $\mathfrak{k}$, where

$$\begin{aligned} \tilde{\mathsf{X}}^k &:= \pi^{-1}(V^k) \times_{\mathfrak{k}} U^k, \\ \tilde{\mathsf{T}}^k &:= V^k \times_{\mathfrak{k}} U^k \\ &= \mathrm{Spec}\left(\mathcal{O}_V(V^k)\left[S_{m,ij}^k, \frac{1}{\det(S_m^k)}, \ m = 0, 1, 2, \ldots, 2n, \right.\right. \\ &\qquad \left.\left. i, j = 1, 2, \ldots, \mathsf{b}_m\right]\right). \end{aligned}$$

It is obtained from $\pi : Y \to V$ and the identity map $U^k \to U^k$. We also define $\alpha^k = \{\alpha_m^k, \ m = 0, 1, \ldots, 2n\}$ by

$$\alpha_m^k := S_m^k \cdot \omega_m^k, \quad m = 0, 1, \ldots, 2n \tag{3.15}$$

and we get a full family $\tilde{\mathsf{X}}^k/\tilde{\mathsf{T}}^k$ of weakly enhanced projective varieties with α^k as global sections of the de Rham cohomology bundle of $\tilde{\mathsf{X}}^k \to \tilde{\mathsf{T}}^k$.

Now, the next step is to cover V with local charts V^k, $k \in I$ and get local charts $\tilde{\mathsf{T}}^k$, $k \in I$ for $\tilde{\mathsf{T}}$, and $\tilde{\mathsf{X}}^k$, $k \in I$ for $\tilde{\mathsf{X}}$, respectively. For each fixed $k \in I$, we have a collection of global sections α^k, $k \in I$ of $H_{\mathrm{dR}}^*(\tilde{\mathsf{X}}^k/\tilde{\mathsf{T}}^k)$. The gluing of $\tilde{\mathsf{T}}^k$'s and $\tilde{\mathsf{X}}^k$ in V^k and $\pi^{-1}(V^k)$ factors is just the usual one coming from the family $Y \to V$. In other factors, it is done by assuming that the global sections α^k in their common domains are equals. More precisely, if in

V^{k_1} and V^{k_2} we have taken global sections ω^{k_1} and ω^{k_2} then in $V^{k_1} \cap V^{k_2}$ we have $\omega_m^{k_1} = B_m^{k_1 k_2 j} \omega_m^{k_2}$ and the gluing in U^{k_1} and U^{k_2} factors is done by

$$S_m^{k_1} B_m^{k_1 k_2} = S_m^{k_2}$$

which amounts to say that $\alpha^{k_1} = \alpha^{k_2}$. The morphisms $\tilde{\mathsf{X}}^k \to \tilde{\mathsf{T}}^k$ and global sections α^k glue to each other to give us $\tilde{\mathsf{X}} \to \tilde{\mathsf{T}}$ and global sections α of $H_{\rm dR}^*(\tilde{\mathsf{X}}/\tilde{\mathsf{T}})$.

The construction of an enhanced family $\mathsf{X} \to \mathsf{T}$ from $\pi : Y \to V$ is similar and it is as follows. In this case $V^k \times_{\mathfrak{k}} U^k$ must be replaced with

$$\mathrm{Spec}\left(\frac{\mathcal{O}_V(V^k)[S^k_{m,ij}, \frac{1}{\det(S^k_m)}, \ m = 0,1,2,\ldots,2n, \ i,j = 1,2,\ldots,\mathsf{b}_m]}{\begin{array}{l}\langle [\alpha^k_{m_1} \cup (\alpha^k_{m_2})^{\mathsf{tr}}] = \sum_{i=1}^{\mathsf{b}_{m_1+m_2}} \Phi_{m_1,m_2,i}\alpha^k_{m_1+m_2,i}, \\ m_1, m_2 = 0,1,2,\ldots,2n\rangle\end{array}}\right).$$

The ideal in the denominator is given by comparing the coefficients of ω_i's in both sides of the equalities written between $\langle$ and $\rangle$. These equalities come from (3.2). In other words, there are many algebraic relations between the entries of S^k_m's and with coefficients in $\mathcal{O}_V(V^k)$ and we have to work modulo these relations. The following theorem is the outcome of the above construction.

Theorem 3.3. *Let $\pi : Y \to V$ be a morphism of projective schemes as in §2.6. We have a commutative diagram*

$$\begin{array}{ccc} \mathsf{X} & \to & \mathsf{T} \\ \downarrow & & \downarrow \\ Y & \to & V \end{array} \tag{3.16}$$

of projective schemes defined over $\mathfrak{k}$ such that

(1) *X is the fiber product of $\mathsf{T} \to V$ and $Y \to V$.*
(2) *$\mathsf{X} \to \mathsf{T}$ is a full family of enhanced projective varieties, and hence, there is an action of G on both X and T which commutes with $\mathsf{X} \to \mathsf{T}$.*
(3) *The universal geometric quotients $\mathsf{X}/\tilde{\mathsf{G}}$ and T/G exists as schemes over $\mathfrak{k}$ and we have isomorphisms $\mathsf{X}/\mathsf{G} = Y$ and $\mathsf{T}/\mathsf{G} = V$ such that*

$$\begin{array}{ccc} \mathsf{X}/\mathsf{G} & \to & \mathsf{T}/\mathsf{G} \\ \| & & \| \\ Y & \to & V \end{array} \tag{3.17}$$

commutes, that is, the induced map $\mathsf{X}/\mathsf{G} \to \mathsf{T}/\mathsf{G}$ is just $Y \to V$.

(4) *The action of* G *on* T *and* X *is free and its orbits are given by the fibers of* $\mathsf{T} \to V$ *and* $\mathsf{X} \to Y$*, respectively.*

The same is true replacing T *with* $\tilde{\mathsf{T}}$ *and enhanced with weakly enhanced.*

Remark 3.2. Sometimes it is more convenient to redefine the matrix S_m^k to be its inverse, and hence, α_m^k in (3.15) is given by $\alpha_m^k = (S_m^k)^{-1}\omega_m$. In this way, the equality $[\langle \alpha_{m,i}^k, \alpha_{m,j}^k \rangle] = \Phi_m$ turns out to be

$$S_m^k \Phi_m (S_m^k)^{\mathsf{tr}} = [\langle \omega_{m,i}^k, \omega_{m,j}^k \rangle].$$

The entries of the left-hand side are quadratic polynomials in the entries of S_m^k and the entries of the right-hand side are in $\mathcal{O}_V(V^k)$.

3.7 Enhanced families with an action of a reductive group

For constructing moduli spaces we need to add a new ingredient to the discussion in §3.6. That is, we let an algebraic group $\mathbf{G}$ (mainly reductive) act on V from the left so that $\mathbf{G}\backslash V$ is a classical moduli of projective varieties X. In this section, we do this and we add more data to Theorem 3.3. A typical example in our mind is the family of hypersurfaces of degree d in $\mathbb{P}^{n+1}$ and the corresponding action of $\mathbf{G} := \mathrm{GL}(n+1)$ induced by the linear action of $\mathbf{G}$ in $\mathbb{P}^{n+1}$. This will be discussed in more details in Chapter 12.

Definition 3.5. An enhanced (or weakly enhanced) family $(\mathsf{X}/\mathsf{T}, \alpha)$ is equipped with a left action of a (reductive) group $\mathbf{G}$ if:

(1) The group $\mathbf{G}$ acts on the morphism $\mathsf{X} \to \mathsf{T}$ of $\mathfrak{k}$-schemes, see Definition 2.15.
(2) The action of G and $\mathbf{G}$ on both X and T are independent from each other, see Definition 2.16.
(3) The induced action of $\mathbf{G}$ on $H_{\mathrm{dR}}^*(\mathsf{X}/\mathsf{T})$ under the isomorphism α in (3.8) is the identity in $H_{\mathrm{dR}}^*(X_0)$ times the action of $\mathbf{G}$ on T. In geometric terms, this means that for all $t \in \mathsf{T}$ and $\mathbf{g} \in \mathbf{G}$, the following diagram commutes:

$$\begin{array}{ccc} H_{\mathrm{dR}}^*(\mathsf{X}_t) & & \\ & \searrow^{\alpha_t} & \\ \mathbf{g}^*\uparrow & & H_{\mathrm{dR}}^*(X_0) \\ & \nearrow_{\alpha_{\mathbf{g}\cdot t}} & \\ H_{\mathrm{dR}}^*(\mathsf{X}_{\mathbf{g}\cdot t}) & & \end{array} \tag{3.18}$$

where we have the isomorphism $\mathsf{X}_t \to \mathsf{X}_{\mathbf{g}\cdot t}$, $x \mapsto \mathbf{g}\cdot x$ and $\mathbf{g}^*$ is the induced map in de Rham cohomologies.

The following is the continuation of Theorem 3.3.

Theorem 3.4. *Let $\pi : Y \to V$ be a family of smooth projective schemes defined over $\mathfrak{k}$ and let $\mathbf{G}$ be a group scheme which acts on Y/V from the left. Let also $\mathsf{X} \to \mathsf{T}$ be as in Theorem 3.3. We have a left action of $\mathbf{G}$ on $(\mathsf{X}/\mathsf{T}, \alpha)$ in the sense of Definition 3.5. Moreover, the action of $\mathbf{G}$ commutes with the four maps in (3.16). The same is also true for the weakly enhanced family $\tilde{\mathsf{X}} \to \tilde{\mathsf{T}}$ in Theorem 3.3.*

Proof. The proof is the continuation of the proof of Theorem 3.3. Let $\mathbf{g} : Y_t \to Y_{\mathbf{g}\cdot t}$ be the isomorphism induced by the action of $\mathbf{g} \in \mathbf{G}$ and

$$\mathbf{g}^* : H^*_{\mathrm{dR}}(Y_{\mathbf{g}\cdot t}) \to H^*_{\mathrm{dR}}(Y_t) \tag{3.19}$$

be the induced map in de Rham cohomologies. A point of T is given by (t, α), where $t \in V$ and α is a basis of $H^*_{\mathrm{dR}}(Y_t)$. The action of $\mathbf{G}$ on $\tilde{\mathsf{T}}$ is given by

$$\mathbf{g}\cdot(t, \alpha) := (\mathbf{g}\cdot t, (\mathbf{g}^*)^{-1}\alpha),$$

see Fig. 3.1. Since $\tilde{\mathsf{X}}$ is the fiber product of $Y \to V$ and $\mathsf{T} \to V$ and the action of $\mathbf{G}$ commutes with both maps, we have the action of $\mathbf{G}$ in $\tilde{\mathsf{X}}$ in a canonical way. By definition, the pairs (X_t, α_t) and $(\mathsf{X}_{\mathbf{g}\cdot t}, \alpha_{\mathbf{g}\cdot t})$ are isomorphic. Note that $(\mathbf{g}^*)^{-1} = (\mathbf{g}^{-1})^*$ is a linear map and so it commutes with the action of G on α and hence we have (2.21).

It is useful to describe the action of $\mathbf{G}$ in local charts. Recall the local chart V^k and sections ω^k_m used in the proof of Theorem 3.3. We write

$$\mathbf{g}^*\omega^k_m = B_{m,\mathbf{g}}\omega^k_m,$$

where $\mathbf{g}^*$ is the map in (3.19) and $B_{m,\mathbf{g}}$ is a $\mathsf{b}_m \times \mathsf{b}_m$ matrix with entries in $\mathcal{O}_{V,t}$. For simplicity, we have assumed that t, $\mathbf{g}\cdot t$ are in the same chart V^k. The action of $\mathbf{g} \in \mathbf{G}$ in $\tilde{\mathsf{X}}^k := \pi^{-1}(V^k) \times U^k$ and $\tilde{\mathsf{T}}^k := V^k \times U^k$ is given by

$$\mathbf{g}\cdot(x, S) = (\mathbf{g}\cdot x, S^k_m B^{-1}_{m,\mathbf{g}}),$$

where x is either in V^k or $\pi^{-1}(V^k)$. This does not depend on the chosen charts $\tilde{\mathsf{T}}^k$ and $\tilde{\mathsf{X}}^k$. □

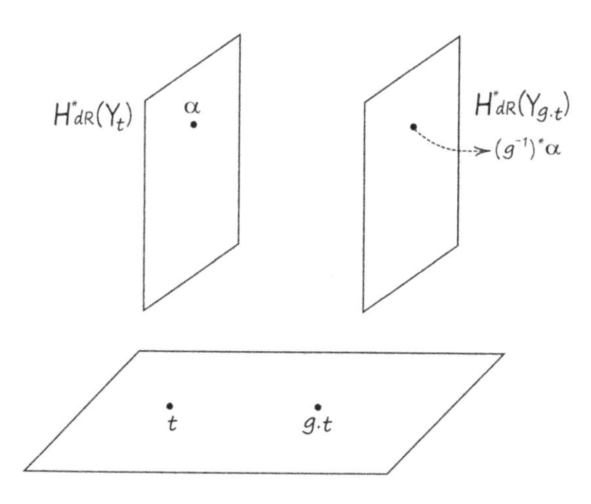

Fig. 3.1. An identification.

For the construction of the moduli of enhanced schemes the following will play a crucial role.

Theorem 3.5. *Let $\pi : Y \to V$ be a family of smooth projective schemes defined over $\mathfrak{k}$ and let $\mathbf{G}$ be a group scheme which acts on Y/V from the left. Let also $\mathsf{X} \to \mathsf{T}$ be as in Theorem 3.3. If a point $t \in V$ is semistable (resp. stable) for the action of $\mathbf{G}$, then all the points in the fiber of $\mathsf{T} \to V$ over t are also semistable (resp. stable) for the action of $\mathbf{G}$.*

Proof. By construction $\pi : \mathsf{T} \to V$ is a finite morphism of $\mathfrak{k}$-schemes, that is, there is a covering of V by open affine subsets $V^k = \text{Spec}(\mathsf{R}_k)$, such that $\pi^{-1}(V^k) = \text{Spec}(\check{\mathsf{R}}_k)$ is affine and $\check{\mathsf{R}}_k$ is a finitely generated R_k-algebra. The theorem follows from Proposition 2.9. □

The scheme T might have points which are not mapped to any stable or semistable point of V, and it is highly recommended to study such points in T without referring to available results in V. We will use Theorem 3.5 in order to construct the moduli of enhanced varieties in cases where the classical moduli spaces are constructed.

3.8 Gauss–Manin connection

Recall the definition of Gauss–Manin connection in §2.15. Let us consider an enhanced family $\mathsf{X} \to \mathsf{T}$ of smooth projective varieties. In this section, we fix $1 \leq m \leq 2n$ and work with the mth de Rham cohomology $H^m(\mathsf{X}/\mathsf{T})$.

We denote by

$$\nabla : H^m_{\rm dR}(\mathsf{X}/\mathsf{T}) \to \Omega^1_{\mathsf{T}} \otimes_{\mathcal{O}_{\mathsf{T}}} H^m_{\rm dR}(\mathsf{X}/\mathsf{T}), \quad m = 0, 1, \ldots, 2n$$

the algebraic Gauss–Manin connection of the family $\mathsf{X} \to \mathsf{T}$. By our definition of enhanced varieties we have automatically global sections $\alpha_{m,i}$, $i = 1, 2, \ldots, \mathsf{b}_m$ of the free $\mathcal{O}_{\mathsf{T}}$-module sheaf $H^m_{\rm dR}(\mathsf{X}/\mathsf{T})$ such that for any closed point $t \in \mathsf{T}$ they form a basis of $H^m(\mathsf{X}_t)$. Let

$$\alpha_m := \left[\alpha_{m,1},\ \alpha_{m,1}, \ldots, \alpha_{m,\mathsf{b}_m}\right]^{\mathsf{tr}}.$$

Definition 3.6. We can write the Gauss–Manin connection in the basis α_m:

$$\nabla(\alpha_m) = \mathsf{A}_m \otimes \alpha_m.$$

Here, A_m is a $\mathsf{b}_m \times \mathsf{b}_m$ matrix with entries which are global sections of Ω^1_{T}. We call it the mth Gauss–Manin connection matrix.

By Griffiths transversality we have

$$\mathsf{A}_m^{i,j} = 0, \quad j \geq i + 2 \tag{3.20}$$

that is, it is of the form

$$\mathsf{A}_m = \begin{pmatrix} * & * & 0 & 0 & 0 \\ * & * & * & 0 & 0 \\ * & * & * & * & 0 \\ * & * & * & * & * \\ * & * & * & * & * \end{pmatrix}$$

(a sample for $m = 4$), where we have used Hodge blocks notation for a matrix, see §2.8. Recall ∇_i's in (2.29). We have

$$\nabla_1 \circ \nabla_0(\alpha_m) = \nabla_1(\mathsf{A}_m \alpha_m) = d\mathsf{A}_m \otimes \alpha_m - \mathsf{A}_m \wedge \nabla \alpha_m = (d\mathsf{A}_m - \mathsf{A}_m \wedge \mathsf{A}_m) \otimes \alpha_m.$$

Since the Gauss–Manin connection is integrable, we have $\nabla_1 \circ \nabla_0 = 0$, and so

$$d\mathsf{A}_m = \mathsf{A}_m \wedge \mathsf{A}_m. \tag{3.21}$$

Proposition 3.6. *Let* g *be an element of* G. *We have*

$$\mathsf{g}^*\mathsf{A}_m = \mathsf{g}_m^{\mathsf{tr}} \cdot \mathsf{A}_m \cdot \mathsf{g}_m^{-\mathsf{tr}},$$

that is, the pull-back of the Gauss–Manin connection matrix A_m *under the isomorphism* $\mathsf{g} : \mathsf{T} \to \mathsf{T}$, $t \mapsto t \bullet \mathsf{g}$ *is* $\mathsf{g}_m^{\mathsf{tr}} \cdot \mathsf{A}_m \cdot \mathsf{g}_m^{-\mathsf{tr}}$.

Proof. Since g is considered to be constant , we have

$$\nabla(\mathsf{g}_m^{\mathsf{tr}}\alpha_m) = \mathsf{g}_m^{\mathsf{tr}}\nabla\alpha_m = (\mathsf{g}_m^{\mathsf{tr}}\mathsf{A}_m\mathsf{g}_m^{-\mathsf{tr}})(\mathsf{g}_m^{\mathsf{tr}}\alpha_m).$$

From another side the pair $(X_{t\bullet \mathsf{g}}, \alpha_m)$ is, by definition, isomorphic to $(X_t, \mathsf{g}_m^{\mathsf{tr}}\alpha_m)$, see (3.11). □

The Gauss–Manin connection and cup product satisfy the following equalities:

$$\nabla(\alpha_1 \cup \alpha_2) = \nabla(\alpha_1) \cup \alpha_2 + \alpha_1 \cup \nabla(\alpha_2), \quad \alpha \in H^{m_i}_{\mathrm{dR}}(\mathsf{X}/\mathsf{T}), \quad i = 1, 2.$$

The polarization $\theta \in H^2_{\mathrm{dR}}(\mathsf{X}/\mathsf{T})$ is flat in the sense that $\nabla\theta = 0$ and hence

$$\nabla(\theta^m) = 0, \quad 0 \le m \le n.$$

Recall the constant matrix Φ_m in (3.1). There are some natural $\mathfrak{k}$-linear relations between the entries of A_m that we introduce them below.

Proposition 3.7. *The Gauss–Manin connection matrix* A_m *satisfy*

$$\mathsf{A}_m\Phi_m + \Phi_m\mathsf{A}_m^{\mathsf{tr}} = 0. \tag{3.22}$$

Proof. The proposition follows after taking ∇ from the equality $\Phi_m = [\langle\alpha_m, \alpha_m^{\mathsf{tr}}\rangle]$. □

Note that by our choice of the basis α, we have $\mathsf{A}_{2n-m} = \mathsf{A}_m$. The whole discussion of this section, except Proposition 3.22, can be done for weakly enhanced families.

Let us describe the Gauss–Manin connection matrix in the local chart $\tilde{\mathsf{T}}^k$ of the variety $\tilde{\mathsf{T}}$ constructed in Theorem 3.3. For simplicity, we remove the upper index k from our notations.

Proposition 3.8. *The* m*th Gauss–Manin connection matrix of* $\tilde{\mathsf{X}}/\tilde{\mathsf{T}}$ *in a local chart* $\tilde{\mathsf{T}}^k$ *constructed in Theorem 3.3 is given by*

$$\tilde{\mathsf{A}}_m = dS_m \cdot S_m^{-1} + S_m \cdot B_m \cdot S_m^{-1},$$

where B_m *is the Gauss–Manin connection matrix in the basis* ω_m.

Proof. This follows from the construction of the global sections α in (3.15) and the Leibniz rule. □

3.9 Gauss–Manin connection and reductive group

Recall our notations of an action of a reductive group $\mathbf{G}$ on the domain and image of the morphism $\mathsf{X} \to \mathsf{T}$ introduced in §3.7.

Proposition 3.9. *Let* $\mathbf{g}$ *be an element of* $\mathbf{G}$. *The pull-back of the Gauss–Manin connection matrix* A_m *under the isomorphism* $\mathsf{T} \to \mathsf{T}$, $t \mapsto \mathbf{g}{\cdot}t$ *is* A_m *itself, that is, the entries of* A_m *are invariant under the action of* $\mathbf{G}$. *In particular, for any* $\mathfrak{g} \in \mathrm{Lie}(\mathbf{G})$ *we have* $\mathsf{A}_{\mathsf{v}_\mathfrak{g}} = 0$.

Proof. This follows from the equality $\nabla\alpha_m = \mathsf{A}_m\alpha_m$ and the fact that for the enhanced family $\mathsf{X} \to \mathsf{T}$, the global sections α are $\mathbf{G}$-invariant. □

As a corollary of Proposition 3.8 we get the following proposition.

Proposition 3.10. *If the geometric quotient* $\check{\mathsf{T}} := \mathbf{G}\backslash\mathsf{T}$ *exists then we have matrices* $\check{\mathsf{A}}_m$ *whose entries are global differential* 1*-forms in* $\check{\mathsf{T}}$, *and such that the pull-back of* $\check{\mathsf{A}}_m$ *under the canonical map* $\mathsf{T} \to \check{\mathsf{T}}$ *is* A_m.

We also call $\check{\mathsf{A}}_m$'s the Gauss–Manin connection matrices of $\mathbf{G}\backslash\mathsf{X} \to \mathbf{G}\backslash\mathsf{T}$, however, note that we do not claim that $\mathbf{G}\backslash\mathsf{X}$ as a scheme over $\mathfrak{k}$ exists. In general, universal moduli spaces are rare, and most of the time we have only coarse moduli spaces. However, in our context of enhanced varieties, it seems that the existence of $\check{\mathsf{T}}$ implies the existence of the corresponding family over $\check{\mathsf{T}}$.

3.10 Marked projective scheme

One of the great, and in the same time simple, discoveries in Hodge theory due to A. Grothendieck and P. Deligne in the sixties was that the Hodge decomposition cannot be defined in the framework of Algebraic Geometry over an arbitrary field, see [Gro66, Del71a, Del74]. However, the Hodge filtration can be defined. In the passage from Hodge decomposition to Hodge filtrations one loses harmonic forms for the sake of defining objects by polynomials. This motivated many other cohomology theories, such as étale and crystalline cohomologies, in Algebraic Geometry. It turns out that for some special varieties the Hodge decomposition is also defined over a base field. These varieties have usually so many automorphisms such that all their periods up to a power of $2\pi i$ factor are algebraic numbers. In this section, we explain this idea. Later, we will give an application of this topic in the codimension of modular foliations.

Definition 3.7. Let X be a projective variety defined over a field $\mathfrak{k}$ of characteristic zero. We say that X has the Hodge decomposition defined over $\bar{\mathfrak{k}}$ if its algebraic de Rham cohomologies over $\bar{\mathfrak{k}}$ can be written as directs sums

$$H^m_{\mathrm{dR}}(X/\bar{\mathfrak{k}}) = H^{m,0} \oplus H^{m-1,1} \oplus \cdots \oplus H^{0,m}$$

for all $m = 0, 1, 2, \ldots, 2n$ such that for any embedding $\bar{\mathfrak{k}} \hookrightarrow \mathbb{C}$ of fields it becomes the Hodge decomposition of the de Rham cohomologies of $X_{\mathbb{C}} := X \times_{\bar{\mathfrak{k}}} \mathbb{C}$.

If X is defined over $\mathfrak{k}$ and enjoys the above property, its Hodge decomposition might be defined over an extension of $\mathfrak{k}$. This is why we have to consider $\bar{\mathfrak{k}}$ in the decomposition. Note that the complex conjugation in de Rham cohomologies which maps $H^{p,q}$ to $H^{q,p}$ depends on the embedding $\bar{\mathfrak{k}} \subset \mathbb{C}$ and we do not (or cannot) insert it inside the above definition.

Recall our convention of Hilbert schemes in §2.11. For the purpose of the present text we need the following property:

Property 3.1. Any irreducible component of a Hilbert scheme of projective varieties has a point X_0 with Hodge decompositions defined over $\bar{\mathfrak{k}}$.

In the above property, we are only considering components of Hilbert schemes whose generic point parametrizes smooth projective varieties. We know that Fermat varieties have the Hodge decomposition defined over $\bar{\mathfrak{k}}$, see Proposition 12.1. The same is expected to be true for CM principally polarized abelian varieties, see 11.3.

Proposition 3.11. *If Property 3.1 is valid then we can choose the matrices $\Phi_{m_1,m_2,i}$ in (3.2) such that the only possibly non-zero Hodge blocks of $\Phi_{m_1,m_2,i}$ are*

$$(\Phi_{m_1,m_2,i})^{i_1,i_2}, \quad (m_1 + m_2 - i_1 - i_2, i_1 + i_2) = \mathrm{Type}(\alpha_{m_1,m_2,i}).$$

These are matrices with one line of possibly non-zero blocks parallel to the anti-diagonal. In particular, Φ_m is anti-diagonal with respect to the Hodge *blocks, that is*

$$\Phi_m = \begin{pmatrix} 0 & 0 & \cdots & 0 & \Phi_m^{0,m} \\ 0 & 0 & \cdots & \Phi_m^{1,m-1} & 0 \\ \vdots & \vdots & \iddots & \vdots & \vdots \\ 0 & \Phi_m^{m-1,1} & 0 & 0 & 0 \\ \Phi_m^{m,0} & 0 & 0 & 0 & 0 \end{pmatrix} \tag{3.23}$$

with $\Phi_m^{m-i,i} = (-1)^m \Phi_m^{i,m-i}$.

Proof. The marked point X_0 is going to be the one with the Hodge decomposition defined over $\bar{\mathfrak{k}}$. We replace $\mathfrak{k}$ with its finite extension such that this property holds. Now, we take a basis of $H^*_{\mathrm{dR}}(X_0)$ compatible with the Hodge decomposition. This basis is automatically compatible with the Hodge filtration. The cup product in this basis gives us the desired format of matrices. □

In practice, we will take the following matrices:

$$\Phi_m = \begin{pmatrix} 0 & 0 & \cdots & 0 & I \\ 0 & 0 & \cdots & I & 0 \\ \vdots & \vdots & \iddots & \vdots & \vdots \\ 0 & I & 0 & 0 & 0 \\ I & 0 & 0 & 0 & 0 \end{pmatrix}, \quad \Phi_m = \begin{pmatrix} 0 & 0 & \cdots & 0 & I \\ 0 & 0 & \cdots & I & 0 \\ \vdots & \vdots & \iddots & \vdots & \vdots \\ 0 & -I & 0 & 0 & 0 \\ -I & 0 & 0 & 0 & 0 \end{pmatrix} \tag{3.24}$$

for m an even and odd number, respectively, where I is the identity matrix of size compatible with our Hodge blocks notations.

Let us consider an enhanced scheme $(\mathsf{X}/\mathsf{T}, \alpha)$, where α is the isomorphism in (3.8). We further assume that there is a point $0 \in \mathsf{T}$ such that the fiber X_0 of $\mathsf{X} \to \mathsf{T}$ over 0 has Hodge decomposition defined over $\bar{\mathfrak{k}}$.

Definition 3.8. If Property 3.1 is valid then we define

$$\mathsf{T}_{\mathrm{real}} := \{(X, \alpha) \in \mathsf{T}(\mathbb{C}) \mid \alpha \text{ is an isomorphism between Hodge decompositions}\}.$$

The set $\mathsf{T}_{\mathrm{real}}$ lives in the complex manifold $\mathsf{T}(\mathbb{C})$ and it is neither algebraic nor complex analytic subset of $\mathsf{T}(\mathbb{C})$. It is a real analytic subset of $\mathsf{T}(\mathbb{C})$.

Proposition 3.12. *For any* $t \in \mathsf{T}(\mathbb{C})$ *there is an element* $\mathsf{g} \in \mathsf{G}(\mathbb{C})$ *such that* $t \bullet \mathsf{g} \in \mathsf{T}_{\mathrm{real}}$.

Proof. This follows from the fact that any two enhanced varieties (X_i, α_i), $i = 1, 2$ with $X_1 = X_2$ are transformed to each other by an action of G. In our proposition one of the enhanced varieties is defined over $\mathbb{C}$ (due to the usage of Hodge decomposition) and so the statement make sense over complex numbers. □

Finally, when X_0 has the Hodge decomposition defined over $\bar{\mathfrak{k}}$ it is natural to define the following algebraic subgroup of G:

$$\check{\mathsf{G}} := \mathrm{Aut}(H^*_{\mathrm{dR}}(X_0), H^{*,*}, \cup, \theta_0) \subset \mathsf{G}. \tag{3.25}$$

3.11 Moduli spaces of enhanced varieties

One of the main motivations for us to introduce Hilbert schemes and the action of reductive groups on such spaces, and developing our main topics with the presence of a reductive group action is that, we wanted to avoid the construction of moduli spaces. The theory developed in the present text will be best seen in the moduli T of projective varieties enhanced with elements in their algebraic de Rham cohomologies as we presented in §3.4. The main reason for this comes from the fact that both classical automorphic functions and topological string partition functions can be realized as regular functions on such moduli spaces. The author is not aware of any literature in mathematics discussing the moduli space T except in his earlier works. A smaller moduli space S in the case of elliptic curves is essentially derived from the two parameter Weierstrass form, however, the moduli T itself appears in the works [Mov08b, Mov12b]. Computations on such moduli space in the case of elliptic curves are done by Katz in the appendix of [Kat76], however he does not define the moduli explicitly. A general definition of such a moduli space for smooth projective varieties has been done in [Mov13] in both contexts of algebraic varieties and Hodge structures. In the case of Calabi–Yau threefolds, explicit computations on T, without defining or constructing it, are done by string theorist in the framework of special geometry, see for instance [CdlO91, CDLOGP91a, Str90, CDF+97, Ali13a]. Its definition in this case is done in [AMSY16]. The special case of mirror quintic goes back to the author's work [Mov17b]. The present section is a continuation of the algebraic geometry part of our previous work [Mov13]. We slightly modify our approach in [Mov13] by considering the whole cohomology ring of a variety, whereas in the mentioned article we have basically considered the middle primitive cohomology of a variety.

The priority in the present text has been to work with projective varieties over a field $\mathfrak{k}$ of characteristic zero. In this way, we have emphasized more geometric questions rather than arithmetic ones. P. Deligne in a personal communication (November 23, 2010) has emphasized that in the case of elliptic curves, the moduli space T is an algebraic stack, and for two invertible, it is actually an algebraic scheme over $\mathrm{Spec}(\mathbb{Z}[1/2])$. This point of view in the case of elliptic curves and abelian varieties is worked out in [Fon21]. A possible applications of this general context might be obtained from a combination of the results in [Ser97] for modular curves and generalizations of these in the framework of our moduli spaces, see [Mov15b].

We first developed the intuitional approach to the moduli space of enhanced projective varieties. Recall the set theoretical description of the moduli space M in §2.9. In this section we explain in geometric terms a bigger moduli space that we would like to construct. For the purpose of this section we consider projective varieties X over an algebraically closed field $\mathfrak{k}$. Recall that we have fixed a smooth projective variety X_0 over $\mathfrak{k}$.

Definition 3.9. The moduli space T is the set of all enhanced projective varieties (X, α), where X is a smooth projective variety over $\mathfrak{k}$ and α is an isomorphism

$$(H^*_{\mathrm{dR}}(X), F^*, \cup, \theta) \overset{\alpha}{\simeq} (H^*_{\mathrm{dR}}(X_0), F^*_0, \cup, \theta_0). \tag{3.26}$$

Two such pairs (X_i, α_i), $i = 1, 2$ are equivalent if we have an isomorphism

$$(H^*_{\mathrm{dR}}(X_1), F^*_1, \cup, \theta_1) \overset{\beta}{\simeq} (H^*_{\mathrm{dR}}(X_2), F^*_2, \cup, \theta_2) \tag{3.27}$$

such that $\alpha_1 = \alpha_2 \circ \beta$.

This definition introduces T as a set. The algebraic group G acts on T from the left in a canonical way

$$\mathsf{T} \times \mathsf{G} \to \mathsf{T}, \ ((X, \alpha), \ \mathsf{g}) \mapsto (X, \mathsf{g} \circ \alpha), \tag{3.28}$$

where $\alpha \circ g$ is the composition of two maps. We transform it into the right action by taking dual of $\mathfrak{k}$-vector spaces as in (3.5) and we denote it by $t \bullet \mathsf{g}$ for $t \in \mathsf{T}$ and $\mathsf{g} \in \mathsf{G}$. We have the isomorphism of sets

$$\mathsf{T}/\mathsf{G} \simeq \mathsf{M}, \ (X, \alpha) \mapsto X$$

obtained by throwing away the structure α. From now on we denote an element of T by $t := (X, \alpha)$ and we make the convention $X_t := X$ and $\alpha_t := \alpha$. Therefore,

$$t = (X_t, \alpha_t).$$

Most of the times, we omit the subscript t as it is clear from the context which X and α we are talking about.

Remark 3.3. There is an alternative way to describe the moduli space T. We fix a basis of $H^*_{\text{dR}}(X_0)$ as in §3.2. The pull-back of this basis by the isomorphism α gives us a basis of $H^*_{\text{dR}}(X)$. We denote it again by $\alpha_{m,i}$. If there is no danger of confusion, we use the notation α both for the isomorphism (3.8) and the set of differential forms $\alpha_{m,i}$. The basis α satisfies the same properties as in §3.2 and moreover the cup product in $H^*_{\text{dR}}(X)$ is constant, that is, if we write (3.2) in this basis the matrices $\Phi_{m_1,m_2,i}$ has constant entries and do not depend on the particular choice of $t \in \mathsf{T}$.

Remark 3.4. Let $\pi : H \to \mathsf{M}$ be the de Rham cohomology bundle over M. The moduli T is not the total space of all choices of basis for the fibers of π with the property (1) and (2) in §3.2. Let $\alpha_{m,i}$ be a basis of $H^*_{\text{dR}}(X)$ as above and let $f \in \text{Aut}(X)$. By definition the pair (X, α) is equivalent to $(X, f^*\alpha)$ and so they represent the same point in T.

We now present the categorical approach to the construction of moduli space of enhanced varieties. We would like to know whether T introduced earlier has a structure of a variety, scheme etc. In the best possible scenario, T might be the underlying scheme of a full family of enhanced varieties. The following property or conjecture has been originally appeared in [Mov13]. Many examples that we discuss in the present text satisfy this property.

Property 3.2. There is a scheme T over $\mathfrak{k}$, an action of G from the right on T and a full family X/T of enhanced projective varieties over $\mathfrak{k}$ which is universal in the following sense:

(1) For any family of enhanced projective varieties Y/S over $\mathfrak{k}$ we have a morphism $Y/S \to \mathsf{X}/\mathsf{T}$ of enhanced varieties.

(2) Any two such morphisms are related to each other by an automorphism of Y/S, that is, there is $a : Y/S \to Y/S$ such that

$$\begin{array}{ccc} Y/S & & \\ & \searrow & \\ \downarrow & & \mathsf{X}/\mathsf{T} \\ & \nearrow & \\ Y/S & & \end{array}$$

commutes.

(3) If Y/S is full then $Y/S \to \mathsf{X}/\mathsf{T}$ respects the action of G.

The main focus of the present text is not to verify Property 3.2 in general, as we are mainly interested in particular examples for which we can verify it by constructing explicit coordinates for T. This includes elliptic curves, curves of arbitrary genus, hypersurfaces, complete intersections, Abelian varieties, K3 surfaces and many types of Calabi–Yau varieties. In all our examples there is a fibration $\bar{\mathsf{X}} \to \bar{\mathsf{T}}$, where $\bar{\mathsf{T}}$ contains T in Property 3.2 as an open set. Using this family we know how enhanced varieties degenerate. For many examples T and $\bar{\mathsf{T}}$ are schemes over $\mathbb{Z}[\frac{1}{N}]$. However, similar to Shimura varieties and many other moduli problems, it is reasonable to think that they are in fact defined over other rings.

The $\mathfrak{k}$-algebra $\mathcal{O}_{\mathsf{T}}$ of regular functions on the moduli T for many different projective varieties, are vast generalizations of algebras of automorphic forms, topological string partition functions and more. In the case of elliptic curves $\mathcal{O}_{\mathsf{T}}$ is isomorphic to the algebra of quasi-modular forms, see [Mov12b], and for mirror quintic Calabi–Yau threefolds $\mathcal{O}_{\mathsf{T}}$ contains elements which encode the Gromov–Witten invariants of a generic quintic, see [Mov17b]. In general, for projective varieties whose period domain is Hermitian symmetric, $\mathcal{O}_{\mathsf{T}}$ is expected to be an algebra of differential automorphic forms and for non-rigid compact Calabi–Yau threefolds, topological partition functions are elements in $\mathcal{O}_{\mathsf{T}}$, see [AMSY16]. Beyond these cases such algebras are not studied at all.

Mumford's geometric invariant theory may be applied in order to find moduli spaces for which Property 3.2 is valid, see [MFK94]. In fact Mumford's article [Mum77] on stability of projective varieties is much more related to Property 3.2. The set of all Chow stable projective varieties $X \subset \mathbb{P}^N_{\bar{\mathfrak{k}}}$ has a canonical action of $\mathrm{SL}(N+1, \bar{\mathfrak{k}})$ and the quotient is the moduli of of projective varieties, let us denote it by M. We throw away singularities of M and those points X in M with an automorphism which acts non-trivially on $H^*_{\mathrm{dR}}(X)$, and call it again M. The vector bundles arising from de Rham cohomologies, Hodge filtrations and $\cup$ product on them, are defined over $\bar{\mathfrak{k}}$. From this we can easily construct an open subset of T which corresponds to enhanced varieties with $X \in \mathsf{M}$, and importantly, it is an algebraic variety. We may also use the Hilbert scheme of projective varieties and an action of a reductive group, see [Vie95]. Mukai's book [Muk03] is more accessible for a general audience.

Apart from the above classical approach for constructing moduli spaces, there is a completely new idea in order to approach Property 3.2. As we will see in §6.11, we can identify many natural vector fields and regular functions in T. In Chapter 13, we will construct such objects in the case of Calabi–Yau varieties. If we take at least one regular function on T then we can differentiate it along vector fields and get more functions. All these new functions on T might give us an embedding of T in some affine variety. For more details of this approach, see Chapter 13.

We might add some other data to X such that the set of its automorphisms becomes finite. For instance, in the case of abelian varieties over $\mathfrak{k}$, we have to fix a point $p \in X(\bar{\mathfrak{k}})$. In general, the verification of the following would be useful for the study of local properties of T.

Property 3.3. Let $X \subset \mathbb{P}^N$ be a smooth polarized projective variety over $\bar{\mathfrak{k}}$ and let G be the group of automorphisms of $X \subset \mathbb{P}^N$ (it preserves the polarization) which induces identity in the de Rham cohomology $H^*_{\mathrm{dR}}(X)$. Then G is a subgroup of the group of automorphisms of all the elements in the moduli M of X.

For abelian varieties the above property follows from classical facts, see for instance Lange–Birkenhake's book [LB92]. In this case G is generated by $x \mapsto x + a, \quad a \in X$ and $x \mapsto -x$. For Calabi–Yau varieties it is proved in Theorem 12 on p. 694 of Todorov's article [Tod03]. In this case G turns out to be finite. From the above property one may conclude that the moduli T is smooth.

We do not want to get stuck in construction of moduli of enhanced varieties because, first, this may distract us from our main objective, which is the properties of modular vector fields and foliations and second, this can be done case by case without using any machinery of moduli spaces, see for instance [Mov12b] for the case of elliptic curves and [Mov17b] for the case of mirror quintic Calabi–Yau threefolds. For this reason, from now on by $\mathsf{X} \to \mathsf{T}$ we mean a full family of enhanced projective varieties. At the best case, when property 3.2 is valid, it is the universal family of enhanced varieties.

3.12 Other moduli spaces

There are other moduli spaces of enhanced projective varieties, and in this paragraph we want to discuss this. Apart from the moduli spaces T used earlier, we have also the followings:

(1) The moduli T_1 of the smooth projective varieties $X/\mathfrak{k}$ equipped with a decomposition

$$H^m_{\mathrm{dR}}(X) = H^{m,0} \oplus H^{m-1,1} \oplus \cdots \oplus H^{1,m-1} \oplus H^{0,m}$$

for all $m = 0, 1, \ldots, 2n$ such that it gives us the Hodge filtration on $H^m_{\mathrm{dR}}(X)$, that is,

$$\begin{aligned} F^m &= H^{m,0} \\ F^{m-1} &= H^{m-1,0} \oplus H^{m-1,1} \\ &\vdots \\ F^1 &= H^{m,0} \oplus H^{m-1,1} \oplus \cdots \oplus H^{1,m-1} \\ F^0 &= H^m_{\mathrm{dR}}(X). \end{aligned}$$

Further,

$$\langle H^{ij}, H^{i'j'} \rangle = 0, \quad i + i' \neq m \text{ or } j + j' \neq m.$$

Note that the Hodge decomposition is intrinsically defined, however, the above decomposition is not intrinsic. The actual Hodge decomposition using harmonic forms gives a point in T_1. We call T_1 the moduli of Hodge decompositions.

(2) The moduli T_2 of the smooth projective varieties X equipped with a decomposition

$$H^m_{\mathrm{dR}}(X) = F^{\frac{m+1}{2}} \oplus W$$

for all odd m. I do not see any canonical way to insert $\langle \cdot, \cdot \rangle$ in the definition of T_2.

(3) Let us fix an element $\mathsf{C}_0 \in H^*_{\mathrm{dR}}(X_0)$ (it can be also a one dimensional subspace of $H^*_{\mathrm{dR}}(X_0)$) and define

$$\mathsf{G}_{\mathsf{C}_0} := \mathrm{Aut}(H^*_{\mathrm{dR}}(X_0), F^*_0, \cup, \theta_0, \mathsf{C}_0), \tag{3.29}$$

which is a subgroup of G. In Chapter 6 we will consider the quotient

$$\mathsf{T}_3 := \mathsf{T}/\mathsf{G}_{\mathsf{C}_0}$$

which can be interpreted as a moduli space in the following way. It is the moduli of triples (X, α, C), where $\mathsf{C} \in H^*_{\mathrm{dR}}(X)$ and α is an isomorphism as in (3.8) such that it sends C to C_0. Two such triples $(X_i, \mathsf{C}_i, \alpha_i)$, $i = 1, 2$ are equivalent if there is β as in (3.27) (which does not necessarily send C_1 to C_2) such that $\mathsf{g} \circ \alpha_1 = \alpha_2 \circ \beta$ for some $\mathsf{g} \in \mathsf{G}_{\mathsf{C}_0}$.

We have canonical surjective maps $\mathsf{T} \to \mathsf{T}_1 \to \mathsf{T}_2$, $\mathsf{T} \to \mathsf{T}_3$.

3.13 Compactifications

Once the moduli space T is constructed, it would be necessary to look for its partial compactifications which describes the degenerations of the projective variety X when it becomes singular. In this direction the following property is valid in many examples.

Property 3.4. There exists a variety $\bar{\mathsf{T}}$ over $\bar{\mathfrak{k}}$ such that

(1) T is an open subset of $\bar{\mathsf{T}}$,
(2) the action of G on T extends to $\bar{\mathsf{T}}$,
(3) and the quotient $\bar{\mathsf{T}}/\mathsf{G}$ is a projective variety (and hence compact).

Note that we do not claim that that there is a compactification of T itself. Apart from T we will consider other moduli spaces such that the problem of full compactification is a reasonable task. One of them is the moduli of Hodge decompositions which is introduced in (3.12). In the case of elliptic curves such a compactification turns out to be the weighted projective space $\mathbb{P}^{1,2,3}$, see §9.8. Another moduli space for which the full compactification seems to be plausible is the moduli space T_3. In the case of elliptic curves and $\mathsf{C}_0 \in H^1_{\mathrm{dR}}(X_0)$ with $\mathsf{C}_0 \notin F^1 H^1_{\mathrm{dR}}(X_0)$, $\mathsf{T}_3 = \mathsf{T}_1$.

Chapter 4

Topology and Periods

M. Picard a donné à ces integrales le nom de périodes; je ne saurais l'en blâmer puisque cette dénomination lui a permis d'exprimer dans un langage plus concis les intéressants résultats auxquels il est parvenu. Mais je crois qu'il serait fâcheux qu'elle s'introduisit définitivement dans la science et qu'elle serait propre à engendrer de nombreuses confusions (H. Poincaré's remarks on the name period used for integrals, see [Poi87, p. 323]).

4.1 Introduction

Singular homologies and cohomologies for algebraic varieties over complex numbers cannot be defined in the framework of algebraic geometry because they do not behave canonically under the automorphisms of the field of complex numbers. As a result, the comparison of the algebraic de Rham cohomology and the singular cohomology leads us to the notion of period, which is again out of the domain of Algebraic Geometry.

In this chapter $X \subset \mathbb{P}^N$ denotes a smooth projective variety of dimension n over $\mathbb{C}$. There is a canonical isomorphism between algebraic de Rham cohomology of X and the usual de Rham cohomology of X defined by C^∞ forms. Therefore, it makes sense to talk about an integrals

$$\int_\delta \alpha \in \mathbb{C}, \quad \delta \in H_m(X,\mathbb{Z}), \quad \alpha \in H^m_{\mathrm{dR}}(X).$$

We denote by $\delta^{\mathsf{pd}} \in H^{2n-m}_{\mathrm{dR}}(X)$, the Poincaré dual of $\delta \in H_m(X,\mathbb{Z})$ defined by the equality

$$\int_\delta \alpha = (2\pi i)^n \cdot \mathrm{Tr}(\alpha \cup \delta^{\mathsf{pd}}) := \int_X \alpha \cup \delta^{\mathsf{pd}}, \quad \forall \alpha \in H^m_{\mathrm{dR}}(X).$$

4.2 Intersections in homologies

We consider $H_*(X,\mathbb{Z})$ as a $\mathbb{Z}$-algebra with the product $\cdot$ of topological cycles

$$H_{m_1}(X,\mathbb{Z}) \times H_{m_2}(X,\mathbb{Z}) \to H_{m_1+m_2-2n}(X,\mathbb{Z})$$
$$(\delta_1,\delta_2) \mapsto \delta_1 \cdot \delta_2.$$

Recall that by definition, the homologies we consider are defined modulo torsions and hence are free. Usually, we take a basis $\delta_{m,i}$ of the $\mathbb{Z}$-algebra $H_*(X,\mathbb{Z})$. Here, for fixed m, $\delta_{m,i}$'s form a basis of $H_m(X,\mathbb{Z})$. We further assume that the intersection forms have a fixed matrix format, that is, if we write

$$\delta_{m_1,i_1} \cdot \delta_{m_2,i_2} = \sum_i c_{m_1,m_2,i_1,i_2,i}\delta_{m_1+m_2-2n,i} \tag{4.1}$$

then the coefficients $c_{m_1,m_2,i_1,i_2,i}$ are fixed. We write (4.1) in the form

$$[\delta_{m_1} \cdot \delta_{m_2}^{\mathsf{tr}}] = \sum_{i=1}^{\mathsf{b}_{m_1}+\mathsf{b}_{m_2}} \Psi_{m_1,m_2,i}\delta_{m_1+m_2-2n,i}, \tag{4.2}$$

where $\Psi_{m_1,m_2,i}$ are constant $\mathsf{b}_{m_1} \times \mathsf{b}_{m_2}$ matrices and

$$\delta_m := \begin{pmatrix} \delta_{m,1} \\ \delta_{m,2} \\ \vdots \\ \delta_{m,\mathsf{b}_m} \end{pmatrix}. \tag{4.3}$$

For $m_1 = m$ and $m_2 = 2n - m$, the intersection of cycles composed with the canonical map $H_0(X,\mathbb{Z}) \cong \mathbb{Z}$, gives us an intersection form

$$\langle \cdot,\cdot \rangle : H_m(X,\mathbb{Z}) \times H_{2n-m}(X,\mathbb{Z}) \to \mathbb{Z}$$

which is non-degenerate. We define

$$\Psi_m := [\langle \delta_{m,i}, \delta_{2n-m,j} \rangle]. \tag{4.4}$$

The polarization of X gives a homology class $[Y] = [\mathbb{P}^{N-1} \cap X] \in H_{2n-2}(X,\mathbb{Z})$ which is Poincaré dual to the topological polarization $u := \frac{1}{2\pi i}\theta$:

$$\theta = (2\pi i) \cdot [Y]^{\mathsf{pd}}.$$

Recall that θ belongs to the algebraic de Rham cohomology $H^2_{\rm dR}(X)$. The self-intersections of $[Y]$ gives us

$$[Y]_m := \underbrace{[Y] \cdot [Y] \cdot \ldots \cdot [Y]}_{(n-\frac{m}{2})\text{-times}} \in H_m(X, \mathbb{Z})$$

for m an even number. In particular, for $m = 0$ the element $[Y]_0 \in H_0(X, \mathbb{Z})$ is $\deg(X)$-times the generator of $H_0(X, \mathbb{Z})$ induced by a point. There is another intersection form which we are going to use:

$$H_m(X, \mathbb{Z}) \times H_m(X, \mathbb{Z}) \to \mathbb{Z}, \quad (\delta_1, \delta_2) \mapsto \delta_1 \cdot \delta_2 \cdot [Y]_{2m}, \quad m = n, n+1, \ldots, 2n. \tag{4.5}$$

The corresponding intersection matrix is denoted by

$$\tilde{\Psi}_m := [\langle \delta_{m,i}, \delta_{m,j} \rangle]. \tag{4.6}$$

By our definition of $\langle \cdot, \cdot \rangle$ in de Rham cohomologies, we have

$$\langle \delta_1, \delta_2 \rangle = (2\pi i)^n \langle \delta_1^{\sf pd}, \delta_2^{\sf pd} \rangle, \quad \forall \delta_1 \in H_m(X, \mathbb{Z}), \ \delta_2 \in H_{2n-m}(X, \mathbb{Z}), \tag{4.7}$$

$$\langle \delta_1, \delta_2 \rangle = (2\pi i)^{2n-m} \langle \delta_1^{\sf pd}, \delta_2^{\sf pd} \rangle, \quad \forall \delta_1, \delta_2 \in H_m(X, \mathbb{Z}). \tag{4.8}$$

Remark 4.1. By hard Lefschetz theorem the map $H_{2n-m}(X, \mathbb{Z}) \to H_m(X, \mathbb{Z})$, $\delta \mapsto \delta \cdot [Y]_{2m}$ is injective, however, it may not be surjective. It becomes an isomorphism only after tensoring with $\mathbb{Q}$. Therefore, we may not be able to take the basis $\delta_{m,i}$ in such a way that the matrices Ψ_m, $\tilde{\Psi}_m$ are equal. This is only valid for $m = n$. Over rational numbers we can take $\delta_m = \delta_{2n-m} \cdot [Y]_{2m}$, $m = n, n+1, \ldots, 2n$ and hence $\Psi_m = \tilde{\Psi}_m$.

4.3 Monodromy group and covering

Let ${\sf X} \to {\sf T}$ be a family of smooth projective varieties over the complex numbers and let X_0 be a fiber of this at $b = 0 \in {\sf T}$ (marked projective variety). We define

$$\begin{aligned}\Gamma_{\mathbb{Z}} &:= \mathrm{Aut}(H_*(X_0, \mathbb{Z}), \ \cdot, \ [Y_0]\ ) \\ &= \{A : H_*(X_0, \mathbb{Z}) \to H_*(X_0, \mathbb{Z}) \mid \mathbb{Z}\text{-linear, respects the homology} \\ &\qquad \text{grading}, \ \forall x, y \in H_*(X_0, \mathbb{Z}), \ Ax \cdot Ay = A(x \cdot y), \ \ A([Y_0]) = [Y_0]\}.\end{aligned} \tag{4.9}$$

It acts on $\delta_{m,j}$'s as a change of basis.

Definition 4.1. We have the monodromy map

$$\pi_1(\mathsf{T},0) \to \Gamma_{\mathbb{Z}}$$

which is a homomorphism from the fundamental group $\pi_1(\mathsf{T},0)$ of T based at 0 to $\Gamma_{\mathbb{Z}}$. Its image $\check{\Gamma}_{\mathbb{Z}}$ is usually called the monodromy group.

We usually write $\Gamma_{\mathbb{Z}}$ in a basis $\delta_{m,i}$, $i = 1, 2, \ldots, \mathsf{b}_m$ and denote its elements by collection of $\mathsf{b}_m \times \mathsf{b}_m$-matrices with entries in $\mathbb{Z}$. We also use $\Gamma_{\mathbb{Z}}$ for the monodromy group $\check{\Gamma}_{\mathbb{Z}}$; being clear in the context which we mean.

Recall the moduli space M of projective varieties introduced in §2.9.

Definition 4.2. Let $\tilde{\mathbb{H}}$ be the moduli of $(X, \delta, [Y])$, where $X \in \mathsf{M}$ is a projective variety, $[Y] \in H_{2n-2}(X, \mathbb{Z})$ is the homology class induced by a hyperplane section and

$$\delta : (H_*(X, \mathbb{Z}), \cdot, [Y]) \cong (H_*(X_0, \mathbb{Z}), \cdot, [Y_0])$$

is an isomorphism of the homology rings sending $[Y]$ to $[Y_0]$. Two such triples $(X_i, \delta_i, [Y_i])$, $i = 1, 2$ are equivalent if we have an isomorphism

$$(H_*(X_1, \mathbb{Z}), \cdot, [Y_1]) \stackrel{\beta}{\simeq} (H_*(X_2, \mathbb{Z}), \cdot, [Y_2]) \tag{4.10}$$

such that $\delta_1 = \delta_2 \circ \beta$. We denote by $\mathbb{H}$ a connected component of $\tilde{\mathbb{H}}$ which contains the triple $(X_0, \delta_0, [Y_0])$, where δ_0 is the identity map, and call it the monodromy covering.

Similar to the case of enhanced varieties, one can replace δ with a basis of $H_*(X, \mathbb{Z})$ with fixed intersection matrices $\Phi_{m_1,m_2,i}$ in §4.2. The group $\Gamma_{\mathbb{Z}}$ acts on $\tilde{\mathbb{H}}$ from the left in a natural way and

$$\Gamma_{\mathbb{Z}} \backslash \mathbb{H} \cong \mathsf{M}, \ (X, \delta) \mapsto X.$$

Let us assume that M is an analytic variety, possibly singular. The set $\tilde{\mathbb{H}}$ has also a canonical structure of an analytic variety, not necessarily connected, and $\Gamma_{\mathbb{Z}}$ acts also on the space of connected components of $\tilde{\mathbb{H}}$. It turns out that the monodromy group $\check{\Gamma}_{\mathbb{Z}} \subset \Gamma_{\mathbb{Z}}$ is the stablizer of the point $\mathbb{H}$ in the space of connected components of $\tilde{\mathbb{H}}$. All connected components of $\tilde{\mathbb{H}}$ are obtained by $\mathbb{H}_\alpha := \alpha(\mathbb{H})$, $\alpha \in \Gamma_{\mathbb{Z}}/\check{\Gamma}_{\mathbb{Z}}$:

$$\tilde{\mathbb{H}} := \bigcup_{\alpha \in \Gamma_{\mathbb{Z}}/\check{\Gamma}_{\mathbb{Z}}} \mathbb{H}_\alpha.$$

The pull-back of holomorphic functions in M by $\mathbb{H} \to \mathsf{M}$ gives us the first class of holomorphic functions on $\mathbb{H}$. Apart from this, we have also other holomorphic functions on $\mathbb{H}$ constructed from periods. This is as follows.

Let us take holomorphic sections $\omega_1, \omega_2, \ldots, \omega_a$ of the Hodge bundle $F^p H^m_{\mathrm{dR}}(X)$, $X \in \mathsf{M}$ such that in an open subset of M they form a basis of $F^p H^m_{\mathrm{dR}}(X)$, or any other subbundle of the cohomology bundle H^m_{dR} over M. The following meromorphic function is trivially independent of the choice of ω_i's:

$$\tau : \quad \mathbb{H} \dashrightarrow \mathbb{C}, \tag{4.11}$$

$$\tau(X, \delta) := \frac{\det[\int_{\delta_{m,k_i}} \omega_j]}{\det[\int_{\delta_{m,l_i}} \omega_j]} \tag{4.12}$$

where we have chosen two subsets δ_{m,k_i}, $i = 1, 2, \ldots, a$ and δ_{m,l_i}, $i = 1, 2, \ldots, a$ of the basis $\delta_{m,i}$, $i = 1, 2, \ldots, \mathsf{b}_m$ of $H_m(X, \mathbb{Z})$, and hence, τ depends on these choices. Note that we regard δ as a choice of a basis for $H_*(X, \mathbb{Z})$. In case X is a Calabi–Yau variety of dimension n and $a := \mathsf{h}^{n,0} = 1$, τ is the mirror map used in physics literature. In the case of elliptic curves we have $\tau := \frac{\int_{\delta_1} \omega}{\int_{\delta_2} \omega}$, where ω is a holomorphic 1-form in X and δ_1, δ_2 is a basis of $H_1(X, \mathbb{Z})$ with $\delta_1 \cdot \delta_2 = -1$. In this case it turns out that τ a biholomorphism between $\mathbb{H}$ and the upper half-plane in $\mathbb{C}$ which is the origin of our notation. One can also get the Poincaré metric on the upper half-plane in a canonical way, see [Mov17b, Appendix B]. For further discussion on $\mathbb{H}$ in the case of mirror quintic see [Mov17b, §4.6].

4.4 Period map

Let $(\mathsf{X}/\mathsf{T}, \alpha)$ be an enhanced family of smooth projective varieties defined over $\mathfrak{k} \subset \mathbb{C}$. We are going to define the period matrix for each $0 \leq m \leq 2n$. Let δ_m be as in (4.3). Each $\delta_{m,i}$ is a continuous family of cycles depending on t. For simplicity, in our notations we have omitted the dependence on t.

Definition 4.3. We integrate α_m^{tr} over δ_m and get the period matrix

$$\mathsf{P}_m = \mathsf{P}_m(t) := (2\pi i)^{-\frac{m}{2}} \cdot \int_{\delta_m} \alpha_m^{\mathsf{tr}} := (2\pi i)^{-\frac{m}{2}} \cdot \left[\int_{\delta_{m,j}} \alpha_{m,i} \right].$$

The entries of P_m are called periods defined over $\mathfrak{k}$. Note that by definition X/T, and hence all $\alpha_{m,i}$'s, are defined over $\mathfrak{k}$.

Recall the constant matrix Φ_m in (3.1) and $\Psi_m, \tilde{\Psi}_m$ in (4.4) and (4.6).

Proposition 4.1. *For $0 \leq m \leq n$ we have*

$$\Phi_m = \mathsf{P}_m^{\mathsf{tr}} \Psi_{2n-m}^{-1} \tilde{\Psi}_{2n-m} \Psi_{2n-m}^{-\mathsf{tr}} \mathsf{P}_m.$$

In particular, for $m = n$ we have

$$\Phi_n = \mathsf{P}_n^{\mathsf{tr}} \Psi_n^{-\mathsf{tr}} \mathsf{P}_n. \tag{4.13}$$

Proof. For simplicity we write the equalities up to $2\pi i$-factors. In order to take care of such factors one has to use (4.7) and (4.8). We have

$$\mathsf{P}_m = \left[\int_{\delta_m} \alpha_m^{\mathsf{tr}} \right] = [\langle \alpha_m^{\mathsf{tr}}, \delta_m^{\mathsf{pd}} \rangle].$$

Let us write α_m in terms of $\delta_{2n-m}^{\mathsf{pd}}$ and let q_m be the $\mathsf{b}_m \times \mathsf{b}_m$ change of basis matrix, that is

$$\alpha_m = \mathsf{q}_m \delta_{2n-m}^{\mathsf{pd}}. \tag{4.14}$$

This q_m is the inverse of q in [Mov13]. We obtain the equality

$$\mathsf{P}_m = \langle (\delta_{2n-m}^{\mathsf{pd}})^{\mathsf{tr}}, \delta_m^{\mathsf{pd}} \rangle] \mathsf{q}_m^{\mathsf{tr}} = [\langle \delta_{2n-m}^{\mathsf{tr}}, \delta_m \rangle] \mathsf{q}_m^{\mathsf{tr}} = \Psi_{2n-m}^{\mathsf{tr}} \mathsf{q}_m^{\mathsf{tr}}.$$

Combining both equalities we get

$$\alpha_m = \mathsf{P}_m^{\mathsf{tr}} \Psi_{2n-m}^{-1} \delta_{2n-m}^{\mathsf{pd}}. \tag{4.15}$$

We have used the fact that the cup product in cohomology is Poincaré dual to intersection of cycles in homology. From another side we use

$$\begin{aligned} \Phi_m &= [\langle \alpha_m, \alpha_m^{\mathsf{tr}} \rangle] \\ &= \mathsf{P}_m^{\mathsf{tr}} \Psi_{2n-m}^{-1} [\langle \delta_{2n-m}^{\mathsf{pd}}, \delta_{2n-m}^{\mathsf{pd},\mathsf{tr}} \rangle] \Psi_{2n-m}^{-\mathsf{tr}} \mathsf{P}_m \\ &= \mathsf{P}_m^{\mathsf{tr}} \Psi_{2n-m}^{-1} \tilde{\Psi}_{2n-m} \Psi_{2n-m}^{-\mathsf{tr}} \mathsf{P}_m. \end{aligned}$$

Note that we have used (2.17). For $m = n$ we have automatically $\tilde{\Psi}_n = \Psi_n$. □

Proposition 4.2. *We have*

$$d\mathsf{P}_m = \mathsf{P}_m \cdot \mathsf{A}_m^{\mathsf{tr}} \tag{4.16}$$

and

$$\mathsf{P}_m(t \bullet \mathsf{g}) = \mathsf{P}_m(t) \cdot \mathsf{g}_m, \quad t \in \mathsf{T}, \quad \mathsf{g} \in \mathsf{G}. \tag{4.17}$$

Proof. This follows after taking ∇ of (4.15) and using $\nabla\delta^{\mathsf{pd}}_{2n-m}=0$. The second equality follows from the definition of period matrix. □

Remark 4.2. A full set of polynomial relations between the entries of all period matrices P_m is obtained in the following way. For $0\leq m_1,m_2\leq 2n$, we write the equality (3.2) in the following format. We define Φ_{m_1,m_2} to be the $1\times \mathsf{b}_{m_1+m_2}$ matrix with the entries $\Phi_{m_1,m_2,i}$, which are matrices, and for simplicity write

$$[\alpha_{m_1}\cup\alpha^{\mathsf{tr}}_{m_2}]=\Phi_{m_1,m_2}\alpha_{m_1+m_2}.$$

We use (4.15) and we get

$$\begin{aligned}&\mathsf{P}^{\mathsf{tr}}_{m_1}\Psi^{-1}_{2n-m_1}[\delta_{2n-m_1}\cdot\delta^{\mathsf{tr}}_{2n-m_2}]\Psi^{-\mathsf{tr}}_{2n-m_2}\mathsf{P}_{m_2}\\&\quad=\Phi_{m_1,m_2}(\mathsf{P}^{\mathsf{tr}}_{m_1+m_2}\Psi^{-1}_{2n-(m_1+m_2)}\delta_{2n-m_1-m_2}).\end{aligned}\tag{4.18}$$

We have used the fact that the cup product in cohomology is Poincaré dual to intersection of cycles in homology. These are not all polynomial relations between periods. For instance, if all the members of the moduli space M have a Hodge cycle varying continuously, then we will get more relations between periods. In general, any algebro-geometric phenomena which occurs for all the members of M might produce algebraic relations between periods.

Remark 4.3. The period map is a local embedding at a point $t\in\mathsf{T}$ if there is no vector v in the tangent space of T at t such that $\mathsf{A}_m(v)=0$ for all m. This follows from (4.16). Once the Gauss–Manin connection matrices A_m are computed, this can be verified computationally. For a concrete example see §9.2.

Chapter 5

Foliations on Schemes

It was a sequence of examples, examples and examples, until you feel that there exists a common feature between all these examples, and then the theorem popped up. At the beginning we had just the announcement but not the proof (C. Camacho, on the invention of Camacho–Sad index theorem in an interview).

5.1 Introduction

In this chapter, we collect all necessary material to deal with (holomorphic) foliations on schemes. We will be mainly concerned with parameter schemes defined in §2.3 or a neighborhood of a point in $\mathbb{C}^n$. A foliation in our context is identified with its module of differential 1-forms and so it is not just the underlying geometric object. This is similar to the passage from a variety to a scheme. This distinguishes our study from the available material in the literature which is mainly concentrated on foliations of dimension or codimension one on complex manifolds. Therefore, most of the material in this chapter cannot be found elsewhere. The Jouanolou's lecture notes [Jou79] may be a good start, however, it does include only codimension one foliations. For an introduction to one dimensional holomorphic foliations, mainly in the local context, the reader is referred to Camacho and Sad's book [CS87] or Loray's book [Lor06]. Lins Neto and Scárdua's book [LNS] and Lins Neto's monograph [Net07] give a nice account of foliations in the projective spaces. Local study of foliations of arbitrary codimension is partially discussed in Medeiros' articles [dM77, dM00] and [CL16]. For a more arithmetically oriented text the reader is referred to Loray, Pereira and Touzet's article [LPT11, LPT13] and Bost's article [Bos01] and references therein. In the C^∞ context, Camacho and Lins Neto's book [CL85] and Godbillons' book [God91] contain many geometric statements on foliations, and depending on applications, one might try to find the corresponding scheme theoretic counterparts.

5.2 Foliations

First, let us recall the notion of a (holomorphic) foliation. The adjective holomorphic is just because of historical reasons and it does not mean that we are working over complex numbers. We may also call them algebraic foliations as they are given by algebraic expressions with coefficients in a ring. Recall our definition of a parameter scheme T in §2.3. In order to keep the intuitional and local aspects of foliations, we will also consider the case $\mathsf{T} = (\mathbb{C}^n, 0)$, that is, T is a small open neighborhood of 0 in $\mathbb{C}^n$. For simplicity we start with the affine case.

Definition 5.1. Let T be an affine parameter scheme over $\mathfrak{k}$. A holomorphic foliation $\mathscr{F}$ in T is given by an $\mathscr{O}_\mathsf{T}$-module $\Omega \subset \Omega^1_\mathsf{T}$ with the integrability condition (or sometimes it is called Frobenius condition). We say that Ω is geometrically integrable if for all $\omega \in \Omega$ there is $f \in \mathscr{O}_\mathsf{T}$, $f \neq 0$ (depending on ω) such that

$$f \cdot d\omega \in \Omega^1_\mathsf{T} \wedge \Omega. \tag{5.1}$$

We say that Ω is algebraically integrable if $d\Omega \subset \Omega^1_\mathsf{T} \wedge \Omega$, that is, in (5.1) we have always $f = 1$.

In a geometric context, Ω induces a foliation $\mathscr{F}(\Omega)$ in T. The $\mathscr{O}_\mathsf{T}$-module Ω may not be free and we consider a set of generators $\omega_i \in \Omega^1_\mathsf{T}$, $i = 1, 2, \ldots, a$ for Ω. The following notations are also common:

$$\mathscr{F}(\omega_1, \omega_2, \ldots, \omega_a), \quad \text{or} \quad \mathscr{F} : \omega_1 = 0,\ \omega_2 = 0,\ \ldots,\ \omega_a = 0.$$

The geometric integrability condition is equivalent to the following. For all $i = 1, 2, \ldots, a$ there are $\omega_{i,j} \in \Omega^1_\mathsf{T}$, $j = 1, 2, \ldots, a$ and $f_i \in \mathscr{O}_\mathsf{T}$, $f_i \neq 0$ such that

$$f_i \cdot d\omega_i = \sum_{j=1}^{a} \omega_{i,j} \wedge \omega_j.$$

For algebraic integrability put all f_i equal to 1.

Remark 5.1. In the literature, see for instance [LPT11, LPT13], one mainly assume that Ω is saturated, that is, the inclusion $\Omega \subset \Omega^1_{\mathsf{T}}$ has non-torsion cokernel. This means that if for some $0 \neq f \in \mathcal{O}_{\mathsf{T}}$ and $\omega \in \Omega^1_{\mathsf{T}}$ we have $f\omega \in \Omega$ then $\omega \in \Omega$. As far as one deals with the geometric aspects of holomorphic foliations, this assumption is quite reasonable. However, our main examples of foliations in Chapter 6 do not satisfy this condition. For a foliation $\mathscr{F}(\Omega)$ one can define another foliation $\mathscr{F}(\check{\Omega})$

$$\check{\Omega} := \{\omega \in \Omega^1_{\mathsf{T}} \mid \exists f \in \mathcal{O}_{\mathsf{T}}, f \neq 0, \quad f\omega \in \Omega\}$$

with the above-mentioned property. Note that by our definition of parameter scheme in §2.3, $\mathcal{O}_{\mathsf{T}}$ has no zero divisors and the integrability of $\check{\Omega}$ follows from the integrability of Ω.

Remark 5.2. The most general definition of a foliation must be done using submodules of k-forms, see for instance [dM77, dM00, CL16]. We are interested only on foliations given by 1-forms as our main examples in Chapter 6 are of this form.

Let $\mathfrak{k}(\mathsf{T})$ be the quotient field of $\mathcal{O}_{\mathsf{T}}$ (the function field of T). We take s elements out of ω_i, $i = 1, 2, \ldots, a$ such that $\omega_1, \omega_2, \ldots, \omega_s$ form a basis of the $\mathfrak{k}(\mathsf{T})$-vector space

$$\tilde{\Omega} := \Omega \otimes_{\mathcal{O}_{\mathsf{T}}} \mathfrak{k}(\mathsf{T}). \tag{5.2}$$

The integrability condition implies that

$$d\omega_i \wedge \omega_1 \wedge \omega_2 \wedge \cdots \wedge \omega_s = 0, \quad i = 1, 2, \ldots, a.$$

Finally, let us make the definition of a foliation in an arbitrary parameter scheme:

Definition 5.2. Let T be a parameter scheme as in §2.3. An integrable system in T is given by a subsheaf $\Omega \subset \Omega^1_{\mathsf{T}}$ of $\mathcal{O}_{\mathsf{T}}$-modules such that there is a covering of T by affine charts such that Ω is integrable in each affine chart.

One can easily verify that the integrability of Ω does not depend on the covering. As before the notation $\mathscr{F}(\Omega)$ is reserved to denote the foliation induced by Ω.

Definition 5.3. We say that two foliations $\mathscr{F}(\Omega_1)$ and $\mathscr{F}(\Omega_2)$ are (algebraically) equal if the corresponding $\mathcal{O}_{\mathsf{T}}$-modules Ω_1 and Ω_2 are equal.

We say that they are geometrically equal if

$$\Omega_1 \otimes_{\mathcal{O}_\mathsf{T}} \mathsf{k}(\mathsf{T}) = \Omega_2 \otimes_{\mathcal{O}_\mathsf{T}} \mathsf{k}(\mathsf{T}).$$

5.3 Rational first integrals

The first example of foliations are those with a first integral.

Definition 5.4. A foliation $\mathscr{F}(\Omega)$ in T has the rational first integral $f \subset \mathsf{k}(\mathsf{T})$ (resp. regular first integral $f \subset \mathsf{k}[\mathsf{T}]$) if the $\mathsf{k}(\mathsf{T})$-vector space $\Omega \otimes_\mathsf{k} \mathsf{k}(\mathsf{T})$ (resp. the $\mathcal{O}_\mathsf{T}$-module Ω) is generated by df.

Note that f is just a set with no sheaf or vector space structure. First integrals have usually poles, that is why we work with both $\mathsf{k}(\mathsf{T})$ and $\mathsf{k}[\mathsf{T}]$. Moreover, for an $\mathcal{O}_\mathsf{T}$-module Ω with a first integral the integrability condition is automatic. In algebraic geometry, the main source of foliations are fibrations.

Definition 5.5. Let $f : \mathsf{T} \to V$ be a morphism of finite type of schemes, that is, there exists a covering of V by open affine subsets V_i such that $f^{-1}(V_i)$ is covered by a finite number of open affine open sets T_{ij} and $\mathsf{k}[\mathsf{T}_{ij}]$ is a finitely generated $\mathsf{k}[V_i]$-algebra. The foliation attached to f is $\mathscr{F}(\Omega)$, where

$$\Omega(\mathsf{T}_{ij}) := \text{The } \mathsf{k}[\mathsf{T}_{ij}]\text{-module generated by } d(g \circ f),\ g \in \mathsf{k}[V_i]. \tag{5.3}$$

A foliation with a regular first integral $f \subset \mathsf{k}[\mathsf{T}]$, where $f = \{f_1, f_2, \ldots, f_n\}$ is the foliation attached to the morphism $\mathsf{T} \to \mathbb{A}^n_\mathsf{k}$ given by $(f_1, f_2, \ldots, f_n)$. Another useful source of foliations comes from group actions discussed in §2.12. We will discuss these foliations in §5.10.

5.4 Leaves

Let T be either a parameter scheme or $(\mathbb{C}^n, 0)$.

Definition 5.6. Let L be a subscheme of T (not necessarily closed) and $\mathscr{I}$ be its sheaf of ideals. Let also $\mathscr{F} = \mathscr{F}(\Omega)$ be a foliation in T. We say that L is a (scheme theoretic) leaf of $\mathscr{F}$ if Ω and $\mathcal{O}_\mathsf{T} \cdot d\mathscr{I}$ projected to $\Omega^1_\mathsf{T}/\mathscr{I}\Omega^1_\mathsf{T}$ and regarded as $\mathcal{O}_\mathsf{T}/\mathscr{I}$-modules are equal. In other words, Ω and $\mathcal{O}_\mathsf{T} d\mathscr{I}$ are equal modulo $\mathscr{I}\Omega^1_\mathsf{T}$.

The following concept seems to be fundamental and non-trivial for singular leaves.

Definition 5.7. For the ideal $\mathscr{I}$ as above we define its integral to be

$$\mathrm{Int}(\mathscr{I}) := \{f \in \mathscr{O}_{\mathsf{T}} \mid df \in \mathscr{I} \cdot \Omega^1_{\mathsf{T}}\}.$$

It can be checked that $\mathrm{Int}(\mathscr{I})$ is an algebra and it is not necessarily an ideal. The definition of $\mathrm{Int}(\mathscr{I})$ is motivated by the following proposition.

Proposition 5.1. *Let L be a subscheme of T and $\mathscr{I}$ be its sheaf of ideals. For any subset $S \subset \mathrm{Int}(\mathscr{I})$ we have a leaf $\check{L}$ whose sheaf of ideals is generated by $\mathscr{I}$ and S.*

Proof. This follows immediately from the definition of a leaf and the integral of an ideal. □

We can repeat the construction in Proposition 5.1 and get more leaves. This phenomenon is present for modular foliations, see §6.10.

By our definition a leaf might be reducible. Therefore, it makes sense to talk about irreducible leaves. Moreover, it might be also reduced. It is useful to write Definition 5.6 in terms of generators. Let $\Omega = \langle \omega_1, \omega_2, \ldots, \omega_a \rangle$ and $\mathscr{I} = \langle f_1, f_2, \ldots, f_s \rangle$. Then L is a leaf of $\mathscr{F}$ if there are $P_{ij}, \check{P}_{ij} \in \mathscr{O}_{\mathsf{T}}$ and $\alpha_{ij}, \check{\alpha}_{ij} \in \Omega^1_{\mathsf{T}}$ such that

$$\begin{bmatrix} \omega_1 \\ \omega_2 \\ \vdots \\ \omega_a \end{bmatrix} = \begin{bmatrix} P_{11} & P_{12} & \cdots & P_{1s} \\ P_{21} & P_{22} & \cdots & P_{2s} \\ \vdots & \vdots & \ddots & \vdots \\ P_{a1} & P_{a2} & \cdots & P_{as} \end{bmatrix} \begin{bmatrix} df_1 \\ df_2 \\ \vdots \\ df_s \end{bmatrix} + \begin{bmatrix} \alpha_{11} & \alpha_{12} & \cdots & \alpha_{1s} \\ \alpha_{21} & \alpha_{22} & \cdots & \alpha_{2s} \\ \vdots & \vdots & \ddots & \vdots \\ \alpha_{a1} & \alpha_{a2} & \cdots & \alpha_{as} \end{bmatrix} \begin{bmatrix} f_1 \\ f_2 \\ \vdots \\ f_s \end{bmatrix}, \tag{5.4}$$

$$\begin{bmatrix} df_1 \\ df_2 \\ \vdots \\ df_s \end{bmatrix} = \begin{bmatrix} \check{P}_{11} & \check{P}_{12} & \cdots & \check{P}_{1a} \\ \check{P}_{21} & \check{P}_{22} & \cdots & \check{P}_{2a} \\ \vdots & \vdots & \ddots & \vdots \\ \check{P}_{s1} & \check{P}_{s2} & \cdots & \check{P}_{sa} \end{bmatrix} \begin{bmatrix} \omega_1 \\ \omega_2 \\ \vdots \\ \omega_a \end{bmatrix} + \begin{bmatrix} \check{\alpha}_{11} & \check{\alpha}_{12} & \cdots & \check{\alpha}_{1s} \\ \check{\alpha}_{21} & \check{\alpha}_{22} & \cdots & \check{\alpha}_{2s} \\ \vdots & \vdots & \ddots & \vdots \\ \check{\alpha}_{s1} & \check{\alpha}_{s2} & \cdots & \check{\alpha}_{ss} \end{bmatrix} \begin{bmatrix} f_1 \\ f_2 \\ \vdots \\ f_s \end{bmatrix}. \tag{5.5}$$

The $\mathscr{O}_{\mathsf{T}}$-module Ω is not necessarily free and L is not necessarily a local complete intersection, therefore, combining the above equalities we only get linear relations between ω_i, df_i and f_i's. In the following by abuse of notation, we will also say that $\{f_1 = f_2 = \cdots = f_s = 0\}$ or $\mathrm{Zero}(f_1, f_2, \ldots, f_s)$ is a leaf of $\mathscr{F}$.

Example 5.1. For a foliation $\mathscr{F} := \mathscr{F}(df_1, df_2, \ldots, df_k)$ with a first integral $f_1, f_2, \ldots, f_k \subset \mathfrak{k}[\mathsf{T}]$ and $c_1, c_2, \ldots, c_k \in \mathfrak{k}$, the subscheme $L := \text{Zero}(f_1 - c_1, f_2 - c_2, \ldots, f_k - c_k) \subset \mathsf{T}$ is a leaf of $\mathscr{F}$. In geometric terms, f_i restricted to L is the constant c_i. In general, the fibers of a fibration $\mathsf{T} \to V$ are leaves of the corresponding foliation. The following question arises in a natural way: Is any leaf of $\mathscr{F}$ necessarily of the above format?

Example 5.2. Let us assume that $\mathscr{F}$ is given by a single holomorphic 1-form ω in $\mathsf{T} := (\mathbb{C}^n, 0)$. For $f \in \mathscr{O}_\mathsf{T}$ with $f(0) = 0$, $\{f = 0\}$ is a leaf of $\mathscr{F}$ if and only if

$$\omega = P \cdot df + f\alpha, \quad \text{for some } P \in \mathscr{O}_\mathsf{T}, \quad \alpha \in \Omega^1_\mathsf{T}, \quad P(0) \neq 0.$$

Therefore, if 0 is a singularity of $\mathscr{F}$ then $f = 0$ must be necessarily singular at 0. We do not have a similar description of the case in which $\mathscr{I}$ is generated by more elements.

Proposition 5.2. *Let* $\mathsf{T} = (\mathbb{C}^n, 0)$ *and consider two subschemes* L_i, $i = 1, 2$ *of* T. *Assume that* $L_1 \cup L_2$ *is a leaf of* $\mathscr{F}$. *Then* $L_1 \backslash L_2$ *and* $L_2 \backslash L_1$ *are leaves of* $\mathscr{F}$.

Note that if $\mathscr{I}_i$, $i = 1, 2$ is the sheaf of ideals of L_i then by definition $\mathscr{I}_1 \cdot \mathscr{I}_2$ is the sheaf of ideals of $L_1 \cup L_2$. Moreover, $L_1 \backslash L_2$ is a subscheme of $\mathsf{T} \backslash L_2$.

Proof. By our hypothesis Ω and $\mathscr{O}_\mathsf{T} d(\mathscr{I}_1 \mathscr{I}_2)$ projected to $\Omega^1_\mathsf{T} / \mathscr{I}_1 \mathscr{I}_2 \Omega^1_\mathsf{T}$ are equal. This implies that Ω and $\mathscr{I}_2 d(\mathscr{I}_1)$ projected to $\Omega^1_\mathsf{T} / \mathscr{I}_1 \Omega^1_\mathsf{T}$ are equal. For a closed point $p \notin L_2$ the stalk of $\mathscr{I}_2$ and $\mathscr{O}_\mathsf{T}$ over p are the same and we conclude the desired statement. □

The following questions arise in a natural way.

(1) Are there two distinct leaves of a foliation $\mathscr{F}$ in $(\mathbb{C}^n, 0)$ passing through 0? Note that "distinct" means that the corresponding ideals $\mathscr{I}_1$ and $\mathscr{I}_2$ are distinct. Even with the stronger condition $\mathscr{I}_1 \subset \mathscr{I}_2$ the question does not seem to be trivial.
(2) For a leaf L of the foliation $\mathscr{F}$, are irreducible components of L also leaves of $\mathscr{F}$? In general, we may write the primary decomposition of the ideal sheaf of L and we may ask whether its primary ideals give new leaves of $\mathscr{F}$.

5.5 Smooth and reduced algebraic leaves

In this section, we consider foliations in $\mathsf{T} := (\mathbb{C}^n, 0)$. Let $L \subset \mathsf{T}$ be a germ of an analytic scheme given by the ideal $\mathscr{I} \subset \mathscr{O}_{\mathsf{T}}$.

Definition 5.8. We say that L is smooth and reduced (of codimension s) if $\mathscr{I} = \langle f_1, f_2, \ldots, f_s \rangle$, and the linear part of f_i's are linearly independent. By holomorphic implicit function theorem this is equivalent to say that in some holomorphic coordinate system $(z_1, z_2, \ldots, z_n)$ in $(\mathbb{C}^n, 0)$ we have $\mathscr{I} = \langle z_1, z_2, \ldots, z_s \rangle$.

Another reformulation of the above definition is to say that L is a smooth local complete intersection. Let $\mathscr{F}(\omega_1, \omega_2, \ldots, \omega_a)$ be a foliation in $\mathsf{T} := (\mathbb{C}^n, 0)$ and let L be a germ of a smooth and reduced leaf of $\mathscr{F}$ of codimension s. By a holomorphic change of coordinates, we can assume that L is given by the ideal $\mathscr{I} = \langle z_1, z_2, \ldots, z_s \rangle$. In the equalities (5.4) and (5.5) we can assume that the matrices P and $\check{P}$ depend only on $z_{s+1}, \ldots, z_n$. Let

$$x := (z_1, z_2, \ldots, z_s), \ y = (z_{s+1}, \ldots, z_n), \ z = (x, y).$$

Replacing (5.4) in (5.5) we get $df - \check{P} \cdot Pdf \in \mathscr{I}\Omega^1_{\mathsf{T}}$ which implies that

$$\check{P}_{s\times a}(y) \cdot P_{a\times s}(y) = I_{s\times s}. \tag{5.6}$$

Evaluating this at $(z_{s+1}, \ldots, z_n) = 0$ we conclude that $s \leq a$.

Proposition 5.3. *The variety $L : x = 0$ is a leaf of $\mathscr{F}(\Omega)$ if and only if $s \leq a$, that is, the codimension of L is less than or equal to the minimum number of generators of Ω, and*

$$\omega_{a\times 1} = P(y)_{a\times s} dx_{s\times 1} + \alpha(x, y)_{a\times s} x_{s\times 1}, \tag{5.7}$$

where we have a matrix $\check{P}_{s\times a}(y)$ with (5.6). In particular, if $\mathscr{F}$ is given by one differential 1-form then smooth and reduced leaves are of codimension 1.

It is natural to ask whether a leaf of $\mathscr{F}$ through a point (if exists) is unique (even in the scheme theoretic sense). We can answer this question only in the case when the leaf is smooth and reduced.

Proposition 5.4. *Let $\mathscr{F}$ be a foliation in $\mathsf{T} := (\mathbb{C}^n, 0)$ with a smooth and reduced leaf L passing through 0. Any other leaf of $\mathscr{F}$ passing through 0 is equal to L.*

Proof. Let us assume that there are two leaves L_i, $i = 1, 2$ of $\mathscr{F}$ passing through 0 and L_1 is smooth and reduced. By a holomorphic change of coordinates, we can assume that L_1 is given by the ideal $\mathscr{I}_1 = \langle z_1, z_2, \ldots, z_r \rangle$. Let $\mathscr{I}_2$ be the ideal of L_2.

We first prove that $\mathscr{I}_2 \subset \mathscr{I}_1$. By definition of a leaf we have

$$d\mathscr{I}_2 \subset \mathscr{I}_2 \Omega^1_{\mathsf{T}} + \mathscr{I}_1 \Omega^1_{\mathsf{T}} + \mathscr{O}_{\mathsf{T}} d\mathscr{I}_1.$$

We restrict the ideal $\mathscr{I}_2$ to $z_1 = z_2 = \cdots = z_r = 0$ and we get an ideal $\check{\mathscr{I}}_2$ in $\mathscr{O}_{\mathbb{C}^{n-m},0}$ with the coordinate system $(z_{r+1}, \ldots, z_n)$ whose zero set contains $0 \in \mathbb{C}^{n-m}$ and it is closed under derivations with respect to variables $z_{r+1}, \ldots, z_n$. By Proposition 5.5, we have $\check{\mathscr{I}}_2 = 0$ and so $\mathscr{I}_2 \subset \mathscr{I}_1$.

Now, let us prove that $\mathscr{I}_1 \subset \mathscr{I}_2$. By definition of a leaf and $\mathscr{I}_2 \subset \mathscr{I}_1$ we have

$$dz_i \in \mathscr{I}_1 \Omega^1_{\mathsf{T}} + \mathscr{O}_{\mathsf{T}} d\mathscr{I}_2, \quad i = 1, 2, \ldots, r.$$

Let $\mathscr{I}_2 = \langle f_1, f_2, \ldots, f_s \rangle$. This implies that the $\mathbb{C}$-vector space of linear parts of $f_1, f_2, \ldots, f_s$ contains $z_1, z_2, \ldots, z_r$. After rearranging and taking linear combinations of f_i's we can assume that the linear parts of $f_1, f_2, \ldots, f_r$ is respectively $z_1, z_2, \ldots, z_r$. We claim that

$$\langle f_1, f_2, \ldots, f_r \rangle = \langle z_1, z_2, \ldots, z_r \rangle. \tag{5.8}$$

This together with $\mathscr{I}_2 \subset \mathscr{I}_1$ implies that $\mathscr{I}_1 = \mathscr{I}_2$. The equality (5.8) follows from the fact that the varieties given by both sides are smooth and of codimension r and one is inside the other one. However, the following argument is better from a computational point of view.

In order to prove (5.8) we use the inverse function theorem for $F : (\mathbb{C}^r, 0) \to (\mathbb{C}^r, 0)$, $F = (f_1, f_2, \ldots, f_r)$ and regard $z_{r+1}, \ldots, z_n$ as parameters or constants. The containment $\subset$ was already proved in the first paragraph and $\supset$ follows from taking the inverse of F. Note that $F(0) = 0$ is independent of the value of $z_{r+1}, \ldots, z_n$ and the derivative of F at 0 is the identity matrix. □

Proposition 5.5. *Let $\mathscr{I} \subset \mathscr{O}_{(\mathbb{C}^n,0)}$ be an ideal such that*

$$\forall f \in \mathscr{I}, \quad i = 1, \ldots, n, \qquad \frac{\partial f}{\partial z_i} \in \mathscr{I}. \tag{5.9}$$

Then either $\mathscr{I} = 0$ or $\mathscr{I} = \mathscr{O}_{(\mathbb{C}^n,0)}$.

Proof. Assume $\mathscr{I} \neq 0$ and so we have a non-zero element $f \in \mathscr{I}$ whose leading term in its Taylor series is $a_0 z_1^{m_1} z_2^{m_2} \cdots a z_n^{m_n}$ with $a_0 \neq 0$. After making the derivations $\frac{\partial^{m_1}}{\partial z_1^{m_1}} \frac{\partial^{m_2}}{\partial z_1^{m_2}} \cdots \frac{\partial^{m_n}}{\partial z_1^{m_n}}$ we get an invertible element in $\mathscr{I}$. □

Proposition 5.6. *Let $\mathscr{F}(\Omega)$ be a foliation in $\mathsf{T} := (\mathbb{C}^n, 0)$ and let L be a germ of a leaf of $\mathscr{F}$ (see Definition 5.6). Then, L is smooth and reduced of codimension s if and only if:*

(1) *we have*

$$\mathrm{ZI}\left(\bigwedge^{s+1}\Omega\right) \subset \mathscr{I} \quad (\text{ or equivalently }) \quad L \subset \mathrm{ZS}\left(\bigwedge^{s+1}\Omega\right),$$

where ZI *and* ZS *are the zero ideal and zero scheme defined in Definition 2.2;*

(2) *we have*

$$\mathrm{ZI}\left(\wedge^{s}\Omega\right) = \mathscr{O}_{\mathsf{T}} \quad (\text{ or equivalently }) \quad \mathrm{ZS}\left(\wedge^{s}\Omega\right) = \emptyset.$$

In other words, there are $\omega_1, \omega_2, \ldots, \omega_s \in \Omega$ such that

$$\mathrm{ZI}\left(\omega_1 \wedge \omega_2 \wedge \cdots \wedge \omega_s\right) = \mathscr{O}_{\mathsf{T}}.$$

In geometric terms, the first item means that the wedge product of Ω, $(s+1)$-times, vanishes at any point of L and the second item means that $\omega_1 \wedge \omega_2 \wedge \cdots \wedge \omega_s$ does not vanish at $0 \in L$, and hence, at any point of a neighborhood of 0 in L. However, note that the first geometric statement is weaker than the scheme theoretic statement used in the proposition.

Proof. Let us prove $\Rightarrow$. The first statement follows from the definition of a leaf. Modulo $\mathscr{I}\Omega^1_{\mathsf{T}}$, Ω is equal to the $\mathscr{O}_{\mathsf{T}}$-module generated by $d\mathscr{I}$ and we know that the $s+1$-times wedge product of the latter is zero. In order to prove the second part we use again the definition of a leaf and we have $df_i = g_i + \omega_i$, $i = 1, 2, \ldots, s$, $g_i \in \mathscr{I}\Omega^1_{\mathsf{T}}$, $\omega_i \in \Omega$. Therefore,

$$df_1 \wedge df_2 \wedge \cdots \wedge df_s - \omega_1 \wedge \omega_2 \wedge \cdots \wedge \omega_s \in \mathscr{I}\Omega^s_{\mathsf{T}}$$

and we get the second statement.

Now, we prove $\Leftarrow$. For ω_i, $i = 1, 2, \ldots, s$ given in the second item of hypothesis we find $f_i \in \mathscr{I}$ and $g_i \in \mathscr{O}_{\mathsf{T}}$ such that $\omega_i - g_i df_i \in \mathscr{I}\Omega^1_{\mathsf{T}}$. We claim that $\mathscr{I} = \langle f_1, f_2, \ldots, f_s \rangle$ and it is smooth and reduced using f_i's. Our hypothesis on ω_i implies that $g_i \in \mathscr{O}^*_{\mathsf{T}}$ and at any point of L the linear part of f_i's are linearly independent. Therefore, it only remains to prove that $\mathscr{I}$ is generated by f_i's. We can make a change of coordinate system in $(\mathbb{C}^n, 0)$ such that $f_i = z_i$, $i = 1, 2, \ldots, s$, where $z = (z_1, z_2, \cdots, z_n)$ is the coordinate system of $(\mathbb{C}^n, 0)$. By our hypothesis the $\mathscr{O}_\mathsf{T}$-modules Ω and $\mathscr{O}_\mathsf{T} \cdot d\mathscr{I}$ are equal modulo $\mathscr{I}\Omega^1_\mathsf{T}$. For any $f \in \mathscr{I}$, we have

$$dz_1 \wedge dz_2 \cdots \wedge dz_s \wedge df = \sum_{i=s+1}^{n} \frac{\partial f}{\partial z_i} dz_1 \wedge dz_2 \cdots \wedge dz_s \wedge dz_i$$

and by our hypothesis $\frac{\partial f}{\partial z_i} \in \mathscr{I}$. This statement is stronger than to say that $\frac{\partial f}{\partial z_i}$ vanishes at any point of L. This will be a crucial point in which we use the scheme theoretic language. Let $\check{\mathscr{I}}$ be the ideal of $\mathscr{O}_{(\mathbb{C}^{n-s},0)}$ obtained by setting $(z_1, z_2, \ldots, z_s) = 0$ in all elements of $\mathscr{I}$ and let $\check{L} = \mathrm{Zero}(\check{\mathscr{I}})$. Our assertion follows from Proposition 5.5 replacing $\mathscr{I}$ of the proposition with $\check{\mathscr{I}}$. □

The "if" part of Proposition 5.6 might be false if we do not use the scheme theoretic notation and it would be interesting to find an example for this. Inspired by Proposition 5.6 we make the following definition.

Definition 5.9. Let T be a parameter scheme over $\mathfrak{k}$ and $\mathscr{F} = \mathscr{F}(\Omega)$ be a foliation of codimension s in T. We consider the inclusions

$$\emptyset \subset \mathrm{ZS}(\Omega) \subset \cdots \subset \mathrm{ZS}(\wedge^i\Omega) \subset \mathrm{ZS}(\wedge^{i+1}\Omega) \subset \cdots. \tag{5.10}$$

The underlying reduced scheme of two schemes in (5.10) might be equal. Neglecting equalities we get

$$\emptyset = \mathsf{T}_{k+1}(\mathbb{C}) \subsetneq \mathsf{T}_k(\mathbb{C}) \subsetneq \cdots \subsetneq \mathsf{T}_1(\mathbb{C}) \subsetneq \mathsf{T}_0(\mathbb{C}) = \mathsf{T}(\mathbb{C}). \tag{5.11}$$

Any consecutive inclusion $\mathsf{T}_i \subsetneq \mathsf{T}_{i+1}$ with T_i as in (5.11) has the property that for some s_i, which only depends on i, $\mathsf{T}_i = \mathrm{ZS}(\wedge^{s_i+1}\Omega)$ and $\mathsf{T}_{i+1} = \mathrm{ZS}(\wedge^{s_i}\Omega)$. We call (5.11) the flag singular locus of $\mathscr{F}(\Omega)$.

Proposition 5.7. *Any smooth and reduced leaf L of $\mathscr{F}$ lies inside $\mathsf{T}_i(\mathbb{C}) \backslash \mathsf{T}_{i+1}(\mathbb{C})$ for some i.*

Proof. This follows from Proposition 5.6. □

Remark 5.3. Something which does not seem to be true is the following. If a leaf L of $\mathscr{F}$ contains a point $p \in \mathsf{T}_i(\mathbb{C}) \backslash \mathsf{T}_{i+1}(\mathbb{C})$ then it must be entirely inside it. By Proposition 5.7 this is true if L is smooth and reduced, but in general it might be false, see Fig. 5.1. This motivates us to define the notion of an essential singularity in Definition 5.12.

5.6 Singular scheme of a foliation

In the literature when one says a foliation, then it is usually non-singular, and so, one has reserved the term singular foliation for those with singularities. In the present text a foliation might have singularities.

Definition 5.10. The codimension of a foliation $\mathscr{F}(\Omega)$ is the smallest integer c such that $\wedge^{c+1}\Omega \subset \Omega_{\mathsf{T}}^{c+1}$ is a torsion sheaf but $\wedge^{c}\Omega$ is not.

Note that $\mathsf{T} = (\mathbb{C}^n, 0)$ has no torsions and so the above definition in this case becomes $\wedge^{c+1}\Omega = 0$ and $\wedge^{c}\Omega \neq 0$.

Definition 5.11. Let T be a parameter scheme and $\mathscr{F}(\Omega)$ be a foliation in T of codimension c. We define

$$\mathrm{Sing}(\mathscr{F}(\Omega)) := \mathrm{ZS}(\wedge^{c}\Omega),$$

where the zero scheme ZS is defined in Definition 2.4, see Fig. 5.1.

It is left to the reader to describe the singular scheme of a foliation attached to a fibration and an action of a group. Inspired by Proposition 5.4, we make the following definition.

Definition 5.12. For a foliation $\mathscr{F}$ on a smooth parameter scheme T, the set of its essential singularities is defined to be

$$\mathrm{ESing}(\mathscr{F}) := \{p \in \mathsf{T}(\mathbb{C}) \mid \text{there is no smooth and reduced leaf of } \mathscr{F} \text{ through } p\}.$$

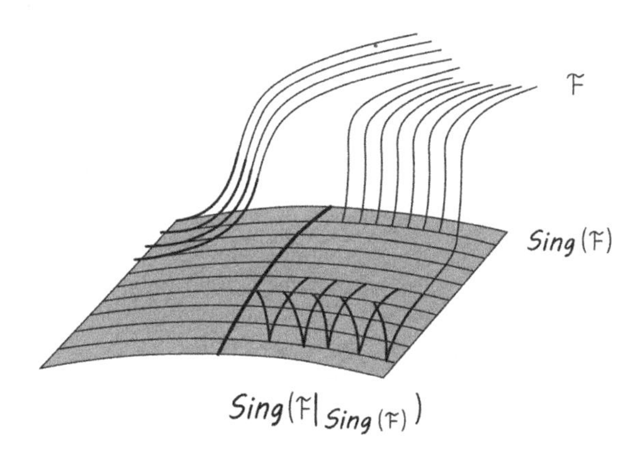

Fig. 5.1. Singular set of a foliation.

Remark 5.4. The set ESing($\mathscr{F}$) does not seem to be an analytic subset of $\mathsf{T}(\mathbb{C})$. It would be natural to find a foliation with non-empty ESing($\mathscr{F}$). In the framework of modular foliations in Chapter 6 the most simple example of a foliation for this purpose seems to be modular foliations attached to either a rank 17 family of $K3$ surfaces described in [CD12], see also [DMWH16], or principally polarized abelian surfaces, see Chapter 13.

5.7 Classical or general leaves

We might develop the theory of algebraic foliations on schemes, however, a foliation in a geometric context makes sense when we are talking about its leaves, and these live only in the complex manifolds. Therefore, in this section we consider the complex and local context $\mathsf{T} = (\mathbb{C}^n, 0)$. The first fundamental theorem in the theory of holomorphic foliations is the following:

Theorem 5.8 (Frobenius theorem). *Let $\mathscr{F}(\Omega)$ be a foliation in $\mathsf{T} = (\mathbb{C}^n, 0)$ and assume that Ω is freely generated by r differential 1-forms and 0 is not a singular point of $\mathscr{F}$. There is a coordinate system $(z_1, z_2, \ldots, z_n)$ in T such that Ω is freely generated as $\mathscr{O}_\mathsf{T}$-module by $dz_1, dz_2, \ldots, dz_r$. In particular,*

$$L : z_1 = \text{const}_1,\ \ z_2 = \text{const}_2, \ldots, z_r = \text{const}_r$$

is a smooth and reduced leaf of $\mathscr{F}$ (Fig. 5.2).

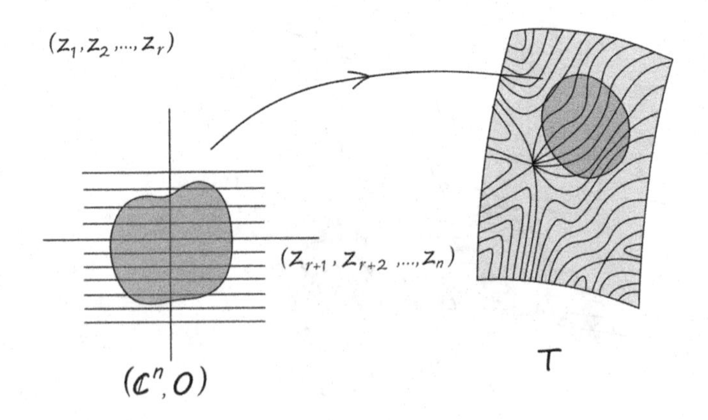

Fig. 5.2. Local charts of a foliation.

Proof. The proof in Camacho and Lins Neto's book [CL85, Appendix 2], can be reproduced easily in the holomorphic context. □

The Frobenius theorem is no more true for points $t \in \mathrm{Sing}(\mathscr{F})$. For a study of foliations in $(\mathbb{C}^2, 0)$ with an isolated singularity at 0 the reader might consult [CS87, Lor06].

Definition 5.13. By a geometric local leaf of a foliation we mean any irreducible component of the underlying analytic variety of the scheme theoretic leaf L in Definition 5.6. The analytic variety L in (5.8) is called a general geometric local leaf of $\mathscr{F}$.

Let us now consider a parameter scheme T over $\mathbb{C}$, a foliation $\mathscr{F}(\Omega)$ in T and a leaf $L \subset \mathsf{T}$ which might be inside $\mathrm{Sing}(\mathscr{F})$. In general, the local analytic variety L is not an open set of an algebraic subvariety of T and its analytic continuation may result in complicated dynamics in the ambient space T (it can become dense in the whole space). We usually remove the adjectives "geometric" and "scheme theoretic"; being clear in the context which leaf we mean.

Definition 5.14. A global leaf L_t of $\mathscr{F}$ passing through $t \in \mathsf{T}\backslash\mathrm{Sing}(\mathscr{F})$ is the union of all local leaves $\check{L}$ which are analytic continuation of the local leaf L passing through t, that is, there is a finite sequence of local leaves $L = L_0, L_1, \ldots, L_s = \check{L}$ with $L_i \cap L_{i+1} \neq \emptyset$. We will use L_t for both local leaf through t and the global leaf as defined above, being clear from the text which we mean.

Chow theorem, which is a part of Serre's GAGA principle [Ser56], says that closed analytic subvarieties of projective varieties are algebraic. This statement is no more true for subvarieties of non-projective varieties and in particular affine varieties. Leaves of modular foliations give us a good class of counterexamples. A leaf of a modular foliation in T may be closed, however, it may have a complicated behavior near the complement of T in its projectivization.

5.8 N-smooth leaves

Let us consider a foliation in $\mathsf{T} := (\mathbb{C}^n, 0)$ and let $L \subset \mathsf{T}$ be a germ of an analytic scheme given by the ideal $\mathscr{I} \subset \mathscr{O}_{\mathsf{T}}$. If both $\mathscr{F}$ and L are given by explicit equations then the property of L to be smooth and reduced boils down to many equalities of formal power series, see for instance

[Mov19, §18.5], and we might check these equalities up to a finite order. This motivates us to define the following concept.

Definition 5.15. For a natural number $N \in \mathbb{N}$, we say that L is N-smooth (and reduced of codimension s) if there are $f_1, f_2, \ldots, f_s \in \mathscr{I}$ such that the linear part of f_i's are linearly independent and

$$\mathscr{I} \subset \langle f_1, f_2, \ldots, f_s \rangle + \mathscr{M}_{\mathbb{C}^n,0}^{N+1},$$

where $\mathscr{M}_{\mathbb{C}^n,0}$ is the maximal ideal of $\mathscr{O}_{\mathbb{C}^n,0}$. In other words, $\mathscr{I} = \langle f_1, f_2, \ldots, f_s \rangle$ modulo power series which vanish at 0 of order $\geq N+1$.

Note that for $k \leq N$ we have $\mathscr{M}_{\mathbb{C}^n,0}^{k+1} \supset \mathscr{M}_{\mathbb{C}^n,0}^{N+1}$ and so N-smoothness implies k-smoothness for all $k \leq N$.

In Proposition 5.10, we will need the following theorem in commutative algebra.

Theorem 5.9 (Krull intersection theorem). *Let I be an ideal in a Noetherian ring R. Suppose either R is a domain and I is a proper ideal; or I is contained in the intersection of all maximal ideals of R (Jacobson radical of R). Then $\bigcap_{n=1}^{\infty} I^n = \{0\}$.*

For a proof see for instance [Cla, Mila].

Proposition 5.10. *Let $\mathscr{I} \subset \mathscr{O}_{\mathbb{C}^n,0}$ be an ideal and $d \in \mathbb{N}$ be a natural number. There is a natural number $N \in \mathbb{N}$, depending only on $\mathscr{I}$ and d with the following property: For any polynomial $P \in \mathbb{C}[z_1, z_2, \ldots, z_n]$ with $\deg(P) \leq d$, if $P \in \mathscr{I} + \mathscr{M}_{\mathbb{C}^n,0}^{N+1}$ then $P \in \mathscr{I}$, that is, if $P \in \mathscr{I}$ modulo power series which vanish at 0 of order $\geq N+1$ then $P \in \mathscr{I}$.*

Proof. The ring $R := \mathscr{O}_{\mathbb{C}^n,0}/\mathscr{I}$ is Noetherian and so by Krull's intersection theorem, see Theorem 5.9, for any proper ideal $I \subset R$ we have $\bigcap_{k=1}^{\infty} I^k = \{0\}$. We use this for the maximal ideal I of R which is is the image of $\mathscr{M}_{\mathbb{C}^n,0}$ in R. From another side, the vector space of polynomials of degree $\leq d$ is finite dimensional, and its projection V in R is also finite dimensional. Since I^k's are also (infinite dimensional) vector spaces, there exists N such that $I^N \cap V = \{0\}$. □

In the case $\mathscr{I}$ is a prime ideal, there is another proof of Proposition 5.10 using Noether's normalization theorem, see [GP07, Theorem 6.2.16], see also [Mov17a]. The advantage of this proof is that it gives also an algorithm how to compute the number N. The author knew of Krull's intersection theorem after a web search by keywords "local ring", "ideal is closed", "Groebner basis for formal power series".

Theorem 5.11. *Let* T *be a smooth parameter scheme over* $\mathbb{C}$ *and let* $\mathscr{F}$ *be an algebraic foliation in* T*. There is an integer* N*, depending on* T *and* $\mathscr{F}$ *such that for all point* $p \in \mathsf{T}$ *and a leaf* L *of* $\mathscr{F}$ *through* p*, if* L *is* N*-smooth then* L *is smooth and reduced.*

Proof. Let $\mathscr{F} = \mathscr{F}(\Omega)$ and $\mathscr{I} = \langle f_1, f_2, \ldots, f_k\rangle$ be the ideal of the leaf L. By definition of a leaf, the $\mathscr{O}_{\mathsf{T},p}$-modules Ω and $\langle df_1, df_2, \ldots, df_k\rangle$ are equal modulo $\mathscr{I} \cdot \Omega^1_{\mathsf{T},p}$. Let us assume that L is N-smooth of codimension $s \leq k$ and the linear part of $f_1, f_2, \ldots, f_s$ are linearly independent. It turns out that the $\mathscr{O}_{\mathsf{T},p}$-modules $\wedge^{s+1}\Omega$ and

$$\left\langle df_{i_1} \wedge df_{i_2} \wedge \cdots \wedge df_{i_{s+1}} \mid \{i_1, i_2, \ldots, i_{s+1}\} \subset \{1, 2, \ldots, k\}\right\rangle \tag{5.12}$$

are equal modulo $\mathscr{I} \cdot \Omega^{s+1}_{\mathsf{T},0}$. The $\mathscr{O}_{\mathsf{T},p}$-module (5.12) is 0 modulo $\mathscr{M}^{N+1}_{\mathsf{T},0} \cdot \Omega^{s+1}_{\mathsf{T},0}$, therefore,

$$\wedge^{s+1}\Omega = 0, \quad \text{modulo } \mathscr{I} \cdot \Omega^{s+1}_{\mathsf{T},0} + \mathscr{M}^{N+1}_{\mathsf{T},0} \cdot \Omega^{s+1}_{\mathsf{T},0}.$$

The $\mathscr{O}_{\mathsf{T},p}$-module $\wedge^{s+1}\Omega$ is generated by algebraic differential forms whose ingredients are polynomials. Therefore, for N big enough depending on Ω, we can use Proposition 5.10 and conclude that (5.12) is in $\mathscr{I} \cdot \Omega^{s+1}_{\mathsf{T},0}$. The rest of the proof is similar to the last part in the proof of Proposition 5.6. Without lose of generality, we can assume that $f_1 = z_1, \ldots, f_s = z_s$ are parts of a coordinate system in (T, p). For any other $f \in \mathscr{I}$ we have $\frac{\partial f}{\partial z_i} \in \mathscr{I}$, $i = s+1, \ldots, n$. After restricting to $z_1 = z_2 = \cdots = z_s = 0$ and applying Proposition 5.5, we conclude that $\mathscr{I} = \langle z_1, z_2, \ldots, z_s\rangle$. □

It would be extremely interesting to give a formula for N in the case of modular foliations attached to families of hypersurfaces. This number seems to be readable from the corresponding Gauss–Manin connection matrix. It would result in the discovery of new components of Hodge loci, for which we might expect that the corresponding Hodge cycles are not algebraic, see [Mov19, Chapter 18].

5.9 Foliations and vector fields

A holomorphic foliation can be also described in terms of vector fields. Recall the preliminaries of vector fields on schemes in §2.4. Let T be a parameter scheme as in §2.3 and consider a finitely generated $\mathscr{O}_{\mathsf{T}}$-module $\Omega \subset \Omega^1_{\mathsf{T}}$ and define the $\mathscr{O}_{\mathsf{T}}$-module

$$\Theta := \{\mathsf{v} \in \Theta_{\mathsf{T}} \mid i_{\mathsf{v}}\Omega = 0\}.$$

Proposition 5.12. *The $\mathscr{O}_{\mathsf{T}}$-module Ω is geometrically integrable if and only if Θ is closed under the Lie bracket.*

Proof. Recall that Ω is geometrically integrable if $\Omega \otimes_{\mathscr{O}_{\mathsf{T}}} \mathfrak{k}(\mathsf{T})$ is integrable. In the following, we take $\omega \in \Omega$ and $\mathsf{v}_1, \mathsf{v}_2 \in \Theta$. First we prove $\Rightarrow$. We have

$$i_{[\mathsf{v}_1,\mathsf{v}_2]}\omega = i_{\mathsf{v}_1} \circ \mathscr{L}_{\mathsf{v}_2}\omega = i_{\mathsf{v}_1} \circ i_{\mathsf{v}_2} d\omega \tag{5.13}$$

for which we have used (2.8). The integrability condition implies that the right-hand side of this is zero.

We now prove $\Leftarrow$. It is enough to prove it for T an affine parameter scheme. Let $\omega_1, \omega_2, \ldots, \omega_a \in \Omega^1_{\mathsf{T}}$ form a basis of the $\mathfrak{k}(\mathsf{T})$-vector space $\Omega^1_{\mathsf{T}} \otimes_{\mathscr{O}_{\mathsf{T}}} \mathfrak{k}(\mathsf{T})$ and for $s \leq a$, $\omega_1, \omega_2, \ldots, \omega_s$ form a basis of $\Omega \otimes_{\mathscr{O}_{\mathsf{T}}} \mathfrak{k}(\mathsf{T})$. For $\omega \in \Omega$ let us write $d\omega := \sum_{i<j} f_{ij}\omega_i \wedge \omega_j$ for some $f_{ij} \in \mathfrak{k}(\mathsf{T})$. It is enough to show that for $s < i < j$ we have $f_{ij} = 0$. By our hypothesis, the left-hand side of (5.13) is zero and hence the right-hand side is zero too. We have

$$0 = i_{\mathsf{v}_1} \circ i_{\mathsf{v}_2} d\omega = i_{\mathsf{v}_1} \circ i_{\mathsf{v}_2} \left(\sum_{s<i<j} f_{ij}\omega_i \wedge \omega_j \right).$$

Taking vector fields v_1 and v_2 dual to ω_i and ω_j we get $f_{ij} = 0$ and hence the integrability condition is satisfied. □

A holomorphic foliation in T is also given by a sub-Lie Algebra Θ of Θ_{T} in which case we denote it by $\mathscr{F}(\Theta)$. This will give us the notion of an algebraically integrable foliation given by vector fields. We may also define geometrically integrable $\mathscr{F}(\Theta)$ in which $\Theta \otimes_{\mathscr{O}_{\mathsf{T}}} \mathfrak{k}(\mathsf{T})$ is closed under the Lie bracket. In both cases, the corresponding set of differential forms is given by

$$\Omega = \{\omega \in \Omega^1_{\mathsf{T}} \mid i_{\mathsf{v}}\omega = 0, \quad \forall \mathsf{v} \in \Theta\}.$$

The definition of a leaf of $\mathscr{F}(\Theta)$ is done using this Ω, however, it might be useful to formulate such a definition without passing through differential forms. For instance, a geometric intuition suggests the following definition.

Definition 5.16. A subscheme L of T is a leaf of $\mathscr{F}(\Theta)$ if the Zariski tangent space of L at any point $t \in L$ is equal to Θ_t.

It is left to the reader to compare this definition with Definition 5.6. Finally, note that two distinct submodules Ω and $\check{\Omega}$ of Ω^1_{T} might have the same $\mathscr{O}_{\mathsf{T}}$-modules of vector fields, that is, $\Theta = \check{\Theta}$, and vice versa. For instance, in the affine line $\mathsf{T} := \mathrm{Spec}(\mathfrak{k}[t])$ and trivial foliations $\Omega = \langle p(t)dt \rangle$ and $\check{\Omega} = \langle \check{p}(t)dt \rangle$, we have $\Theta = \check{\Theta} = \{0\}$. In a similar way, for $\Theta = \langle p(t)\frac{\partial}{\partial t} \rangle$, $\check{\Theta} = \langle \check{p}(t)\frac{\partial}{\partial t} \rangle$ we have $\Omega = \check{\Omega} = \{0\}$.

5.10 Algebraic groups and foliations

Let T be a parameter scheme and G be an algebraic group acting on T from the right, all defined over an algebraically closed field $\mathfrak{k}$. Recall our notations in §2.14 and notice that we have to adapt our notations to the right action of groups.

Definition 5.17. We define

$$i(\mathrm{Lie}(\mathsf{G})) := \{\mathsf{v}_{\mathfrak{g}} \mid \mathfrak{g} \in \mathrm{Lie}(\mathsf{G})\} \subset H^0(\mathsf{T}, \Theta_{\mathsf{T}}), \tag{5.14}$$

which is the image of the map i in (2.24). It is equipped with the Lie bracket of vector fields. We denote by $\mathscr{F}(\mathsf{G})$ the foliation in T induced by $i(\mathrm{Lie}(\mathsf{G}))$ in T and call it the foliation induced by the action of G on T.

Note that $[\mathsf{v}_{\mathfrak{g}_1}, \mathsf{v}_{\mathfrak{g}_2}] = \mathsf{v}_{[\mathfrak{g}_1, \mathfrak{g}_2]}$ for $\mathfrak{g}_1, \mathfrak{g}_2 \in \mathrm{Lie}(\mathsf{G})$ and so $i(\mathrm{Lie}(\mathsf{G}))$ is closed under the Lie bracket. The following problems arise in a natural way:

(1) Describe the leaves of $\mathscr{F}(i(\mathrm{Lie}(\mathsf{G})))$ in terms of the action of G. In particular, describe smooth and reduced leaves. Recall that we have made a purely algebraic definition of a leaf, see Definition 5.6, and the geometric intuition might not be enough for this purpose.
(2) What is the singular locus $\mathrm{Sing}(\mathscr{F}(i(\mathrm{Lie}(\mathsf{G}))))$?
(3) Are all the points in $\mathsf{T}\backslash\mathrm{Sing}(\mathscr{F}(i(\mathrm{Lie}(\mathsf{G}))))$ stable in the sense of geometric invariant theory?
(4) What is the flag singular locus of the foliation $\mathscr{F}(i(\mathrm{Lie}(\mathsf{G})))$ defined in (5.11)?

Answering these questions would require an intensive study of geometric invariant theory, see [MFK94]. For $t \in \mathsf{T}$ the orbit $t \bullet \mathsf{G}$ seems to be an algebraic leaf of $\mathscr{F}(i(\mathrm{Lie}(\mathsf{G})))$ in the sense of Definition 5.16. This is true at least in the smooth part of the orbit $t \bullet \mathsf{G}$, because all the vector field

$\mathsf{v}_\mathfrak{g} \in i(\mathrm{Lie}(\mathsf{G}))$ are tangent to a smooth point of $t \bullet \mathsf{G}$ and they generate the tangent space of $t \bullet \mathsf{G}$ at that point.

Definition 5.18. A holomorphic foliation $\mathscr{F}(\Omega)$ in T is algebraically invariant under the action of G if the induced action in Ω^1_T sends Ω to itself. It is called geometrically invariant if the induced action on Ω^1_T sends Ω to some submodule $\Omega_1 \subset \Omega^1_\mathsf{T}$ such that the foliations $\mathscr{F}(\Omega)$ and $\mathscr{F}(\Omega_1)$ are geometrically equal (see Definition 5.3).

The geometry behind the above definition is as follows.

Proposition 5.13. *For $\mathsf{k} = \mathbb{C}$, a foliation $\mathscr{F}$ is geometrically invariant under the action of G if and only if for all $\mathsf{g} \in \mathsf{G}$, the action map $i_\mathsf{g} : \mathsf{T} \to \mathsf{T}$, $t \mapsto t \bullet \mathsf{g}$ sends a general local leaf of $\mathscr{F}$ to another general local leaf.*

Proof. By definition i_g sends the tangent space of $\mathscr{F}$ at $t \in \mathsf{T}$ to the tangent space of T at $t \bullet \mathsf{g}$. The result follows from the uniqueness of a leaf passing through a regular point. □

In other words, G acts on the space of leaves of $\mathscr{F}$. For a leaf L of $\mathscr{F}$, we may expect that $L \bullet \mathsf{G}$'s are leaves of another foliation. This is in fact the case.

Proposition 5.14. *Let $\mathscr{F}$ be a foliation in T given by the $\mathscr{O}_\mathsf{T}$-module $\mathsf{R} \subset \Theta_\mathsf{T}$ and let us assume that $\mathscr{F}$ is geometrically invariant under the action of G. Then $(\mathsf{R} + i(\mathrm{Lie}(\mathsf{G}))) \otimes_{\mathscr{O}_\mathsf{T}} \mathsf{k}(\mathsf{T})$ is closed under the Lie bracket and hence induces a holomorphic foliation $\tilde{\mathscr{F}}$ in T. Further, for $\mathsf{k} = \mathbb{C}$ we have*

(1) *All the leaves of $\mathscr{F}$ are inside the leaves of $\tilde{\mathscr{F}}$.*
(2) *For any point $t \in \mathsf{T}$, we have*

$$\tilde{L}_t := L_t \bullet \mathsf{G}, \tag{5.15}$$

where L_t (respectively, $\tilde{L}_t$) is the leaf of $\mathscr{F}$ (respectively, $\tilde{\mathscr{F}}$) through t.

Proof. Only the first part is non-trivial. For this we do not have a purely algebraic argument. Therefore, we use Lefschetz principle and assume that $\mathsf{k} = \mathbb{C}$. In fact, it seems that $\mathsf{R} + i(\mathrm{Lie}(\mathsf{G}))$ itself is closed under the Lie bracket, however, our transcendental proof below does provide this statement if R is a free $\mathscr{O}_\mathsf{T}$-module. It is enough to prove the statement locally around a point $t \in \mathsf{T}$ which is not a singularity of $\mathscr{F}$. There is a holomorphic coordinate system $(z_1, z_2) \in (\mathbb{C}^{n-r} \times \mathbb{C}^r, 0)$ around t such that $\mathscr{F}$ in these coordinates is given by z_2 = constant. Therefore, the space of leaves of $\mathscr{F}$ is given by $z_2 \in (\mathbb{C}^r, 0)$, and by our assumption, a neighborhood of the identity in G, say it $(\mathsf{G}, 1)$, acts on z_2 coordinate. By further holomorphic coordinate change we can replace the z_2-coordinate with $(z_2, z_3) \in \mathbb{C}^{r-a} \times \mathbb{C}^a$ such that the induced foliation by the action of $(\mathsf{G}, 1)$ in the space (z_2, z_3) of leaves of $\mathscr{F}$ is given by z_3 =constant. Now, a leaf of the foliation $\tilde{\mathscr{F}}$ is given by z_3 =constant. □

As an immediate corollary of the above proposition we have the following proposition.

Proposition 5.15. *If a leaf of $\mathscr{F}$ through t is algebraic, then the leaf of $\tilde{\mathscr{F}}$ through t is also algebraic.*

Proof. This follows from the equality (5.15). □

5.11 $\mathscr{F}$-invariant schemes

For a foliation $\mathscr{F}$ in T the notion of $\mathscr{F}$-invariant subschemes of T are as natural as algebraic leaves of $\mathscr{F}$.

Definition 5.19. Let us consider the sheaf of ideals $\mathscr{I} \subset \mathscr{O}_{\mathsf{T}}$ defining the subscheme $V := \text{Zero}(\mathscr{I})$ and a foliation $\mathscr{F}(\Omega)$. We say that V is algebraically $\mathscr{F}(\Omega)$-invariant if for all $f \in \mathscr{I}$, df is in the $\mathscr{O}_{\mathsf{T}}$-module generated by Ω and $\mathscr{I} \cdot \Omega^1_{\mathsf{T}}$:

$$d\mathscr{I} \subset \mathscr{I} \cdot \Omega^1_{\mathsf{T}} + \Omega. \tag{5.16}$$

If in a local chart we take generators $\mathscr{I} = \langle f_1, f_2, \ldots, f_s \rangle$, $\Omega = \langle \omega_1, \omega_2, \ldots, \omega_a \rangle$ for both $\mathscr{O}_{\mathsf{T}}$-modules $\mathscr{I}$ and Ω, then our definition is equivalent to the equality (5.5). Note that a subscheme V of T is a leaf of $\mathscr{F}$ if it is algebraically $\mathscr{F}$-invariant and $\Omega \subset \mathscr{I} \cdot \Omega^1_{\mathsf{T}} + \mathscr{O}_{\mathsf{T}} \cdot d\mathscr{I}$.

Proposition 5.16. *Let $\mathscr{F}(\Omega)$ be of codimension c. If V is algebraically $\mathscr{F}(\Omega)$-invariant, then*

$$d\mathscr{I} \wedge \underbrace{\Omega \wedge \Omega \wedge \cdots \wedge \Omega}_{c \text{ times}} \subset \mathscr{I}\Omega^{c+1}_{\mathsf{T}}. \tag{5.17}$$

Proof. This follows from the definition of the codimension of a foliation in Definition (5.10) and (5.16). □

It is desirable to find a definition of $\mathscr{F}$-invariant subscheme V of T which corresponds to a geometric concept. For instance, we may ask that in (5.5) one has to multiply df_i with some $g_i \notin \mathscr{I}$. The correct definition must be done depending on applications and we content ourselves to the following:

Definition 5.20. We say that V is geometrically $\mathscr{F}$-invariant if we have (5.17).

5.12 Transcendental numbers vs. variables

One of the basic ideas in Deligne's proof of absolute Hodge cycles for abelian varieties is the interchange between a transcendental number and a variable. Foliations give us a convenient machinery to determine the situations in which a transcendental number is replaced with a variable and the variable, in turn, is replaced by any number. In Hodge theory such transcendental numbers are periods of algebraic varieties defined over $\bar{\mathbb{Q}}$. Since the integration is not an algebraic operation, the above philosophy cannot be applied directly. For this purpose we use foliations. In this section, we explain a theorem which does this job for us.

Let $\tilde{\mathfrak{k}} = \overline{\mathfrak{k}(z_1, z_2, \ldots, z_n)} \subset \mathbb{C}$ be an algebraically closed field of finite transcendence degree over $\mathfrak{k}$.

Definition 5.21. The Zariski closure Z of $(z_1, z_2, \ldots, z_n) \in \mathbb{C}^n$ over $\bar{\mathfrak{k}}$ is the zero set (or the spectrum) of the ideal

$$\left\{P \in \bar{\mathfrak{k}}[x_1, x_2, \ldots, x_n] \mid P(z_1, z_2, \ldots, z_n) = 0\right\}.$$

It is easy to see that Z is an irreducible variety over $\bar{\mathfrak{k}}$. The following is a classical statement.

Proposition 5.17. *Let Z be the Zariski closure of $(z_1, z_2, \ldots, z_n) \in \mathbb{C}^n$ over $\bar{\mathfrak{k}}$. The dimension of Z is the transcendence degree of $\tilde{\mathfrak{k}}$ over $\bar{\mathfrak{k}}$.*

See for instance Lemma 1.7 in Deligne's lecture notes in [DMOS82]. The following theorem enables us to regard a transcendental number as a variable. A basic idea is that if in the middle of an algebraic structure, where only polynomials are used, if you find a transcendental number like π or e then you can replace it with a variable and hence by any other number in the base field.

Theorem 5.18. *Let $\tilde{\mathfrak{k}}$ be an algebraically closed field extension of $\bar{\mathfrak{k}}$. Let V be variety over $\bar{\mathfrak{k}}$ and let $\mathscr{F}$ be an algebraic foliation on $V_{\tilde{\mathfrak{k}}} := V \otimes_{\bar{\mathfrak{k}}} \tilde{\mathfrak{k}}$ defined over $\tilde{\mathfrak{k}}$. Let also $Y \subset V_{\tilde{\mathfrak{k}}}$ be an algebraic leaf of $\mathscr{F}$ defined over $\tilde{\mathfrak{k}}$. There are a variety Z, a foliation $\tilde{\mathscr{F}}$ on $V \times Z$ and an algebraic subset $\tilde{Y}$ of $V \times Z$, all of them defined over $\bar{\mathfrak{k}}$, and a point $a \in Z(\tilde{\mathfrak{k}})$ such that*

(1) *The fibers of the projection* $\pi : V \times Z \to Z$ *are* $\tilde{\mathscr{F}}$*-invariant.*
(2) $\tilde{Y}$ *is* $\tilde{\mathscr{F}}$*-invariant and the leaves of* $\tilde{\mathscr{F}}$ *in* $\tilde{Y}$ *are fibers of* $\pi|_{\tilde{Y}}$.
(3) *The triple* $(V \times \{a\}, \tilde{\mathscr{F}}|_{V\times\{a\}}, \tilde{Y} \cap (V \times \{a\}))$ *is isomorphic over* $\tilde{\mathfrak{k}}$ *to* $(V_{\tilde{\mathfrak{k}}}, \mathscr{F}, Y)$.

Moreover, if $\mathscr{F}$ *is defined over* $\bar{\mathfrak{k}}$ *then all the pairs* $(V \times \{b\}, \tilde{\mathscr{F}}|_{V\times\{b\}})$, $b \in Z(\tilde{\mathfrak{k}})$ *are isomorphic to* $(V_{\tilde{\mathfrak{k}}}, \mathscr{F})$ *under the map* $V \times \{b\} \to V$ *induced by the identity.*

Proof. First note that in the definition of $\mathscr{F}$ and Y we have used a finite number of coefficients and so we can assume that the field $\tilde{\mathfrak{k}}$ is of finite transcendence degree over $\bar{\mathfrak{k}}$, that is, $\tilde{\mathfrak{k}} = \overline{\bar{\mathfrak{k}}(z_1, z_2, \ldots, z_n)}$. Let Z be the affine variety as in Proposition 5.17. In a local chart U of V, the foliation $\mathscr{F}$ is given by differential 1-forms ω_i, $i = 1, 2, \ldots, a$. We can regard ω_i as a differential form on $U \times Z$. Looking in this way they may not give a foliation in $U \times Z$, as the integrability may fail. However, the set

$$\{\omega_i, \quad i = 1, 2, \ldots, a, \ df, \ f \in \mathscr{O}_Z(Z)\} \tag{5.18}$$

satisfies the integrability condition. The foliation $\tilde{\mathscr{F}}$ in the local chart $U \times Z$ of $V \times Z$ is given by the differential forms (5.18). The algebraic set Y can be regarded as an algebraic set $\tilde{Y}$ in $V \times Z$ in a canonical way. The first statement follows from the fact that the defining differential forms of $\tilde{\mathscr{F}}$ contain all df, $f \in \mathscr{O}_Z(Z)$. The second statement is a direct consequence of the fact that Y is $\mathscr{F}$-invariant. The point a in the third statement is the point $(z_1, z_2, \ldots, z_n) \in Z(\tilde{\mathfrak{k}})$ (Fig. 5.3). □

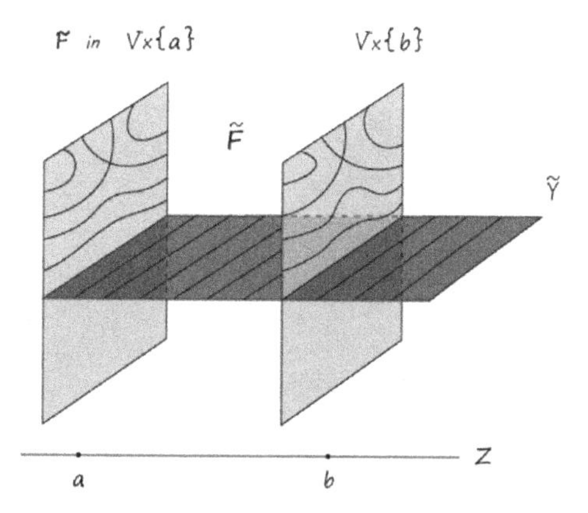

Fig. 5.3. Leaves and transcendental numbers.

Chapter 6

Modular Foliations

I have found a friend in you who views my labours sympathetically. ... I am already a half starving man. To preserve my brains I want food and this is my first consideration. Any sympathetic letter from you will be helpful to me here to get a scholarship either from the university or from the government (S. Ramanujan in his letter to G.H. Hardy, see [OR16]).

6.1 Introduction

In this chapter, we introduce the notion of a modular foliation in the parameter space T of an enhanced family of projective varieties. In order to motivate the reader we will start considering foliations over the field of complex numbers, that is, $\mathfrak{k} = \mathbb{C}$, and hence, we will freely use periods. However, as far as we are not talking about non-algebraic leaves we can work with foliations on a variety over a field $\mathfrak{k}$ of characteristic zero or even schemes. A very fascinating and simple fact about foliations is that many statements about them, which are apparently talking about transcendental aspects, can be translated into purely algebraic statements. Since we are going to work with a fixed (co)homology, we sometimes drop the subindex m from our notations. Our main example in this chapter has been the case of elliptic curves and their product, and so, it is strongly recommended that the reader read this chapter together with Chapters 9 and 10 simultaneously. For the case of Abelian varieties (respectively, hypersurfaces) the reader is recommended to read Chapter 11 (respectively, Chapter 12) simultaneously. In this chapter, δ denotes a continuous family of cycles as in §6.3 and so it must not be confused with a basis $\delta = \delta_m$ of a homology as in Chapter 4.

6.2 A connection matrix

What we need in this chapter is the Gauss–Manin connection matrix A in §3.8. The reader who is not familiar with the concept of Gauss–Manin connections is recommended to follow the content of this chapter for an arbitrary connection matrix as follows. We will only borrow the terminology used in Chapter 3.

Let T be a parameter scheme over the field $\mathfrak{k}$ as in §2.3. Let also b, $\mathsf{h}^{m,0}, \mathsf{h}^{m-1,1}, \ldots, \mathsf{h}^{1,m-1}, \mathsf{h}^{0,m}$ be natural numbers such that $\mathsf{h}^{i,j} = \mathsf{h}^{j,i}$ and $\mathsf{b} = \mathsf{h}^{m,0} + \mathsf{h}^{m-1,1} + \cdots + \mathsf{h}^{1,m-1} + \mathsf{h}^{0,m}$. We consider a $\mathsf{b} \times \mathsf{b}$ matrix A whose entries are (global) differential 1-forms (with no pole) in T and it is of the form

$$\mathsf{A} = [\mathsf{A}^{ij}] = \begin{pmatrix} \mathsf{A}^{00} & \mathsf{A}^{01} & 0 & 0 & \cdots & 0 \\ \mathsf{A}^{10} & \mathsf{A}^{11} & \mathsf{A}^{12} & 0 & \cdots & 0 \\ \mathsf{A}^{20} & \mathsf{A}^{21} & \mathsf{A}^{22} & \mathsf{A}^{23} & \cdots & 0 \\ \vdots & \vdots & \vdots & \ddots & \vdots & \vdots \\ \mathsf{A}^{m0} & \mathsf{A}^{m1} & \mathsf{A}^{m2} & \mathsf{A}^{m3} & \cdots & \mathsf{A}^{mm} \end{pmatrix},$$

where we have used the Hodge block notation for matrices introduced in §2.8. We further assume the following:

$$d\mathsf{A} = \mathsf{A} \wedge \mathsf{A} \tag{6.1}$$

and for some constant $(-1)^m$-symmetric matrix Φ

$$\mathsf{A}\Phi + \Phi\mathsf{A}^{\mathsf{tr}} = 0. \tag{6.2}$$

We can imitate further the Gauss–Manin connection matrix, putting more structures in this abstract context, such as the action of two algebraic groups G and $\mathbf{G}$, etc. Note that it would be also reasonable to consider the collection of matrices A_m, $m = 0, 1, \ldots, 2n$ instead of a single one. For a small open set U (in the usual topology) of the underlying analytic variety $\mathsf{T}(\mathbb{C})$, we can find a $\mathsf{b} \times \mathsf{b}$ matrix P with entries which are holomorphic functions in U and

$$d\mathsf{P} = \mathsf{P} \cdot \mathsf{A}^{\mathsf{tr}}. \tag{6.3}$$

The rows of P form a basis of the vector space of holomorphic solutions of $dY = Y \cdot \mathsf{A}^{\mathsf{tr}}$ with Y as $1 \times \mathsf{b}$ matrix. We will use the notation of a period matrix for P as in Chapter 4, however, note that there is no family X/T.

6.3 The loci of constant periods

For a moment assume that $\mathfrak{k} = \mathbb{C}$ and let us consider an enhanced family X/T and a continuous family of cycles

$$\delta = \{\delta_t\}_{t\in(\mathsf{T},t_0)}, \qquad \delta_t \in H_m(X_t, \mathbb{C}).$$

We consider such a family with coefficients in $\mathbb{C}$ but later we will work with coefficients in $\mathbb{Q}$ and $\mathbb{Z}$. Let

$$\mathsf{C} = \begin{pmatrix} \mathsf{C}^0 \\ \mathsf{C}^1 \\ \vdots \\ \mathsf{C}^m \end{pmatrix}$$

be a $\mathsf{b} \times 1$ constant matrix. We call it a (constant) period vector. We are interested in the locus

$$L_\delta := \left\{ t \in (\mathsf{T}, t_0) \,\middle|\, \int_\delta \alpha = \mathsf{C} \right\}, \tag{6.4}$$

where the entries of $\alpha = [\alpha_1, \alpha_2, \ldots, \alpha_{\mathsf{b}}]^{\mathsf{tr}}$ are global sections of $H^m_{\rm dR}(\mathsf{X}/\mathsf{T})$ coming automatically from the definition of an enhanced family. We consider L_δ as an analytic scheme. Its defining ideal is generated by b holomorphic functions which are the entries of $\int_\delta \alpha - \mathsf{C}$.

Proposition 6.1. *The $\mathscr{O}_{\mathsf{T}}$-module generated by the differential forms in the entries of $\mathsf{A} \cdot \mathsf{C}$ is algebraically integrable in the sense of Definition 5.1, and hence, it induces a holomorphic foliation $\mathscr{F}(\mathsf{C})$ in T. The analytic scheme L_δ in (6.4) is a local leaf of $\mathscr{F}(\mathsf{C})$ in the sense of Definition 5.6.*

Proof. Note that L_δ might be singular and for different δ's it might have different codimensions. We have

$$d(\mathsf{A} \cdot \mathsf{C}) = d\mathsf{A} \cdot \mathsf{C} = \mathsf{A} \wedge (\mathsf{A} \cdot \mathsf{C}),$$

which proves that the entries of $\mathsf{A} \cdot \mathsf{C}$ gives us an algebraically integrable submodule of differential forms in T. Let

$$f := \int_\delta \alpha - \mathsf{C}. \tag{6.5}$$

We have

$$df = \mathsf{A} \cdot \mathsf{C} + \mathsf{A} \cdot f, \tag{6.6}$$

which follows from

$$df = \int_\delta \nabla\alpha = \mathsf{A}\cdot\int_\delta \alpha = \mathsf{A}\cdot\mathsf{C} + \mathsf{A}\cdot f.$$

This implies the last statement. □

Definition 6.1. We denote by $\mathscr{F}(\mathsf{C})$ the holomorphic foliation given by the entries of $\mathsf{A}\cdot\mathsf{C}$ and call it a modular foliation (associate to the period vector C and the enhanced family X/T).

For the definition of a modular foliation we do not need the assumption $\mathfrak{k} = \mathbb{C}$. Therefore, we can talk about a modular foliation $\mathscr{F}(\mathsf{C})$ in T which is defined over $\mathfrak{k}$ under the condition that the enhanced family $\mathsf{X}\to\mathsf{T}$ and the constant vector C are defined over $\mathfrak{k}$. Recall the period map from §4.4.

Proposition 6.2. *The local leaves of the modular foliation $\mathscr{F}(\mathsf{C})$ are given by*

$$L_{\tilde{\mathsf{C}}} : \mathsf{P}^{\mathsf{tr}}\tilde{\mathsf{C}} = \mathsf{C}, \tag{6.7}$$

for a constant vector $\tilde{\mathsf{C}} \in \mathbb{C}^{\mathsf{b}}$. In other words, the local holomorphic first integral of $\mathscr{F}(\mathsf{C})$ is given by the set of entries of $\mathsf{P}^{-\mathsf{tr}}\mathsf{C}$.

Proof. Let δ_m be as in (4.3). We just write $\delta = \delta_m^{\mathsf{tr}}\cdot\tilde{\mathsf{C}}$, where $\tilde{\mathsf{C}}$ is a constant $\mathsf{b}\times 1$-matrix with coefficients in $\mathbb{C}$ and δ is given in (4.3), and integrate α over this equality. The second statement follows from the first one or from the equalities

$$d\mathsf{P}^{-\mathsf{tr}} = -\mathsf{P}^{-\mathsf{tr}}\cdot d\mathsf{P}^{\mathsf{tr}}\cdot\mathsf{P}^{-\mathsf{tr}} = -\mathsf{P}^{-\mathsf{tr}}\cdot\mathsf{A}\cdot\mathsf{P}^{\mathsf{tr}}\cdot\mathsf{P}^{-\mathsf{tr}} = -\mathsf{P}^{-\mathsf{tr}}\cdot\mathsf{A},$$

multiplied with C from the right. □

Hopefully, the three different notations L_t, the leaf of $\mathscr{F}(\mathsf{C})$ passing through t, L_δ defined in (6.4) and $L_{\tilde{\mathsf{C}}}$ as above, will not produce any confusion. We have used (6.7) and we have introduced the notion of a modular foliation in a generalized period domain in the sense of the reference [Mov13].

The notion of a modular foliation in the present text is slightly different from the same notion introduced in [Mov11b]. In the context of the present text, the modular foliation of [Mov11b] is defined in the following way.

Definition 6.2. Let C be a period vector with entries in $\mathfrak{k}$. The modular foliation $\tilde{\mathscr{F}}(\mathsf{C})$ is given by the differential forms $\omega_1, \omega_2, \ldots, \omega_{\mathsf{b}}$, where

$$\nabla\left(\alpha^{\mathsf{tr}}\mathsf{C}\right) = \sum_{i=1}^{\mathsf{b}} \omega_i \otimes \alpha_i.$$

The leaves of $\tilde{\mathscr{F}}(\mathsf{C})$ are given by the constant loci of the periods of

$$\alpha^{\mathsf{tr}}\mathsf{C} = \sum_{i=1}^{\mathsf{b}} \mathsf{C}_i \cdot \alpha_i,$$

that is, the constant loci of the entries of $\mathsf{P}\cdot\mathsf{C}$. Note that C_i's are the entries of C, whereas C^i's are Hodge blocks of C. The two notions of modular foliations $\tilde{\mathscr{F}}(\mathsf{C})$ and $\mathscr{F}(\mathsf{C})$ are the same if we interchange the role of the Gauss–Manin connection by its dual, that is, what is a modular foliation in the first sense and derived from the Gauss–Manin connection matrix A is modular in the second sense and derived from the dual A^{tr} of the Gauss–Manin connection matrix and vice versa. Note that the dual of the Gauss–Manin connection does not satisfy the Griffiths transversality.

The modular foliation $\mathscr{F}(\mathsf{C})$ may have singularities in T, that is, points t_0 where the wedge of all the entries of $\mathsf{A}\cdot\mathsf{C}$ vanishes. For curiosity, one may try to characterize such a t_0 in terms of a property of the projective variety X_{t_0}. Throughout the present text we will avoid $\mathrm{Sing}(\mathscr{F}(\mathsf{C}))$.

Remark 6.1. The reader who does not like taking basis of vector spaces and like a more intrinsic presentation of objects may identify C with a $\mathfrak{k}$-linear map

$$\mathsf{C} : H^m(X_0/\mathfrak{k}) \to \mathfrak{k}.$$

This maps $\alpha_{m,i}$ to the ith entry of C_i. In this way, C is interpreted as an element in the mth homology of $X/\mathfrak{k}$. Let $\check{\delta}_t \in H^m(X_t, \mathbb{C})$ be obtained after taking Poincaré dual of δ_t and then taking its pull-back by the isomorphism $L^{n-m} : H^m(X_t, \mathbb{C}) \to H^{2n-m}(\mathsf{X}_t, \mathbb{C})$ in the Hard Lefschetz theorem. For simplicity, we use δ instead of $\check{\delta}$. A leaf L_δ of $\mathscr{F}(\mathsf{C})$ is the loci of parameters $t \in \mathsf{T}$ such that

(1) δ is a holomorphic flat section of $H^m(\mathsf{X}/\mathsf{T})$ and
(2) the image of δ under the composition

$$H^*_{\rm dR}(X/\mathsf{T}) \overset{\alpha}{\cong} H^*_{\rm dR}(X_0) \otimes_{\mathfrak{k}} \mathscr{O}_\mathsf{T} \overset{\mathsf{C}\otimes \mathrm{Id}}{\cong} \mathscr{O}_\mathsf{T} \tag{6.8}$$

maps to $\mathfrak{k} \subset \mathscr{O}_\mathsf{T}$, that is, its derivation with respect to any vector field in T is zero.

Finally, the following definition seems to be useful in many instances.

Definition 6.3. A leaf L of the modular foliation $\mathscr{F}(\mathsf{C})$ is homologically defined over $\mathbb{Q}$ if the corresponding family of continuous cycles $\delta = \delta_t$, $t \in (\mathsf{T}, t_0)$ with $L = L_\delta$ is in $H_m(\mathsf{X}_t, \mathbb{Q})$, that is, it lives in homologies with rational coefficients. It is weak homologically defined over $\mathbb{Q}$ if $c \cdot \delta_t$, for some constant $c \in \mathbb{C}$, is in $H_m(\mathsf{X}_t, \mathbb{Q})$.

Note that in the above definition L might be a transcendental leaf. In case, L is an algebraic leaf, we have also the field of definition of L which *a priori* has nothing to do with the above definition. Moreover, for a period vector C, the modular foliation $\mathscr{F}(\mathsf{C})$ might not have any leaf defined over $\mathbb{Q}$. We do not know any example of such a foliation. The notion of a weak homologically defined leaf over $\mathbb{Q}$ is inspired by the notion of a weak absolute Hodge cycle introduced in Chapter 7.

6.4 Algebraic groups and modular foliations

Let X/T be a full family of enhanced smooth projective varieties. In this section, we describe the relation between the algebraic group G which acts on T and the modular foliation $\mathscr{F}(\mathsf{C})$ which lives in T.

Proposition 6.3. *For an element* g *of the algebraic group* G*, the isomorphism* $\mathsf{T} \to \mathsf{T}$, $t \mapsto t \bullet \mathsf{g}$ *maps* $\mathscr{F}(\mathsf{g}^{-\mathsf{tr}}\mathsf{C})$ *to* $\mathscr{F}(\mathsf{C})$*. In the complex context, this means that it maps a leaf of* $\mathscr{F}(\mathsf{g}^{-\mathsf{tr}}\mathsf{C})$ *to a leaf of* $\mathscr{F}(\mathsf{C})$*.*

Proof. Using Proposition 3.6 we know that the pull-back of AC is $\mathsf{g}^{\mathsf{tr}}\mathsf{A}\mathsf{g}^{-\mathsf{tr}}\mathsf{C}$. We know also that foliations induced by $\mathsf{g}^{\mathsf{tr}}\mathsf{A}\mathsf{g}^{-\mathsf{tr}}\mathsf{C}$ and $\mathsf{A}\mathsf{g}^{-\mathsf{tr}}\mathsf{C}$ are the same. Therefore, the pull-back of $\mathscr{F}(\mathsf{C})$ is $\mathscr{F}(\mathsf{g}^{-\mathsf{tr}}\mathsf{C})$. For $\mathfrak{k} = \mathbb{C}$ the leaves of $\mathscr{F}(\mathsf{C})$ are described in Proposition 6.1. Using (3.11) we have

$$\int_{\delta_{t\bullet \mathsf{g}}} \alpha = \int_{\delta_t} \mathsf{g}^{\mathsf{tr}}\alpha = \mathsf{g}^{\mathsf{tr}} \int_{\delta_t} \alpha = \mathsf{g}^{\mathsf{tr}}(\mathsf{g}^{-\mathsf{tr}}\mathsf{C}) = \mathsf{C}, \tag{6.9}$$

which implies the desired statement. Note that the action of $\mathsf{g} \in \mathsf{G}$ in X induces an isomorphism $\mathsf{X}_t \cong \mathsf{X}_{t\bullet\mathsf{g}}$ and $\delta_{t\bullet\mathsf{g}}$ is the push-forward of δ_t by this isomorphism. □

Let

$$\mathrm{Stab}(\mathsf{G}, \mathsf{C}) := \{\mathsf{g} \in \mathsf{G} \mid \mathsf{g}^{\mathrm{tr}}\mathsf{C} = \mathsf{C}\} \tag{6.10}$$

be the stabilizer of C and let $\mathrm{Stab}(\mathsf{G}, \mathsf{C})_0$ be the identity component, that is, the connected component of $\mathrm{Stab}(\mathsf{G}, \mathsf{C})$ that contains the identity element.

Proposition 6.4. *For a point $t \in \mathsf{T}$, the set $t \bullet \mathrm{Stab}(\mathsf{G}, \mathsf{C})_0$ is contained in the leaf of $\mathscr{F}(\mathsf{C})$ through t.*

Proof. For $\mathfrak{k} = \mathbb{C}$, this follows from (6.9). □

Definition 6.4. A modular foliation $\mathscr{F} = \mathscr{F}(\mathsf{C})$ is called trivial if the leaves of $\mathscr{F}$ in T are all of the form $t \bullet \mathrm{Stab}(\mathsf{G}, \mathsf{C})_0$, $t \in \mathsf{T}$, and hence, all the leaves are algebraic.

A strategy to prove that a foliation is trivial is by showing that it induces a zero dimensional foliation restricted to the leaves of $\mathscr{F}(2)$. The foliation $\mathscr{F}(2)$ will be introduced in §6.10. For a period vector of Hodge type, we may try to prove that the differential forms in the matrix $\mathsf{A}_{\frac{n}{2}-1,\frac{n}{2}}\mathsf{C}_{\frac{n}{2}}$ restricted to the leaves of $\mathscr{F}(2)$ are linearly independent.

Let us now consider a full enhanced family X/T with an action of a reductive group as in §3.7. By Proposition 3.9 the entries of the Gauss–Manin connection matrix A are invariant under the action of $\mathbf{G}$. This implies that the modular foliation $\mathscr{F}(\mathsf{C})$ is invariant under $\mathbf{G}$ and the orbits of $\mathbf{G}$ are inside the leaves of $\mathscr{F}(\mathsf{C})$. In this case, we might modify Definition 6.4 as follows.

Definition 6.5. A modular foliation $\mathscr{F} = \mathscr{F}(\mathsf{C})$ attached to a full enhanced family with an action of a reductive group is called trivial if the leaves of $\mathscr{F}$ in T are all of the form

$$\mathbf{G} \cdot t \bullet \mathrm{Stab}(\mathsf{G}, \mathsf{C})_0, \quad t \in \mathsf{T}.$$

Note that by our assumption, the reductive group $\mathbf{G}$ is irreducible. It might happen that a modular foliation $\mathscr{F}$ is trivial in $\mathsf{T}\backslash\mathrm{Sing}(\mathscr{F})$ but not in T and it would be interesting to construct an explicit example for this. For a period vector of Hodge type C, trivial modular foliations are related to isolated Hodge cycles that we will discuss them in §7.4.

6.5 A character

In our study of modular foliations, it turns out that the algebraic group $\mathsf{Stab}(\mathsf{G},\mathsf{C})$ defined in §6.4 can be replaced with

$$\mathsf{G}_\mathsf{C} := \{\mathsf{g} \in \mathsf{G} \mid \mathsf{g}^{\mathsf{tr}}\mathsf{C} = \lambda(\mathsf{g})\mathsf{C}, \text{ for some } \lambda(g) \in \mathfrak{k}^*\} \tag{6.11}$$

which turns out to be more useful. It follows from the definition that G_C is a subgroup of G and

$$\lambda : \mathsf{G}_\mathsf{C} \to \mathbb{G}_m$$

is a character of G_C. Here, $\mathbb{G}_m = (\mathfrak{k}^*, \cdot)$ is the multiplicative group of $\mathfrak{k}$. We will also use $\mathbb{G}_a = (\mathfrak{k}, +)$ for the additive group of $\mathfrak{k}$. The kernel of λ is the algebraic group defined in (6.10).

Proposition 6.5. *The algebraic group* G_C *maps* $\mathscr{F}(\mathsf{C})$ *to itself. Moreover, it maps the leaf* $L_{\lambda(\mathsf{g})\check{\mathsf{C}}}$ *to* $L_{\check{\mathsf{C}}}$.

Proof. It follows from the equalities

$$(\mathsf{Pg})^{-\mathsf{tr}}\mathsf{C} = \lambda(\mathsf{g})^{-1}\mathsf{P}^{-\mathsf{tr}}\mathsf{C}$$

for $\mathsf{g} \in \mathsf{G}_\mathsf{C}$. □

Definition 6.6. We say that a modular foliation $\mathscr{F}(\mathsf{C})$ has a non-trivial character if λ is non-trivial, that is, its image is not a finite subgroup of $\mathbb{G}_m$.

The universal family of enhanced elliptic curves and the corresponding modular foliation given by Ramanujan vector field has a non-trivial character, see Chapter 9. However, in the case of product of two elliptic curves, the modular foliation has a trivial character, see Chapter 10. This simple fact will affect the way we study the dynamics of modular foliations. We use Proposition 5.14 with the group action G_C and conclude that the foliation $\mathscr{F}(\mathsf{C})$ is inside another foliation of bigger dimension.

Definition 6.7. We define $\mathscr{F}(\mathsf{C},\lambda)$ to be the foliation $\tilde{\mathscr{F}}$ in Proposition 5.14 constructed from $\mathscr{F}(\mathsf{C})$ and the action of G_C in T.

Note that if the modular foliation $\mathscr{F}(\mathsf{C})$ has trivial character then

$$\mathscr{F}(\mathsf{C},\lambda) = \mathscr{F}(\mathsf{C}).$$

The property of a modular foliation having a trivial or non-trivial character is reflected directly in its dynamics. In fact, this is the main reason behind the definition of such a property.

6.6 Space of leaves

In this section, we only deal with the notion of a geometric leaf as in Definition 5.13. Therefore, we discard first the scheme structure of L and then we replace it with one of its irreducible components. In particular, the local leaves of the modular foliation $\mathscr{F}(\mathsf{C})$ are irreducible components of $L_{\tilde{\mathsf{C}}}$ given in (6.7) and this format is obtained after fixing a basis

$$\delta_{m,i} \in H_m(X_t, \mathbb{Z}), \qquad i = 1, 2, \ldots, \mathsf{b}, \qquad t \in (\mathsf{T}, t_0).$$

Recall that the discrete group $\Gamma_{\mathbb{Z}}$ in §4.2 acts on the basis $\delta_{m,i}$ of $H_m(X, \mathbb{Z})$ as an operation of basis change. The foliation $\mathscr{F}(\mathsf{C})$ may have singularities in T, and if t_0 is not a singularity of $\mathscr{F}(\mathsf{C})$ then $L_{\tilde{\mathsf{C}}}$ is smooth and hence it is a local leaf in the sense which is explained at the beginning of this section.

Proposition 6.6. *Let $L_1 \subset L_{\tilde{\mathsf{C}}_1} \subset (\mathsf{T}, t_1)$ and $L_2 \subset L_{\tilde{\mathsf{C}}_2} \subset (\mathsf{T}, t_2)$ be two local leaves. If they lie in a global leaf L of $\mathscr{F}(\mathsf{C})$ (see Definition 5.14), then there is $A \in \Gamma_{\mathbb{Z}}$ such that $A\tilde{\mathsf{C}}_1 = \tilde{\mathsf{C}}_2$.*

Proof. If $L_1 = L_2$ then $\mathsf{P}^{-\mathsf{tr}}\mathsf{C}$ evaluated on this set gives us both $\tilde{\mathsf{C}}_1$ and $\tilde{\mathsf{C}}_2$ and hence these are equal. Otherwise, we take a path from t_1 to t_2 in L and consider the analytic continuation of $\mathsf{P}^{-\mathsf{tr}}\mathsf{C}$ along this path. Note that both $L_{\tilde{\mathsf{C}}_1}$ and $L_{\tilde{\mathsf{C}}_2}$ are defined fixing the same basis $\delta_{m,i}$. □

There is no any reason why the converse of Proposition 6.6 must be true, that is, two analytic sets $L_{\tilde{\mathsf{C}}}$ and $L_{A\tilde{\mathsf{C}}} \subset (\mathsf{T}, t_0)$ for some $A \in \Gamma_{\mathbb{Z}}$ might not have any common connected component. Moreover, we might have two local leaves in different open subsets of T and with the same $\tilde{\mathsf{C}}$ such that they do not lie in a global leaf of $\mathscr{F}(\mathsf{C})$. Even in the local context $L_{\tilde{\mathsf{C}}}$ might have many irreducible components with different codimensions. For elliptic curves (see §9.10) we will actually show that the converse of Proposition 6.6 is true, however, this is a very particular case. In general, we expect that the set $\bigcup_{A\in\Gamma_{\mathbb{Z}}} L_{A\tilde{\mathsf{C}}}$ consists of finitely many leaves of $\mathscr{F}(\mathsf{C})$. This is actually the case when the generalized period map is a biholomorphism (global Torelli in our context). We will show this property for elliptic curves, K3 surfaces, cubic fourfolds and abelian varieties, see §8.12.

Definition 6.8. Let us define

$$\mathscr{L}(\mathsf{C}) := \{\tilde{\mathsf{C}} \in \mathbb{C}^{\mathsf{b}} \mid L_{\tilde{\mathsf{C}}} \neq \emptyset, \quad \text{for some } t_0 \in \mathsf{T} \text{ and a choice of basis } \delta_{m,i}\}. \tag{6.12}$$

This is invariant under $\Gamma_{\mathbb{Z}}$-action from the left. The quotient $\Gamma_{\mathbb{Z}}\backslash\mathscr{L}(\mathsf{C})$ is called the space of leaves of $\mathscr{F}(\mathsf{C})$.

The set $\mathscr{L}(\mathsf{C})$ is the image of the multi-valued function $\mathsf{P}^{-\mathsf{tr}}\mathsf{C}$ from T to $\mathbb{C}^{\mathsf{b}}$. This multivalued function may be a disjoint union of infinitely many multivalued functions. This is due to the fact that the monodromy group inside $\Gamma_{\mathbb{Z}}$ may have infinite index. In $\mathscr{L}(\mathsf{C})$ we consider the topology induced by the usual topology of $\mathbb{C}^{\mathsf{b}}$. For some examples such as elliptic curves, the set $\mathscr{L}(\mathsf{C})$ is an open set in $\mathbb{C}^{\mathsf{b}}$. However, the space of leaves may not have any kind of good topology. For instance, we cannot talk about a global leaf being near to another one, because locally one leaf may be near to another one, however, their analytic continuation may result in dense subsets of the ambient space. We use Theorem 6.6 and find that for the modular foliation $\mathscr{F}(\mathsf{C})$ with a non-trivial character λ, the space of leaves of the foliation $\mathscr{F}(\mathsf{C},\lambda)$ is given by

$$\Gamma_{\mathbb{Z}}\backslash\mathscr{L}(\mathsf{C})/\mathbb{G}_m.$$

The advantage of $\mathscr{F}(\mathsf{C},\lambda)$ is that $\mathscr{L}(\mathsf{C})/\mathbb{G}_m \subset \mathbb{P}^{\mathsf{b}-1}$ and $\mathbb{P}^{\mathsf{b}-1}$ is compact. For the following proposition, we need to mention that we are working with the mth (co)homology, and so, we write the (co)homology subscripts if they are different from m. Recall the matrices Φ_m and Ψ_m in (3.1) and (4.4).

Proposition 6.7. *Let $\mathscr{F}(\mathsf{C})$ be a modular foliation. We have*

$$\mathscr{L}(\mathsf{C}) \subset \{\tilde{\mathsf{C}} \in \mathbb{C}^{\mathsf{b}} \mid \tilde{\mathsf{C}}^{\mathsf{tr}}\Psi_{2n-m}^{\mathsf{tr}}\tilde{\Psi}_{2n-m}^{-1}\Psi_{2n-m}\tilde{\mathsf{C}} = \mathsf{C}^{\mathsf{tr}}\Phi_m^{-1}\mathsf{C}\}. \tag{6.13}$$

In particular, if the number $\mathsf{C}^{\mathsf{tr}}\Phi_m^{-1}\mathsf{C}$ is not zero then the modular foliation $\mathscr{F}(\mathsf{C})$ has a trivial character and $\mathscr{L}(\mathsf{C})$ is not invariant under $\mathbb{G}_m$-action on $\mathbb{C}^{\mathsf{b}}$ given by $(a,\tilde{\mathsf{C}}) \mapsto a\tilde{\mathsf{C}}$ for $a \in \mathbb{G}_m$ and $\tilde{\mathsf{C}} \in \mathbb{C}^{\mathsf{b}}$.

Proof. The affirmation (6.13) follows from $\mathsf{P}^{\mathsf{tr}}\tilde{\mathsf{C}} = \mathsf{C}$ and Proposition 4.1. If $\mathscr{F}(\mathsf{C})$ has a non-trivial character then by Proposition 6.5, $\mathscr{L}(\mathsf{C})$ is closed under the $\mathbb{G}_m$ action on $\mathbb{C}^{\mathsf{b}}$. This is in contradiction with (6.13) and $\mathsf{C}^{\mathsf{tr}}\Phi_m^{-1}\mathsf{C} \neq 0$. □

Remark 6.2. For δ as in the beginning of §6.3, the equality (6.13) is equivalent to

$$\langle\delta_t,\delta_t\rangle = \mathsf{C}^{\mathsf{tr}}\Phi_m^{-1}\mathsf{C},$$

where $\langle\cdot,\cdot\rangle$ is defined in (4.5).

6.7 Moduli of modular foliations

For a constant $a \in \mathsf{k}$, $a \neq 0$ the foliation induced by $\mathscr{F}(\mathsf{C})$ and $\mathscr{F}(a\mathsf{C})$ are the same. We use Proposition 6.3 and we know that the foliations $\mathscr{F}(\mathsf{g}^{\mathsf{tr}}\mathsf{C})$ for all $\mathsf{g} \in \mathsf{G}$ are isomorphic.

Definition 6.9. We define

$$\mathsf{MF}_m := \mathsf{G}\backslash\mathbb{P}^{\mathsf{b}-1}$$

to be the moduli of modular foliations in mth cohomologies. Here, the action is defined by

$$(\mathsf{g}, \mathsf{C}) \mapsto \mathsf{g}^{\mathsf{tr}}\mathsf{C}, \quad \mathsf{g} \in \mathsf{G}, \quad \mathsf{C} \in \mathbb{P}^{\mathsf{b}-1}.$$

For the study of Hodge cycles, we also need the vector

$$\mathsf{C} = \begin{pmatrix} 0 \\ \vdots \\ 0 \\ \mathsf{C}^{\frac{m}{2}} \\ \vdots \\ \mathsf{C}^{m} \end{pmatrix} \in \mathsf{k}^{\mathsf{b}_m} \tag{6.14}$$

or

$$\mathsf{C} = \begin{pmatrix} 0 \\ \vdots \\ 0 \\ \mathsf{C}^{\frac{m}{2}} \\ 0 \\ \vdots \\ 0 \end{pmatrix} \in \mathsf{k}^{\mathsf{b}_m}. \tag{6.15}$$

We call it a period vector of Hodge type. The following property in many examples turns out to be true.

Property 6.1. For any C of the form (6.14) there is $\mathsf{g} \in \mathsf{G}$ such that $\mathsf{g}^{\mathsf{tr}}\mathsf{C}$ is of the form (6.15), that is only the middle Hodge block might be non-zero.

It seems that Property 6.1 has some relation with Property 3.1 and it might be verified using Proposition 3.12. We identify $\mathbb{A}_{\mathsf{k}}^{\mathsf{h}^{\frac{m}{2},\frac{m}{2}}} \subset \mathbb{A}_{\mathsf{k}}^{\mathsf{b}_m}$ by the space of vectors (6.15). Let $\mathrm{Stab}(\mathsf{G}, \mathbb{A}_{\mathfrak{k}}^{\mathsf{h}^{\frac{m}{2},\frac{m}{2}}})$ be the stabilizer group of the set $\mathbb{A}_{\mathfrak{k}}^{\mathsf{h}^{\frac{m}{2},\frac{m}{2}}}$.

Definition 6.10. We call the quotient

$$\mathrm{Stab}(\mathsf{G}, \mathbb{A}_{\mathfrak{k}}^{\mathsf{h}^{\frac{m}{2},\frac{m}{2}}})\backslash\mathbb{P}_{\mathfrak{k}}^{\mathsf{h}^{\frac{m}{2},\frac{m}{2}}-1}$$

the moduli of modular foliations of Hodge type.

This moduli space in many interesting cases such as hypersurfaces turns out to have just one point. For an example see Chapter 10.

For a modular foliation $\mathscr{F}(\mathsf{C})$ attached to an enhanced family X/T and a constant period vector C, we expect that an $\mathscr{F}(\mathsf{C})$-invariant algebraic subset V of T has a geometric meaning in the sense that the fibers X_t, $t \in V$ enjoy a particular algebraic structure, such as the existence of an algebraic cycle, such that those t outside V do not enjoy such a structure. At this point, it might be too early to formulate a precise conjectural statement. In order to develop this idea further we have worked out two examples, the case of elliptic curves and the case of product of two elliptic curves, see Chapters 9 and 10. In the first case, we prove that any algebraic invariant set for the modular foliation given by the Ramanujan vector field is necessarily contained in the discriminant locus. In the second case, we construct a modular foliation and we prove that the only algebraic leaves outside the discriminant locus are those derived from modular curves.

6.8 Modular foliations and the Lefschetz decomposition

Let C be as in §6.3. Since α_m is also compatible with the Lefschetz decomposition of $H^m_{\mathrm{dR}}(X_t)$, we may rearrange the entries of α_m and write

$$\alpha_m = \left[\alpha_m^0, \alpha_{m-2}^1, \ldots, \alpha_{m-2[\frac{m}{2}]}^{[\frac{m}{2}]}\right]^{\mathsf{tr}}, \tag{6.16}$$

where α^q_{m-2q} is a basis of $H^{m-2q}(X_t)_0$ or its image in $H^m(X_t)$ under the cup product with θ^q; being clear in the text which we mean. For instance, in the above equality we have used the second interpretation. In a similar way we write

$$\mathsf{C}=\begin{pmatrix}\mathsf{C}^0_m\\ \mathsf{C}^1_{m-2}\\ \vdots\\ \mathsf{C}^{[\frac{m}{2}]}_{m-2[\frac{m}{2}]}\end{pmatrix}=\begin{pmatrix}\mathsf{C}^0_m\\ 0\\ \vdots\\ 0\end{pmatrix}+\begin{pmatrix}0\\ \mathsf{C}^1_{m-2}\\ \vdots\\ 0\end{pmatrix}+\cdots+\begin{pmatrix}0\\ 0\\ \vdots\\ \mathsf{C}^{[\frac{m}{2}]}_{m-2[\frac{m}{2}]}\end{pmatrix}$$

and call it the decomposition of C into primitive pieces. By abuse of notation, we also write

$$\mathsf{C}=\mathsf{C}^0_m+\mathsf{C}^1_{m-2}+\cdots+\mathsf{C}^{[\frac{m}{2}]}_{m-2[\frac{m}{2}]}.$$

The vectors C^q_{m-2q} are called primitive (constant) period vectors.

Proposition 6.8. *We have*

$$\mathscr{F}(\mathsf{C}):=\bigcap_{q=0}^{[\frac{m}{2}]}\mathscr{F}(\mathsf{C}^q_{m-2q}).$$

Proof. In the algebraic context the proof is as follows. The Gauss–Manin connection respects the Lefschetz decomposition, and so, the Gauss–Manin connection matrix in the basis (6.16) is a block diagonal $\mathsf{A}=\mathrm{diag}(\mathsf{A}_m,\mathsf{A}_{m-2},\ldots)$. This implies that the set of entries of $\mathsf{A}\cdot\mathsf{C}$ is a disjoint union of $\mathsf{A}_{m-2q}\mathsf{C}_{m-2q}$, $q=0,1,\ldots,[\frac{m}{2}]$. By definition this is exactly the desired statement.

In the geometric context the proof is as follows. Let us take a continuous family of cycles $\delta_t\in H_m(X_t,\mathbb{C})$ and write its Lefschetz decomposition

$$\delta_t=\sum_{q=0}^{[\frac{m}{2}]}\delta_{t,m-2q},$$

where $\delta_{t,m-2q}\in H_{m-2q}(\mathsf{X}_t,\mathbb{C})$ is a continuous family of primitive cycles. We have the following identity of local leaves:

$$L=\bigcap_{q=0}^{[\frac{m}{2}]}L_{\delta_{t,m-2q}} \tag{6.17}$$

which follows from

$$\int_{\delta_{t,q}} \alpha_{m-2p} = 0, \quad \text{if } p \neq q. \qquad \square$$

6.9 Some other foliations

The integrability of the Gauss–Manin connection gives us other interesting foliations in T. The equality $d\mathsf{A} = \mathsf{A} \wedge \mathsf{A}$ breaks into

$$d\mathsf{A}^{i,j} = \mathsf{A}^{i,0} \wedge \mathsf{A}^{0,j} + \mathsf{A}^{i,1} \wedge \mathsf{A}^{1,j} + \cdots + \mathsf{A}^{i} \wedge \mathsf{A}^{m,j}$$

and so we have the following definition.

Definition 6.11. For $a = 1, 2, \ldots$, the entries of the following matrices form an integrable $\mathscr{O}_{\mathsf{T}}$-module

$$\mathsf{A}^{i,j}, \quad j - i < a - 1.$$

We denote by $\mathscr{F}(a)$ the corresponding holomorphic foliation in T. For $a = 0, -1, -2, \ldots$, we denote by $\mathscr{F}(a)$ the foliation induced by the entries of

$$\mathsf{A}^{i,j}, \quad j - i > a.$$

For any vector field v_2 in T tangent to $\mathscr{F}(2)$, we have

$$\mathsf{A}_{\mathsf{v}_2} = \begin{pmatrix} 0 & * & 0 & 0 & 0 & 0 \\ 0 & 0 & * & 0 & 0 & 0 \\ 0 & 0 & 0 & * & 0 & 0 \\ 0 & 0 & 0 & 0 & * & 0 \\ 0 & 0 & 0 & 0 & 0 & * \\ 0 & 0 & 0 & 0 & 0 & 0 \end{pmatrix}, \tag{6.18}$$

where $\mathsf{A}_{\mathsf{v}_2}$ is the contraction of the entries of A along the vector field v_2, and for any vector field v_1 in T tangent to $\mathscr{F}(1)$ we have

$$\mathsf{A}_{\mathsf{v}_1} = \begin{pmatrix} * & * & 0 & 0 & 0 & 0 \\ 0 & * & * & 0 & 0 & 0 \\ 0 & 0 & * & * & 0 & 0 \\ 0 & 0 & 0 & * & * & 0 \\ 0 & 0 & 0 & 0 & * & * \\ 0 & 0 & 0 & 0 & 0 & * \end{pmatrix} \tag{6.19}$$

(samples for $m = 5$). Any vector field v_a tangent to the leaves of $\mathscr{F}(a)$, $a = 0, -1, -2, -3, -4$ satisfies

$$\mathsf{A}_{\mathsf{v}_0} = \begin{pmatrix} * & 0 & 0 & 0 & 0 \\ * & * & 0 & 0 & 0 \\ * & * & * & 0 & 0 \\ * & * & * & * & 0 \\ * & * & * & * & * \end{pmatrix}, \quad \mathsf{A}_{\mathsf{v}_{-1}} = \begin{pmatrix} 0 & 0 & 0 & 0 & 0 \\ * & 0 & 0 & 0 & 0 \\ * & * & 0 & 0 & 0 \\ * & * & * & 0 & 0 \\ * & * & * & * & 0 \end{pmatrix}, \quad \mathsf{A}_{\mathsf{v}_{-2}} = \begin{pmatrix} 0 & 0 & 0 & 0 & 0 \\ 0 & 0 & 0 & 0 & 0 \\ * & 0 & 0 & 0 & 0 \\ * & * & 0 & 0 & 0 \\ * & * & * & 0 & 0 \end{pmatrix},$$

$$\mathsf{A}_{\mathsf{v}_{-3}} = \begin{pmatrix} 0 & 0 & 0 & 0 & 0 \\ 0 & 0 & 0 & 0 & 0 \\ 0 & 0 & 0 & 0 & 0 \\ * & 0 & 0 & 0 & 0 \\ * & * & 0 & 0 & 0 \end{pmatrix}, \quad \mathsf{A}_{\mathsf{v}_{-4}} = \begin{pmatrix} 0 & 0 & 0 & 0 & 0 \\ 0 & 0 & 0 & 0 & 0 \\ 0 & 0 & 0 & 0 & 0 \\ 0 & 0 & 0 & 0 & 0 \\ * & 0 & 0 & 0 & 0 \end{pmatrix}$$

(samples for $m = 4$). Note that the foliation $\mathscr{F}(0)$ is given by the entries of

$$\mathsf{A}^{i,i+1}, \quad i = 0, 1, 2, \ldots, m-1. \tag{6.20}$$

Proposition 6.9. *We have*

$$\mathscr{F}(a) = \mathscr{F}(b), \quad a, b \geq 3$$

and along the leaves of this foliation the period matrix P *is constant.*

Proof. By Griffiths transversality we have $\mathsf{A}^{i,j} = 0,\ j - i \geq 2$ and so for $a \geq 3$ the foliation $\mathscr{F}(a)$ is induced by all the entries of A. □

So far, we worked with a fixed m and so we removed the subscript m from our notations. For the following definition, we need m to be back, and so we write $\mathsf{A}_m = \mathsf{A}$, $\mathscr{F}_m(a) = \mathscr{F}(a)$, etc. Note that for $m = 0, 2n$, A_m is the 1×1 zero matrix and hence the corresponding objects are trivial.

Definition 6.12. We define

$$\mathscr{F}(a) = \bigcap_{m=0}^{2n} \mathscr{F}_m(a), \quad a \in \mathbb{Z}. \tag{6.21}$$

Since in the generalized period domain the Griffiths transversality is not necessarily integrable, the foliations $\mathscr{F}(a)$ in that context are non-trivial even for $a \geq 3$, see Chapter 8. Let $\mathscr{F}(\mathsf{G})$ be the holomorphic foliation attached to the action of G on T.

Proposition 6.10. *We have*

$$\mathscr{F}(\mathsf{G}) \subset \mathscr{F}(0).$$

In geometric terms, this means that the leaves of the foliation $\mathscr{F}(\mathsf{G})$ *are contained in the leaves of* $\mathscr{F}(0)$*, that is, for any vector field* v *tangent to* $\mathscr{F}(\mathsf{G})$*,* v *is also tangent to all* $\mathscr{F}(0)$*.*

We will prove this proposition later in Proposition 6.21. From Proposition 3.6, the fact that g^{tr} is block lower triangular and Griffiths transversality data in (3.20), it follows that the $\mathscr{O}_{\mathsf{T}}$-module generated by (6.20) is invariant under the action of G. However, this does not imply Proposition 6.10. We may suspect that the leaves of $\mathscr{F}(\mathsf{G})$ are given by the intersection of the leaves of $\mathscr{F}(0)$'s.

Conjecture 6.1. *We have*

$$\mathscr{F}(\mathsf{G}) = \mathscr{F}(0).$$

The foliation $\mathscr{F}(3)$ has to do with repeated elements in X/T. Let us consider a leaf L of this foliation. The Gauss–Manin connection matrix A_m restricted to L is identically zero for all m, and so all the period matrices P_m restricted to L are constants, see Proposition 6.9. In particular, for any two points $t_1, t_2 \in \mathsf{T}$ the varieties X_{t_1} and X_{t_2} have the same Hodge structures. We get the following proposition.

Proposition 6.11. *Let* X/T *be an enhanced family and assume that the foliation* $\mathscr{F}(3)$ *has a non-zero dimensional leaf* L*. Then all* (X_t, α_t)*,* $t \in L$ *have the same period matrix, and consequently, their Hodge structures are isomorphic.*

It might be of interest to see whether (X_t, α_t), $t \in L$ are all isomorphic to each other. If not, this would give a counterexample to the local Torelli theorem.

Property 6.2. Let X/T be a universal enhanced family. Then all the leaves of $\mathscr{F}(3)$ are zero dimensionals, that is, they are points.

Proposition 6.12. *Let* X/T *be an enhanced family with the action of a reductive group from the left described in §3.7. We have*

$$\mathscr{F}(\mathbf{G}) \subset \mathscr{F}(3).$$

Proof. This follows from the fact that the Gauss–Manin connection matrix A is invariant under the action of $\mathbf{G}$, see Proposition 3.6. □

6.10 The foliation $\mathscr{F}(2)$

The foliation $\mathscr{F}(2)$ plays an important role through the present text. In the case of elliptic curves it turns out to be given by the Ramanujan differential equation. Just in this case we have verified the following property.

Property 6.3. For a universal family of enhanced projective varieties X/T (if it exists), the foliation $\mathscr{F}(2)$ has no algebraic leaf.

We give a plausible argument for the above property using Deligne's global invariant cycle theorem. If $\mathscr{F}(2)$ has an algebraic leaf then we have a proper smooth family of projective varieties $Y \to V$ with global sections $\alpha_{m,\imath}$, $\imath = 1, 2, \ldots,$ of $H^m_{\rm dR}(Y/V)$ which form a basis and further the Gauss–Manin connection written in this basis has the format (6.18), that is, if we write $\alpha_m := [\alpha^0, \alpha^1, \ldots, \alpha^m]$, where α^i's are the Hodge blocks of α_m, then

$$\nabla(\alpha^i) = \mathsf{A}^{i,i+1}\alpha^{i+1}, \quad i = 0, 1, \ldots, m$$

for $\mathsf{h}^{m-i,i} \times \mathsf{h}^{m-i-1,i+1}$ matrices $\mathsf{A}^{i,i+1}$ with entries in $\Omega_V(V)$. In particular, for $i = m$ we have $\nabla(\alpha^i) = 0$ and so all the entries of α^m are flat sections. Let $\bar{Y} \to \bar{V}$ be any smooth projective compactification of $Y \to V$. Deligne's global invariant cycle theorem (see [Del71a, Theorem 4.1.1, "théorèm de la partie fixe"; Del68, Voi07, Voi13, Mov19, Chapter 6]) implies that the restriction map

$$f_{\mathbb{Q}} : H^m(\bar{Y}, \mathbb{Q}) \to H^m(Y_0, \mathbb{Q})$$

is a morphism of Hodge structures and its image is equal to the set of monodromy invariant cycles, therefore, the entries of α^m lies in its image. This implies that $H^{0,m}$ and hence $H^{m,0}$ is in the image of $f_{\mathbb{C}}$. It seems to me that $f_{\mathbb{Q}}$ is surjective and hence all the elements of $H^m(Y_0, \mathbb{Q})$ are invariant under monodromy. We conclude that all the Hodge structures $H^m(Y_b, \mathbb{Q})$, $b \in V$ are isomorphic, and hence, if we assume the local Torelli for X/T we get a contradiction.

It must be remarked that a natural place to define the foliation $\mathscr{F}(2)$ is the moduli of Hodge decompositions defined in §3.12. In this space, a vector field tangent to $\mathscr{F}(2)$ has the property that when it is composed with the Gauss–Manin connection, it sends the $H^{p,q}$ to $H^{p-1,q+1}$. For the description of $\mathscr{F}(2)$ in the case of elliptic curves, see §9.7.

Let $(L, 0) \cong (\mathbb{C}^N, 0)$ be a local leaf of $\mathscr{F}(2)$ and let $\mathscr{F}(\mathsf{C})$ be a modular foliation with C a period vector of Hodge type as in (6.15). Let also

$$f := \left(\int_\delta \alpha - \mathsf{C}\right)\Big|_L = \begin{pmatrix} f^0 \\ f^1 \\ \vdots \\ f^m \end{pmatrix},$$

which is introduced in §6.3. We assume that $f(0) = 0$ and so the loci of constant periods crosses $0 \in L$. The differential equation (6.6) restricted to L breaks into pieces:

$$\begin{cases} df^0 = \mathsf{A}^{01} f^1 \\ df^1 = \mathsf{A}^{12} f^2 \\ \quad \vdots \\ df^{\frac{m}{2}-1} = \mathsf{A}^{\frac{m}{2}-1,\frac{m}{2}} f^{\frac{m}{2}} + \mathsf{A}^{\frac{m}{2}-1,\frac{m}{2}} \mathsf{C}^{\frac{m}{2}} \\ \quad \vdots \\ df^{m-1} = \mathsf{A}^{m-1,m} f^m \\ df^m = 0. \end{cases} \tag{6.22}$$

Note that the Hodge blocks $\mathsf{A}^{i,i+1}$ of the Gauss–Manin connection matrix in the above equalities must be restricted to L.

Proposition 6.13. *We have the equality*

$$\mathsf{A}^{\frac{m}{2}-1,\frac{m}{2}} \mathsf{C}^{\frac{m}{2}}\Big|_L = df^{\frac{m}{2}-1} \tag{6.23}$$

and this gives us the foliation $\mathscr{F}(\mathsf{C})$ restricted to L.

Proof. The only non-zero block for $\mathsf{A} \cdot \mathsf{C}$ is the left-hand side of (6.23). This implies the last statement. The proof of the equality (6.23) is as follows. The loci of constant periods is given by $f^0 = 0,\ f^1 = 0, \dots, f^m = 0$. This and the last equality in (6.22) imply that f^m is identically zero. Using other equations we conclude that $f^m, f^{m-1}, \dots, f^{\frac{m}{2}}$ are all identically zero. Therefore, the last $\frac{m}{2}$ equalities in (6.22) are trivial equalities $0 = 0$. The $\frac{m}{2}$th equation becomes (6.23). □

Proposition 6.14. *The Zariski tangent space of L is the zero set of the linear part of $f^{\frac{m}{2}-1}$.*

Proof. The first $\frac{m}{2}$ equalities

$$df^i = \mathsf{A}^{i,i+1} f^{i+1}, \ i = 0, 1, \ldots, \frac{m}{2} - 1$$

imply that the linear part of f^i, $i = 0, 1, \ldots, \frac{m}{2} - 1$ at 0 is zero. $\square$

The derivatives of f^i's with respect to z_j's are in the ideal generated by f^{i+1}, however, this does not imply the entries of f^i are in the ideal generated by the entries of f^{i+1} (this has inspired us to write down Definition 5.7). We can conclude this if, for instance, there is a holomorphic vector field v_i in L such that

$$df^i(\mathsf{v}_i) = B^i f^i, \quad B_i \in \mathrm{Mat}(\mathsf{h}^{n-i,i} \times \mathsf{h}^{n-i,i}, \mathscr{O}_{L,0}), \quad \det(B^i) \neq 0, \quad \text{at } 0$$

and hence, L is given by $f^{\frac{m}{2}-1} = 0$. This seems to be related to a class of singularities called generalized quasi-homogeneous singularities. Let g be a holomorphic function in $(\mathbb{C}^N, 0)$. The germ of the singularity $\{g = 0\}$ is called a generalized quasi-homogeneous singularity if there is a holomorphic vector field in $(\mathbb{C}^N, 0)$ such that $dg(\mathsf{v}) = g$, in other words, g is in its Jacobian ideal. For instance, for a weighted homogeneous polynomial g, the singularity $\{g = 0\}$ has such a property.

6.11 Modular vector fields

Explicit computations of Gauss–Manin connection produce huge polynomial expressions with big coefficients and this makes difficult to study modular foliations using explicit expressions, see for instance [Mov11b, Mov19] and the author's webpage for samples of such computations. However, there are vector fields in the parameter space T of enhanced families which helps us to understand the dynamics and arithmetic of modular foliations and they have rather simple expressions. Examples of such vector fields are due to Darboux, Halphen and Ramanujan. We call them modular vector fields. The main reason for this naming is that they live in the moduli of enhanced projective varieties, and in many interesting cases such as elliptic curves, abelian and Calabi–Yau varieties, the pull-back of such vector fields by the special map transforms them into derivations with respect to parameters of the underlying moduli space.

Let X/T be an enhanced family and $\mathsf{v} \in \Theta_\mathsf{T}$. We denote by $\mathsf{A}_{m,\mathsf{v}}$ the composition of the entries of A_m with v. The entries of $\mathsf{A}_{m,\mathsf{v}}$ are regular functions in T.

Proposition 6.15. *A vector field* v *in* T *is tangent to the modular foliation* $\mathscr{F}_m(\mathsf{C})$ *if*

$$\mathsf{A}_{m,\mathsf{v}} \cdot \mathsf{C} = 0.$$

Proof. This follows from the fact that a modular foliation is given by the entries of $\mathsf{A}_m \cdot \mathsf{C}$. □

The Gauss–Manin connection matrix A_m is usually huge, however, we can describe many vector fields v in T such that $\mathsf{A}_{m,\mathsf{v}}$ is simple and in many cases it is a constant matrix.

Definition 6.13. We define $\mathfrak{M}_m(\mathsf{X}/\mathsf{T})$ to be the set of vector fields $\mathsf{v} \in \Theta_\mathsf{T}$ such that the Gauss–Manin connection matrix of X/T on the mth cohomology is of the form

$$\mathsf{Y}_{m,\mathsf{v}} := \mathsf{A}_{m,\mathsf{v}} = \begin{pmatrix} 0 & \mathsf{Y}^{01}_{m,\mathsf{v}} & 0 & \cdots & 0 \\ 0 & 0 & \mathsf{Y}^{12}_{m,\mathsf{v}} & \cdots & 0 \\ \vdots & \vdots & \vdots & \ddots & \vdots \\ 0 & 0 & 0 & \cdots & \mathsf{Y}^{m-1,m}_{m,\mathsf{v}} \\ 0 & 0 & 0 & \cdots & 0 \end{pmatrix}, \tag{6.24}$$

where $\mathsf{Y}^{i-1,i}_{m,\mathsf{v}}$ is a $\mathsf{h}^{m-i+1,i-1} \times \mathsf{h}^{m-i,i}$ matrix with entries in $\mathscr{O}_\mathsf{T}$. These are called Yukawa couplings. This is equivalent to say that

$$\nabla_\mathsf{v} \alpha^i_m = \mathsf{Y}^{i,i+1}_{m,\mathsf{v}} \alpha^{i+1}_m, \quad i = 0, 1, 2, \ldots, m. \tag{6.25}$$

An element of $\mathfrak{M}_m(X/\mathsf{T})$ is called a modular vector field.

The $\mathscr{O}_\mathsf{T}$-module $\mathfrak{M}_m(\mathsf{X}/\mathsf{T})$ is finitely generated and the maps

$$\mathsf{Y}^{i-1,i}_{m,\cdot} : \mathfrak{M}_m(\mathsf{X}/\mathsf{T}) \to \mathrm{Mat}(\mathsf{h}^{m-i+1,i-1} \times \mathsf{h}^{m-i,i}, \mathscr{O}_\mathsf{T}) \tag{6.26}$$

are $\mathscr{O}_\mathsf{T}$-linear. Modular vector fields are algebraic incarnation of derivations with respect to quotients of periods. The main motivation for the matrix format (6.24) in the definition of a modular vector field comes from various sources. The case of elliptic curves and corresponding Ramanujan vector field was observed in [Mov08b, Mov12b]. In fact a similar observation for the Halphen differential equation is still true, see [Mov12a]. In the case of mirror quintic [Mov17b], one gets for the first time a matrix (6.24) with non-constant entries. Such a matrix format appears in topological string theory and in particular in the context of special geometry, see for instance [CdlOGP91b, GMP95, CDF+93, Ali13b].

Proposition 6.16. *The matrices* $\mathsf{Y}^{i-1,i}_{m,\mathsf{v}}$ *satisfy the equalities*

$$\mathsf{Y}^{i-1,i}_{m,\mathsf{v}}\,\Phi^{i,m-i}_m + \Phi^{i-1,m-i+1}_m\left(\mathsf{Y}^{m-i,m+1-i}_{m,\mathsf{v}}\right)^{\mathsf{tr}} = 0, \quad i = 1, 2, \ldots, m. \tag{6.27}$$

Proof. The proof follows from the equality (3.22) for the Gauss–Manin connection matrix. Recall that $\Phi^{i,m-i}$ is a $\mathsf{h}^{m-i,i} \times \mathsf{h}^{i,m-i}$ matrix. □

Note that $(\Phi^{i,m-i}_m)^{\mathsf{tr}} = (-1)^m\,\Phi^{m-i,i}_m$ and so (6.27) can be written as

$$\mathsf{Y}^{i-1,i}_{m,\mathsf{v}}\,\Phi^{i,m-i}_m = (-1)^{m-1}\left(\mathsf{Y}^{m-i,m+1-i}_{m,\mathsf{v}}\,\Phi^{m-i+1,i-1}_m\right)^{\mathsf{tr}}. \tag{6.28}$$

In particular, for an odd number m and $i = \frac{m+1}{2}$, the matrix $\Phi^{\frac{m+1}{2},\frac{m-1}{2}}_m$ is antisymmetric, and

$$\mathsf{Y}^{\frac{m-1}{2},\frac{m+1}{2}}_{m,\mathsf{v}}\,\Phi^{\frac{m+1}{2},\frac{m-1}{2}}_m \text{ is symmetric.} \tag{6.29}$$

Proposition 6.17. *For* $\mathsf{v}_1, \mathsf{v}_2 \in \Theta_\mathsf{T}$ *we have*

$$\mathsf{A}_{m,[\mathsf{v}_1,\mathsf{v}_2]} = [\mathsf{A}_{m,\mathsf{v}_2}, \mathsf{A}_{m,\mathsf{v}_1}] + \mathsf{v}_1(\mathsf{A}_{m,\mathsf{v}_2}) - \mathsf{v}_2(\mathsf{A}_{m,\mathsf{v}_1}). \tag{6.30}$$

Proof. For two vector fields $\mathsf{v}_1, \mathsf{v}_2 \in \Theta_\mathsf{T}$ with the Gauss–Manin connection matrices A_1 and A_2 we have

$$\begin{aligned}
\nabla_{[\mathsf{v}_1,\mathsf{v}_2]}\alpha_m &= \nabla_{\mathsf{v}_1} \circ \nabla_{\mathsf{v}_2}\alpha_m - \nabla_{\mathsf{v}_2} \circ \nabla_{\mathsf{v}_1}\alpha_m \\
&= \nabla_{\mathsf{v}_1}\mathsf{A}_2\alpha_m - \nabla_{\mathsf{v}_2}\mathsf{A}_1\alpha_m \\
&= ([\mathsf{A}_2, \mathsf{A}_1] + \mathsf{v}_1(\mathsf{A}_2) - \mathsf{v}_2(\mathsf{A}_1))\alpha_m.
\end{aligned}$$

□

Proposition 6.18. *The* $\mathcal{O}_\mathsf{T}$*-module* $\mathfrak{M}_m(\mathsf{X}/\mathsf{T})$ *is closed under the Lie bracket and for* $\mathsf{v}_1, \mathsf{v}_2 \in \mathfrak{M}_m(X/\mathsf{T})$ *we have*

$$\mathsf{Y}^{i-1,i}_{m,[\mathsf{v}_1,\mathsf{v}_2]} = \mathsf{v}_1(\mathsf{Y}^{i-1,i}_{m,\mathsf{v}_2}) - \mathsf{v}_2(\mathsf{Y}^{i-1,i}_{m,\mathsf{v}_1}), \tag{6.31}$$

$$\mathsf{Y}^{i-1,i}_{m,\mathsf{v}_1}\mathsf{Y}^{i,i+1}_{m,\mathsf{v}_2} = \mathsf{Y}^{i-1,i}_{m,\mathsf{v}_2}\mathsf{Y}^{i,i+1}_{m,\mathsf{v}_1}. \tag{6.32}$$

Proof. This follows from (6.30). □

When two modular vector fields $\mathsf{v}_1, \mathsf{v}_2 \in \mathfrak{M}_m(X/\mathsf{T})$ commutes, that is, $[\mathsf{v}_1, \mathsf{v}_2] = 0$ then we have

$$\mathsf{v}_1(\mathsf{Y}^{i-1,i}_{m,\mathsf{v}_2}) = \mathsf{v}_2(\mathsf{Y}^{i-1,i}_{m,\mathsf{v}_1}). \tag{6.33}$$

One may think that there is a matrix $\mathsf{Y}^{i-1,i}_m$ independent of the vector field v_1 and v_2 such that

$$\mathsf{Y}^{i-1,i}_{m,\mathsf{v}_j} = \mathsf{v}_j(\mathsf{Y}^{i-1,i}_m), \quad j = 1, 2.$$

This can be done in the context of enhanced varieties over complex numbers, where $\mathsf{Y}^{i-1,i}_m$ have holomorphic entries. This discussion is inspired from the case of Calabi–Yau threefolds in [AMSY16] and more details will be given in Chapters 13 and 8. Let us define

$$\mathfrak{M}(\mathsf{X}/\mathsf{T}) := \bigcap_{m=1}^{2n-1} \mathfrak{M}_m(\mathsf{X}/\mathsf{T}). \tag{6.34}$$

Recall the definition of foliations $\mathscr{F}(a)$ in §6.9.

Proposition 6.19. *We have*

$$\mathscr{F}_m(2) = \mathscr{F}(\mathfrak{M}_m(\mathsf{X}/\mathsf{T})),$$

and hence

$$\mathscr{F}(2) = \mathscr{F}(\mathfrak{M}(\mathsf{X}/\mathsf{T}))$$

that is, the foliation $\mathscr{F}(2)$ is given by modular vector fields.

Proof. This follows from the definition of a modular vector field and $\mathscr{F}(2)$. □

The $\mathscr{O}_\mathsf{T}$-module $\mathfrak{M}(\mathsf{X}/\mathsf{T})$ is finitely generated and we would like to find a particular basis of this $\mathscr{O}_\mathsf{T}$-module. Recall that $\mathsf{M} = \mathsf{T}/\mathsf{G}$ is the classical moduli space.

Property 6.4. Let X/T be a universal enhanced family. The $\mathscr{O}_\mathsf{T}$-module $\mathfrak{M}(\mathsf{X}/\mathsf{T})$ is free and it has a basis

$$\mathsf{v}_i, \quad i = 0, 1, 2, \ldots, \dim(\mathsf{T}/\mathsf{G})$$

with $[\mathsf{v}_i, \mathsf{v}_j] = 0$ for all i, j.

We will verify this property in the case of principally polarized abelian varieties, K3 surfaces and many other particular cases. If all the Hodge numbers of the middle cohomology are equal to one, then the $\mathscr{O}_{\mathsf{T}}$-module is of rank at most one and this property is valid, see [Nik15, MN18]. Modular vector fields are not enough in order to generate the $\mathscr{O}_{\mathsf{T}}$-module Θ_{T}. In the following section, we are going to discuss the missing vector fields.

Recall the fundamental vector field map in Definition 2.21 and the $\mathfrak{k}$-vector space $i(\mathrm{Lie}(\mathsf{G})) \subset \Theta_{\mathsf{T}}$ obtained by the action of G on T and defined in (5.14). For many examples, including elliptic curves and their products, principally polarized abelian varieties, the following property is valid.

Property 6.5. Let X/T be a universal enhanced family. We have

$$\Theta_{\mathsf{T}} = \mathfrak{M}(\mathsf{X}/\mathsf{T}) \oplus i(\mathrm{Lie}(\mathsf{G})) \otimes_{\mathfrak{k}} \mathscr{O}_{\mathsf{T}}$$

that is, any vector field in T is a $\mathscr{O}_{\mathsf{T}}$-linear combination of modular vector fields and vector fields coming from the action of G.

The fact that in the above equality we have a direct sum follows from the format of the Gauss–Manin connection composed with the elements of $\mathfrak{M}(\mathsf{X}/\mathsf{T})$ and $i(\mathrm{Lie}(\mathsf{G}))$ and our hypothesis on the enhanced family X/T in Property 6.2. However, it is not clear why any vector field in T must be an $\mathscr{O}_{\mathsf{T}}$-linear combination of the elements in $\mathfrak{M}(\mathsf{X}/\mathsf{T})$ and $i(\mathrm{Lie}(\mathsf{G}))$.

Let us consider the action of a reductive group $\mathbf{G}$ for the enhanced family X/T as in §3.7. By Proposition 3.9 we know that $i(\mathrm{Lie}(\mathbf{G})) \subset \mathfrak{M}(\mathsf{X}/\mathsf{T})$ and $\mathsf{Y}^{i-1,i}_{m,\mathsf{v}_{\mathfrak{g}}} = 0$ for all $\mathfrak{g} \in \mathrm{Lie}(\mathbf{G})$, where i is the fundamental vector field map. Moreover, the action of $\mathbf{G}$ on Θ_{T}, $(\mathbf{g}, \mathsf{v}) \mapsto \mathbf{g}_*\mathsf{v}$, induces an action on $\mathfrak{M}(\mathsf{X}/\mathsf{T})$.

Proposition 6.20. *For a modular vector field $\mathsf{v} \in \mathfrak{M}(\mathsf{X}/\mathsf{T})$ which is invariant under the action of $\mathbf{G}$, that is, $\mathbf{g}_*\mathsf{v} = \mathsf{v}$ for all $\mathbf{g} \in \mathbf{G}$, we have*

$$[\mathsf{v}, \mathsf{v}_{\mathfrak{g}}] = 0, \quad \forall \mathfrak{g} \in \mathrm{Lie}(\mathbf{G}), \tag{6.35}$$

and the entries of $\mathsf{Y}^{i-1,i}_{m,\mathsf{v}}$ are constant along the orbits of $\mathbf{G}$ in T.

Proof. The proof of the first statement is as follows. Let $A : \mathsf{T} \to \mathsf{T}$, $A(t) = \mathbf{g}{\cdot}t$. The equality $\mathbf{g}_*\mathsf{v} = \mathsf{v}$ implies that for any differential 1-form $\omega \in \Omega^1_{\mathsf{T}}$ we have $\omega(\mathsf{v})(\mathbf{g}{\cdot}t) = A^*(\omega(\mathsf{v})) = (A^*\omega)(\mathsf{v})$. For $\omega = df$, where $f \in \mathscr{O}_{\mathsf{T}}$ is a regular function, we get

$$df(\mathsf{v})(\mathbf{g}{\cdot}t) = d(f(\mathbf{g}{\cdot}t))(\mathsf{v}),$$

where both d's are derivations in T. Now, we take the derivation of this equality with respect to $\mathbf{g}$, evaluate it over the vector field $\mathfrak{g} \in \Theta_{\mathsf{G}}$ and use Proposition 2.13 to get the result. By (6.31) we get $\mathsf{v}_{\mathfrak{g}}(\mathsf{Y}^{i-1,i}_{m,\mathsf{v}}) = 0$ for all $\mathfrak{g} \in \mathrm{Lie}(\mathbf{G})$ which implies the second statement. □

Proposition 6.20 tells us that the entries of $\mathsf{Y}^{i-1,i}_{m,\mathsf{v}}$ give us functions on the quotient space $\mathbf{G}\backslash\mathsf{T}$. These functions might be used in order to construct moduli spaces. For further discussion in this direction, see Chapter 13.

6.12 Constant vector fields

Based on the study of many particular examples, we find another class of important vector fields in the parameter space T of a full enhanced family $\mathsf{X} \to \mathsf{T}$. In the following, the elements of $\mathfrak{k}$ are called constants.

Definition 6.14. We define

$$\mathscr{C}(\mathsf{X}/\mathsf{T}) := \{\mathsf{v} \in \Theta_{\mathsf{T}} \mid \mathsf{A}_{m,\mathsf{v}} \text{ is constant for all } m = 1, 2, \ldots, 2n-1\}$$

and we call an element $\mathsf{v} \in \Theta_{\mathsf{T}}$ a constant vector field in T (relative to the Gauss–Manin connection of the enhanced family X/T).

Proposition 6.21. *Let X/T be a full family of enhanced varieties. The vector fields $\mathsf{v}_{\mathfrak{g}}$, $\mathfrak{g} \in \mathrm{Lie}(\mathsf{G})$ defined in (2.24) are constant and so $i(\mathrm{Lie}(\mathsf{G})) \subset \mathscr{C}(\mathsf{X}/\mathsf{T})$, where $i(\mathrm{Lie}(\mathsf{G}))$ is defined in (5.14). Further, the mth Gauss–Manin connection matrix composed with $\mathsf{v}_{\mathfrak{g}}$ satisfies:*

$$\mathsf{A}_{m,\mathsf{v}_{\mathfrak{g}}} = \mathfrak{g}^{\mathsf{tr}}_{m},$$

and hence it is block lower triangular matrix.

Proof. We have a proof of the above proposition in the complex context and using periods. It follows from (4.16) and (4.17). We take the differential of (4.17) with respect to g, evaluate it at $\mathfrak{g} \in \mathrm{Lie}(\mathsf{G})$, and use Proposition 2.13 in the holomorphic context and get the equality:

$$d\mathsf{P}_m(t)(\mathsf{v}_{\mathfrak{g}}) = \mathsf{P}_m \cdot \mathfrak{g}_m.$$

Now we substitute the above equality in (4.16) and get the desired result. □

The set $\mathscr{C}(\mathsf{X}/\mathsf{T})$ of constant vector fields is a $\mathfrak{k}$-vector space and not an $\mathscr{O}_{\mathsf{T}}$-module.

Proposition 6.22. *The space of constant vector fields $\mathscr{C}(\mathsf{X}/\mathsf{T})$ is closed under the Lie bracket. The leaves of the corresponding foliation contain the orbits of the action of G on T.*

Proof. This is a consequence of Proposition 6.17. □

Definition 6.15. We call $\mathscr{F}(\mathscr{C}(\mathsf{X}/\mathsf{T}))$ the constant foliation.

For elliptic curves, abelian varieties and in general for varieties such that their moduli in a natural way are isomorphic to a quotient of a Hermitian symmetric domain by a discrete group, the space of vector fields in T has a basis of constant vector fields and hence, the constant foliation has only one leaf which is the whole space. Therefore, a constant foliation measures how far classical moduli of projective varieties are from the mentioned property. For further discussion see §6.13.

6.13 Constant Gauss–Manin connection

The following class of varieties provide many interesting modular vector fields and foliations.

Definition 6.16. Let X/T be a full family of enhanced varieties. We say that X/T has a constant Gauss–Manin connection if the $\mathscr{O}_{\mathsf{T}}$-module Θ_{T} has a basis v_i, $i = 0, 1, 2, \ldots, n$ with constant vector fields. In other words, we have

$$\Theta_{\mathsf{T}} = \mathscr{C}(\mathsf{X}/\mathsf{T}) \otimes_{\mathfrak{k}} \mathscr{O}_{\mathsf{T}}.$$

In Chapter 11, we will prove that the full family of principally polarized Abelian varieties has a constant Gauss–Manin connection. The same statement for the family of quartics in $\mathbb{P}^3$ (K3 surfaces) will be proved in Chapter 13. Particular examples of elliptic curves and product of elliptic curves are the origin of the above definition. Motivated by all these examples we further claim the following conjecture.

Conjecture 6.2. *If the moduli of enhanced varieties of a fixed topological type X has a constant Gauss–Manin connection then the classical moduli space M of X is biholomorphic in a natural way to some $\Gamma_{\mathbb{Z}}\backslash D$, where D is a Hermitian symmetric domain and $\Gamma_{\mathbb{Z}}$ is a discrete group acting on D.*

Of course, we have to define rigorously what means "in a natural way". The converse of this conjecture seems to be wrong and the case of Calabi–Yau threefolds is a candidate for this claim, see §13.10. For Abelian varieties and K3 surfaces D is the Griffith period domain parameterizing polarized Hodge structures in a fixed $H^m(X,\mathbb{Z})_0$, see for instance [Gri70]. Note that in Griffiths' formulation we fix a (primitive cohomology), whereas in our formulation in Chapter 8 we work with all cohomologies $H^*(X,\mathbb{Z})$, and so our version of the Griffiths period domain is finer. One may expect that D in Conjecture 6.2 is always the Griffiths period domain, however, this does not seem to be the case. The reader is referred to [Hel01, Milb] for preliminaries on Hermitian symmetric domains.

6.14 Constructing modular vector fields

Recall the construction of the full enhanced family X/T in §3.6 and the notations used there. We have the $\mathscr{O}_\mathsf{T}$-module $\mathfrak{M}(\mathsf{X}/\mathsf{T})$ and we will also consider it as an $\mathscr{O}_V$-module. For this we take the pull-back of functions in V under $\mathsf{T} \to V$ and then we perform the usual multiplication of functions and vector fields in T.

Theorem 6.23. *There is an isomorphism*

$$f : \Theta_V \to \mathfrak{M}(\mathsf{X}/\mathsf{T}) \tag{6.36}$$

of $\mathscr{O}_V$-modules such that the following conditions hold.

(1) *Under the canonical map $\mathsf{T} \to V$ $f(\mathsf{v})$ in T is mapped to v in V.*
(2) *For $\mathsf{v}_1, \mathsf{v}_2 \in \Theta_V$ we have $[f(\mathsf{v}_1), f(\mathsf{v}_2)] = f([\mathsf{v}_1, \mathsf{v}_2])$.*
(3) *The $\mathscr{O}_\mathsf{T}$-module $\mathfrak{M}(\mathsf{X}/\mathsf{T})$ is free of rank* $\dim(V)$.

Proof. The main idea of the proof is taken from [MN18, Mov17b]. For simplicity, we will drop the subindex m (cohomology grading) and the upper index k (chart index) from our notations. We first define $\mathfrak{M}(\tilde{\mathsf{X}}/\tilde{\mathsf{T}})$ in a similar way as we did it for $\mathfrak{M}(\mathsf{X}/\mathsf{T})$ in Definition 6.13, and prove that this $\mathscr{O}_{\tilde{\mathsf{T}}}$-module is free. We are looking for vector fields $\mathsf{v} \in \mathfrak{M}(\tilde{\mathsf{X}}/\tilde{\mathsf{T}})$. Since the Gauss–Manin connection matrix $\tilde{\mathsf{A}}$ in a local chart is of the form in Proposition 3.8, we conclude that

$$\dot{S} = \mathsf{Y}_\mathsf{v} \cdot S - S \cdot B_\mathsf{v}, \tag{6.37}$$

where dot over a quantity means its derivation along v. In Hodge block notation and for $m = 4$ this is of the format:

$$\begin{pmatrix} \dot{S}^{00} & 0 & 0 & 0 & 0 \\ \dot{S}^{10} & \dot{S}^{11} & 0 & 0 & 0 \\ \dot{S}^{20} & \dot{S}^{21} & \dot{S}^{22} & 0 & 0 \\ \dot{S}^{30} & \dot{S}^{31} & \dot{S}^{32} & \dot{S}^{33} & 0 \\ \dot{S}^{40} & \dot{S}^{41} & \dot{S}^{42} & \dot{S}^{43} & \dot{S}^{44} \end{pmatrix} = \begin{pmatrix} 0 & \mathsf{Y}_{\mathsf{v}}^{01} & 0 & 0 & 0 \\ 0 & 0 & \mathsf{Y}_{\mathsf{v}}^{12} & 0 & 0 \\ 0 & 0 & 0 & \mathsf{Y}_{\mathsf{v}}^{23} & 0 \\ 0 & 0 & 0 & 0 & \mathsf{Y}_{\mathsf{v}}^{34} \\ 0 & 0 & 0 & 0 & 0 \end{pmatrix} \begin{pmatrix} S^{00} & 0 & 0 & 0 & 0 \\ S^{10} & S^{11} & 0 & 0 & 0 \\ S^{20} & S^{21} & S^{22} & 0 & 0 \\ S^{30} & S^{31} & S^{32} & S^{33} & 0 \\ S^{40} & S^{41} & S^{42} & S^{43} & S^{44} \end{pmatrix}$$

$$- \begin{pmatrix} S^{00} & 0 & 0 & 0 & 0 \\ S^{10} & S^{11} & 0 & 0 & 0 \\ S^{20} & S^{21} & S^{22} & 0 & 0 \\ S^{30} & S^{31} & S^{32} & S^{33} & 0 \\ S^{40} & S^{41} & S^{42} & S^{43} & S^{44} \end{pmatrix} \begin{pmatrix} B_{\mathsf{v}}^{00} & B_{\mathsf{v}}^{01} & 0 & 0 & 0 \\ B_{\mathsf{v}}^{10} & B_{\mathsf{v}}^{11} & B_{\mathsf{v}}^{12} & 0 & 0 \\ B_{\mathsf{v}}^{20} & B_{\mathsf{v}}^{21} & B_{\mathsf{v}}^{22} & B_{\mathsf{v}}^{23} & 0 \\ B_{\mathsf{v}}^{30} & B_{\mathsf{v}}^{31} & B_{\mathsf{v}}^{32} & B_{\mathsf{v}}^{33} & B_{\mathsf{v}}^{34} \\ B_{\mathsf{v}}^{40} & B_{\mathsf{v}}^{41} & B_{\mathsf{v}}^{42} & B_{\mathsf{v}}^{43} & B_{\mathsf{v}}^{44} \end{pmatrix}.$$

The equalities in (6.37) for Hodge blocks M^{ij}, $i \leq j$ can be regarded as the definition of the vector field v for the variable S_{ij} which is the (i, j)th entry of S. We define v_{ij} to be the quantity in the (i, j)th entry of the right-hand side of (6.37), and so, the (i, j)-entry of this equation is $\dot{S}_{ij} = \mathsf{v}_{ij}$. The equalities in (6.37) for Hodge blocks $M^{ij}, j \geq i + 2$ are just $0 = 0$. For Hodge blocks $M^{i,i+1}$, we get the following equalities:

$$0 = \mathsf{Y}_{\mathsf{v}}^{i,i+1} S^{i+1,i+1} - S^{i,i} B_{\mathsf{v}}^{i,i+1},$$

or equivalently

$$\mathsf{Y}_{\mathsf{v}}^{i,i+1} = S^{i,i} B_{\mathsf{v}}^{i,i+1} \left(S^{i+1,i+1}\right)^{-1}. \tag{6.38}$$

This means that we can define v for variables in V an arbitrary quantity and (6.38) can be regarded as the definition of Y_{v}. From now on we discard the usage of v as a vector field in T and use it as a fixed vector field v in V. We substitutes $\mathsf{Y}_{\mathsf{v}}^{i,i+1}$'s in v_{ij} in order to get expressions of v_{ij} in terms of S_{ij} and regular functions in V. The conclusion is that $\mathfrak{M}(\tilde{\mathsf{X}}/\tilde{\mathsf{T}})$ in a local chart is generated by

$$f(\mathsf{v}) := \mathsf{v} + \check{\mathsf{v}}, \quad \text{where } \check{\mathsf{v}} = \sum_{i,j} \mathsf{v}_{ij} \frac{\partial}{\partial S_{ij}}, \tag{6.39}$$

and S_{ij} runs through all (i, j) entries of S with $i \geq j$. Note that $\check{\mathsf{v}}$ depends on v. If we take a basis v_k, $k = 1, 2, 3, \ldots$ of sections of Θ_V in a local chart (if necessary, we take such a local chart smaller) then we have a basis $f(\mathsf{v}_k)$ of $\mathfrak{M}(\tilde{\mathsf{X}}/\tilde{\mathsf{T}})$.

Now, we prove that $\mathfrak{M}(\mathsf{X}/\mathsf{T})$ itself is free of rank $\dim(V)$. Recall the function $f : \tilde{\mathsf{T}} \to \mathbb{A}_{\mathfrak{k}}^{s}$ defined in (3.14) and the fact that T is a fiber of f.

It is enough to prove that any $\mathsf{v} \in \mathfrak{M}(\tilde{\mathsf{X}}/\tilde{\mathsf{T}})$ is tangent to the fibers of f. The morphism f is made of many pieces. It is more instructive to first consider the pieces as in $f = (g_1, g_2, \ldots, g_{2n-1}, \ldots)$, where $g_m = [\langle \alpha_i, \alpha_j \rangle]$ and $\langle \cdot, \cdot \rangle$ is defined in (2.13).

We drop the subindex of g_m and write $g = g_m$. Let $\Omega := [\langle \omega_i, \omega_j \rangle]$ and hence $f(t) = S\Omega S^{\mathrm{tr}}$. Below, the dot over a quantity means derivation along v.

$$
\begin{aligned}
\dot{\overbrace{[\langle \alpha_i, \alpha_j \rangle]}} &= \dot{\overbrace{(S\Omega S^{\mathrm{tr}})}} \\
&= \dot{S}\Omega S^{\mathrm{tr}} + S\dot{\Omega}S^{\mathrm{tr}} + S\Omega\dot{S}^{\mathrm{tr}} \\
&= (\mathsf{Y}_{\mathsf{v}}S - SB)\Omega S^{\mathrm{tr}} + S(B\Omega + \Omega B^{\mathrm{tr}})S^{\mathrm{tr}} + S\Omega(S^{\mathrm{tr}}\mathsf{Y}_{\mathsf{v}}^{\mathrm{tr}} - B^{\mathrm{tr}}S^{\mathrm{tr}}) \\
&= \mathsf{Y}_{\mathsf{v}}\Phi + \Phi\mathsf{Y}_{\mathsf{v}}^{\mathrm{tr}} \\
&\overset{?}{=} 0.
\end{aligned}
$$

The only non-trivial equality is the last one. It is equivalent to

$$
\mathsf{Y}_{\mathsf{v}}^{i,i+1}\Phi^{i+1,m-i-1} + \Phi^{i,m-i}(\mathsf{Y}_{\mathsf{v}}^{m-i-1,m-i})^{\mathrm{tr}} = 0, \tag{6.40}
$$

which means that we have to prove Proposition 6.16 for v. In order to prove this equality, we write the equality $S\Omega S^{\mathrm{tr}} = \Phi$ in the format $S\Omega = \Phi S^{-\mathrm{tr}}$ and we get

$$
S^{i,i}\Omega^{i,m-i} = \Phi^{i,m-i}(S^{m-i,m-i})^{-\mathrm{tr}}, \tag{6.41}
$$

or equivalently

$$
S^{m-i,m-i} = (\Phi^{i,m-i})^{\mathrm{tr}}(S^{i,i})^{-\mathrm{tr}}(\Omega^{i,m-i})^{-\mathrm{tr}}. \tag{6.42}
$$

We substitute $\mathsf{Y}_{\mathsf{v}}^{i,i+1}$'s defined in (6.38) in the left-hand side of (6.40) and then substitute $S^{m-i,m-i}$'s from (6.42). Now the equality (6.40) follows from

$$
0 = B^{i,i+1}\Omega^{i+1,m-i-1} + \Omega^{i,m-i}(B^{m-i-1,m-i})^{\mathrm{tr}},
$$

which is the equality corresponding to the Hodge block $M^{i,m-i-1}$ in $d\Omega = B \cdot \Omega + \Omega \cdot B^{\mathrm{tr}}$.

In general, we have to do the following computation. We write the lower index m of cohomologies, however, we drop the upper index k of charts.

Let $\Omega := [\omega_{m_1} \cup \omega_{m_2}^{\mathsf{tr}}]$ which has entries in $H_{\mathrm{dR}}^{m_1+m_2}(Y/V)$.

$$\begin{aligned}
\overbrace{[\alpha_{m_1} \cup \alpha_{m_2}^{\mathsf{tr}}]}^{\cdot} &= \overbrace{(S_{m_1}[\omega_{m_1} \cup \omega_{m_2}^{\mathsf{tr}}]S_{m_2}^{\mathsf{tr}})}^{\cdot} \\
&= \dot{S}_{m_1}\Omega S_{m_2}^{\mathsf{tr}} + S_{m_1}\dot{\Omega}S_{m_2}^{\mathsf{tr}} + S_{m_1}\Omega\dot{S}_{m_2}^{\mathsf{tr}} \\
&= (\mathsf{Y}_{m_1,\mathsf{v}}S_{m_1} - S_{m_1}B_{m_1})\Omega S_{m_2}^{\mathsf{tr}} + S_{m_1}(B_{m_1}\Omega + \Omega B_{m_2}^{\mathsf{tr}})S_{m_2}^{\mathsf{tr}} \\
&\quad + S_{m_1}\Omega(S_{m_2}^{\mathsf{tr}}\mathsf{Y}_{m_2,\mathsf{v}}^{\mathsf{tr}} - B_{m_2}^{\mathsf{tr}}S_{m_2}^{\mathsf{tr}}) \\
&= \mathsf{Y}_{m_1,\mathsf{v}}S_{m_1}\Omega S_{m_2}^{\mathsf{tr}} + S_{m_1}\Omega S_{m_2}^{\mathsf{tr}}\mathsf{Y}_{m_2,\mathsf{v}}^{\mathsf{tr}} \\
&= \mathsf{Y}_{m_1,\mathsf{v}}[\alpha_{m_1} \cup \alpha_{m_2}^{\mathsf{tr}}] + [\alpha_{m_1} \cup \alpha_{m_2}^{\mathsf{tr}}]\mathsf{Y}_{m_2,\mathsf{v}}^{\mathsf{tr}} \\
&= \sum_{i=1}^{\mathsf{b}_{m_1+m_2}} \left(\mathsf{Y}_{m_1,\mathsf{v}}\Phi_{m_1,m_2,i} + \Phi_{m_1,m_2,i}\mathsf{Y}_{m_2,\mathsf{v}}^{\mathsf{tr}}\right)\alpha_{m_1+m_2,i} = 0
\end{aligned}$$

and in a similar way we argue the last equality. □

As an immediate corollary of Theorem 6.23 we have the following proposition.

Proposition 6.24. *The foliation* $\mathscr{F}(2)$ *is of dimension* $\dim(V)$ *and has no singularities in* T.

Proof. This is because $\mathscr{F}(2)$ is given by $\mathfrak{M}(\mathsf{X}/\mathsf{T})$. □

6.15 Modular vector fields and IVHS

Let us consider the infinitesimal variation of Hodge structures in §2.16 for an enhanced family X/T. The advantage with enhanced families is that all $\mathfrak{k}$-vector spaces in IVHS comes with a basis, except for $\mathbf{T}_t\mathsf{T}$, for which we also take a basis $\frac{\partial}{\partial t_i}$, $i = 1, 2, \ldots, r$. Note that these are not yet vector fields in T and we do not take a coordinate system $(t_1, t_2, \ldots, t_r)$ around the point $t \in \mathsf{T}$. We denote by $dt_1, dt_2, \ldots, dt_r \in (\Omega_{\mathsf{T}}^1)_t := (\mathbf{T}_t\mathsf{T})^\vee$ the dual elements, that is, $dt_i(\frac{\partial}{\partial t_j}) = 1$ if $i = j$ and $= 0$ otherwise. Recall the Gauss–Manin connection matrix $\mathsf{A} = \mathsf{A}_m$ of X/T and its Hodge block format $\mathsf{A} = [\mathsf{A}^{ij}]$.

Proposition 6.25. *If we write the Gauss–Manin connection matrix at the point* $t \in \mathsf{T}$

$$\mathsf{A}_t^{i,i-1} = \sum_{j=1}^{\tilde{r}} B_j^{i,i-1} dt_j, \tag{6.43}$$

then $B_j^{i,i-1}$ *is the* $\mathsf{h}^{m-i,i} \times \mathsf{h}^{m-i-1,i+1}$ *matrix of* $\delta_i(\frac{\partial}{\partial t_j})$ *written in the standard basis.*

Proof. This follows from the construction of IVHS from the Gauss–Manin connection. □

As a corollary we have the following proposition.

Proposition 6.26. *Let* $\mathsf{v} \in \mathfrak{M}(\mathsf{X}/\mathsf{T})$ *be a modular vector field. The matrices* $\mathsf{Y}_{m,\mathsf{v}}^{i,i+1}$ *are the data of infinitesimal variation of Hodge structures* $\delta_{m,k}$ *in* (2.32) *written in the standard basis.*

For the first examples of modular vector fields, we have a uniqueness property which is due to further constrains on regular functions in the entries of $\mathsf{Y}_{\mathsf{v}}^{i,i+1}$'s. This has also to do with choosing a natural basis of the $\mathscr{O}_\mathsf{T}$-module $\mathfrak{M}(\mathsf{X}/\mathsf{T})$. We were not able to formulate a general procedure leading to such a basis. Just in the case of Abelian and Calabi–Yau varieties we were able to find natural generalizations of Ramanujan and Darboux-Halphen vector fields.

Chapter 7

Hodge Cycles and Loci

One may ask whether imposing a certain Hodge class upon a generic member of an algebraic family of polarized algebraic varieties amounts to an algebraic condition upon the parameters (A. Weil in [Wei77].

7.1 Introduction

In this chapter, we are going to relate modular foliations to Hodge loci and weak absolute Hodge cycles. We introduce a holomorphic foliation $\mathscr{F}(\mathsf{C})$ in a larger parameter space T attached to families of enhanced projective varieties. Irreducible components of the Hodge locus in T are algebraic leaves of the foliation $\mathscr{F}(\mathsf{C})$. Under the hypothesis that these are all the algebraic leaves, we get the fact that such algebraic leaves are defined over the algebraic closure of the base field and that Hodge classes are weak absolute in the sense of Voisin. These are also two consequences of the Hodge conjecture. We study such foliations using modular vector fields which are natural generalizations of the vector fields due to Darboux, Halphen and Ramanujan. For an expository account on Hodge loci the reader is referred to [Voi13, CS11] and the references therein. The reader who wants to see the content of this chapter in a concrete example is invited to read Chapter 10 and the article [Mov18]. Throughout the chapter, all varieties and enhanced families are defined over a subfield $\mathfrak{k}$ of $\mathbb{C}$. The algebraic closure of $\mathfrak{k}$ is denoted by $\bar{\mathfrak{k}}$ and $\tilde{\bar{\mathfrak{k}}}$ is a field extension of $\bar{\mathfrak{k}}$ by some transcendental numbers.

7.2 Cattani–Deligne–Kaplan theorem

A Hodge locus in our context turns out to be an algebraic leaf of a modular foliation and this is the main motivations for the present chapter. After Lefschetz $(1,1)$-theorem, without doubt, the theorem of Cattani, Deligne and Kaplan in [CDK95] is the strongest evidence to the Hodge conjecture.

Definition 7.1. Let Y be a smooth projective variety. A Hodge class is any element in the intersection of the rational cohomology $H^m(Y,\mathbb{Q}) \subset H^m_{\mathrm{dR}}(Y)$ and $F^{\frac{m}{2}} \subset H^m_{\mathrm{dR}}(Y)$, where $F^{\frac{m}{2}} = F^{\frac{m}{2}} H^m_{\mathrm{dR}}(Y)$ is the $\frac{m}{2}$-th piece of the Hodge filtration of $H^m_{\mathrm{dR}}(Y)$.

Therefore, the $\mathbb{Q}$-vector space of Hodge classes is simply the intersection $H^m(Y,\mathbb{Q}) \cap H^{\frac{m}{2},\frac{m}{2}} = H^m(Y,\mathbb{Q}) \cap F^{\frac{m}{2}}$. Now, let $Y \to V$ be a family of smooth complex projective varieties ($Y \subset \mathbb{P}^N \times V$ and $Y \to V$ is obtained by projection on the second coordinate).

Definition 7.2. Let $F^{\frac{m}{2}} H^m_{\mathrm{dR}}(Y/V)$ be the vector bundle of $F^{\frac{m}{2}}$ pieces of the Hodge filtration of $H^m_{\mathrm{dR}}(Y_t)$, $t \in V$. The locus of Hodge classes is the subset of $F^{\frac{m}{2}} H^m_{\mathrm{dR}}(Y/V)$ containing all Hodge classes.

Note that $F^{\frac{m}{2}} H^m_{\mathrm{dR}}(Y/V)$ is an algebraic bundle, however, the locus of Hodge classes is a union of local analytic varieties. Now, we define the Hodge locus in V itself.

Definition 7.3. The projection of the locus of Hodge classes under $F^{\frac{m}{2}} H^m_{\mathrm{dR}}(Y/V) \to V$ is called the Hodge locus in V. An irreducible component H of the Hodge locus in a (usual) neighborhood of a point $t_0 \in V$ is characterized in the following way. It is an irreducible closed analytic subvariety of (V, t_0) with a continuous family of Hodge classes $\delta_t \in H^m(Y_t,\mathbb{Q}) \cap H^{\frac{m}{2},\frac{m}{2}}$ in varieties $Y_t, t \in H$ such that for points t in a dense open subset of H, the monodromy of δ_t to a point in a neighborhood (in the usual topology of V) of t and outside H is no more a Hodge class.

One of the main goals of the present text is to develop the theory of modular foliations as much as possible and at the end to give a systematic proof (or a counterexample) for the following consequence of the Hodge conjecture.

Conjecture 7.1. *Let $Y \to V$ be a family of smooth projective varieties defined over a field $\mathfrak{k} \subset \mathbb{C}$. All the irreducible components of the locus of Hodge classes are algebraic subsets of $F^{\frac{m}{2}} H^m_{\mathrm{dR}}(Y/V)$ defined over the algebraic closure of $\mathfrak{k}$.*

In particular, the components of the Hodge locus in V are also algebraic. The algebraicity statement has been proved by Cattani, Deligne and Kaplan.

Theorem 7.1 (Cattani–Deligne–Kaplan [CDK95]). *The irreducible components of the locus of Hodge classes in $F^{\frac{m}{2}}H^m_{\mathrm{dR}}(Y/V)$ are algebraic sets.*

The main ingredient of their proof is Schmid's nilpotent orbit theorem in [Sch73] together with some results in [CKS86]. All these are purely transcendental methods in algebraic geometry, and hence, their proof does not give any light into the second part of the Conjecture 7.1, that is, any component of the locus of Hodge classes is defined over the algebraic closure of the base field $\mathfrak{k}$.

The algebraicity statement for the locus of Hodge classes is slightly stronger than the same statement for the Hodge locus. Let us explain this. We take an irreducible component H of the Hodge locus. Above each point $t \in H$ we have a Hodge class β and the above theorem implies that the action of the monodromy representation $\pi_1(H,t) \to \mathrm{Aut}(H^m(Y_t,\mathbb{Q}))$ on β produces a finite number of cohomological classes (which are again Hodge classes). This topological fact does not follow just from the algebraicity of H. This will be used in Theorem 7.2 which is the adaptation of Theorem 7.1 to our context of modular foliations. We are working with Hodge cycles which lives in homologies in comparison with Hodge classes which live in cohomologies. Both notions are related to each other by Poincaré duality.

7.3 Hodge cycles and enhanced families

Consider an enhanced family $\mathsf{X} \to \mathsf{T}$. For a given family of projective varieties $Y \to V$ we can use the methods introduced in §3.6 and construct an enhanced family.

Definition 7.4. A cycle $\delta_0 \in H_m(\mathsf{X}_0, \mathbb{Z})$ is called Hodge if

$$\int_{\delta_0} \alpha_{m,i} = 0, \quad i = 1, 2, \ldots, \mathsf{h}^{\frac{m}{2}+1},$$

where $\mathsf{h}^{\frac{m}{2}+1} := \mathsf{h}^{m,0} + \mathsf{h}^{m-1,1} + \cdots + \mathsf{h}^{\frac{m}{2}+1,\frac{m}{2}-1}$. Recall that the differential forms $\alpha_{m,i}$, $i = 1, 2, \ldots, \mathsf{h}^{\frac{m}{2}+1}$ form a basis $F^{\frac{m}{2}+1}H^m_{\mathrm{dR}}(\mathsf{X}_t)$ for all $t \in \mathsf{T}$.

Using Poincaré duality, a Hodge cycle is mapped to Hodge class in Definition 7.1. It follows that the period vector of a Hodge cycle, that is

$$\mathsf{C} := \int_{\delta_t} \alpha_m,$$

is of the format (6.14). We have called C a period vector of Hodge type. If the Hodge decomposition of X_0 is defined over $\bar{\mathfrak{k}}$ and α_m is compatible with the Hodge decomposition of X_0 then C is actually of the format (6.15).

Definition 7.5. Let $\delta = \{\delta_t\}_{t\in(\mathsf{T},0)}$, $\delta_t \in H_m(\mathsf{X}_t, \mathbb{Q})$ be a continuous family of cycles. We call

$$L_\delta := \left\{ t \in \mathsf{T} \,\middle|\, \int_{\delta_t} \alpha_m = \mathsf{C},\ \delta_t \text{ is obtained by a monodromy of } \delta_0 \right\} \quad (7.1)$$

the (global) locus of Hodge cycles with constant periods. It is sometimes useful to replace T with a neighborhood of 0 in T, and call it the local locus of Hodge cycles with constant periods.

From now on by Hodge locus in the parameter space T of an enhanced family X/T, we mean the locus of Hodge cycles with constant periods.

Remark 7.1. Let X/T be an enhanced family constructed from Y/V as in §3.6. A Hodge locus with constant periods in T is projected to a Hodge locus in V under the canonical projection $\mathsf{T} \to V$, and the resulting map is not necessarily surjective.

The following theorem is a consequence of Theorem 7.1 and the regularity of the Gauss–Manin connection.

Theorem 7.2. *The set $L_\delta \subset \mathsf{T}$ is algebraic, and hence, it has finitely many components.*

Proof. Let $\tilde{\alpha}_m$ be the submatrix of α_m containing the first $\frac{m}{2}$ Hodge blocks (corresponding to zero blocks in (6.14)). The Cattani–Deligne–Kaplan theorem implies that the local irreducible components of $\int_{\delta_t} \tilde{\alpha}_m = 0$ are in fact algebraic. Let us take one component of this locus, say $H \subset \mathsf{T}$. We prove that restricted to H the entries of $\int_{\delta_t} \alpha_m$ are algebraic over the field of rational functions of H. This would be enough to prove that the locus $\int_{\delta_t} \alpha_m = \mathsf{C}$ is algebraic. We use again Theorem 7.1, and in particular the algebraicity statement in the Hodge bundle $F^{\frac{m}{2}} H^m_{\mathrm{dR}}(\mathsf{X}/\mathsf{T})$, and conclude that the monodromy of δ_t results in finitely many cycles at each fiber. This implies that the entries of $\int_{\delta_t} \alpha_m$ take finite number of values. In particular, they satisfy polynomial equations with coefficients which are holomorphic (one valued) functions in H. The regularity of the Gauss–Manin connection, see for instance Griffiths expository article [Gri70, p. 237], implies that such one valued functions are in fact regular functions on H. □

Note that the locus (7.1) with t near to 0 and δ_t obtained by a monodromy in a neighborhood of 0 is an analytic subset of $(\mathsf{T}, 0)$ and so it has finitely many components. In particular, Theorem 7.2 implies that they are parts of finitely many algebraic sets in T. Apparently, such an algebraic set is of codimension $\mathsf{h}^{\frac{m}{2}+1}$, however, its codimension is less than or equal $\mathsf{h}^{\frac{m}{2}+1,\frac{m}{2}-1}$, see for instance Voisin's book [Voi03, Proposition 5.14] or [Mov19, Theorem 16.2]. Recall that L_δ's are leaves of the foliation $\mathscr{F}(\mathsf{C})$ defined in §6.3. Recall also the definition of $L_\delta = L_{\tilde{\mathsf{C}}}$ in (6.4) and (6.7).

Theorem 7.3. *If $\tilde{\mathsf{C}} \in \mathbb{P}^{\mathsf{b}-1}(\mathbb{Q})$ (equivalently if δ_t up to multiplication by a constant is in $H_m(X_t, \mathbb{Q})$) then all the components of the leaf $L_{\tilde{\mathsf{C}}} = L_\delta$ of $\mathscr{F}(\mathsf{C})$ are algebraic subsets of T.*

Proof. From Proposition 6.2 we know that $\mathscr{F}(\mathsf{C})$ has the local holomorphic first integral $f := \mathsf{P}^{-\mathsf{tr}}\mathsf{C}$, and $L_{\tilde{\mathsf{C}}}$ is the inverse image of $\tilde{\mathsf{C}}$ under f. In this way, this theorem is just Theorem 7.2 in different words. □

Next, we note that not all period vectors of the form (6.14) arise from Hodge cycles. The following proposition is inspired after reading [Voi07, Remark 1.2].

Proposition 7.4. *Let (X, α) be an enhanced variety. For m an even number and for a period vector*

$$\mathsf{C} := \frac{1}{(2\pi i)^{\frac{m}{2}}} \int_\delta \alpha_m$$

arising from a primitive Hodge cycle $\delta \in H_m(X, \mathbb{Q})_0$, we have

$$\langle \delta, \delta \rangle = \mathsf{C}^{\mathsf{tr}} \Phi_m^{-\mathsf{tr}} \mathsf{C} > 0, \tag{7.2}$$

where Φ_m is the matrix in (3.1). In particular, if X is a fiber of an enhanced family then the corresponding modular foliation $\mathscr{F}(\mathsf{C})$ has a trivial character in the sense of Definition 6.6.

Proof. We prove the first statement. We take the Poincaré dual of δ and use the hard Lefschetz theorem to obtain $\tilde{\delta}^{\mathsf{pd}} \in H^m(X,\mathbb{Q})_0$, that is, $\tilde{\delta}^{\mathsf{pd}} \cup u^{n-m} = \delta^{\mathsf{pd}}$. Here, $u \in H^2(X,\mathbb{Z})$ is the cohomology class of a hyperplane section of X (topological polarization). We write $\tilde{\delta}^{\mathsf{pd}} = \alpha_m^{\mathsf{tr}} B$ and get

$$\begin{aligned} \mathsf{C} &= \frac{1}{(2\pi i)^{\frac{m}{2}}} \int_\delta \alpha_m \\ &= \frac{1}{(2\pi i)^{\frac{m}{2}}} \left(\int_X \alpha_m \cup \alpha_m^{\mathsf{tr}} \cup u^{n-m} \right) B \\ &= (2\pi i)^{\frac{m}{2}} \cdot \Phi_m \cdot B. \end{aligned}$$

Note that $\theta = 2\pi i \cdot u$ is the algebraic polarization. Therefore,

$$\begin{aligned} \langle \delta, \delta \rangle = \int_X \tilde{\delta}^{\mathsf{pd}} \wedge \tilde{\delta}^{\mathsf{pd}} \wedge u^{n-m} &= (2\pi i)^m \langle \tilde{\delta}^{\mathsf{pd}}, \tilde{\delta}^{\mathsf{pd}} \rangle \\ &= (2\pi i)^m B^{\mathsf{tr}} \langle \alpha_m, \alpha_m^{\mathsf{tr}} \rangle B \\ &= (2\pi i)^m B^{\mathsf{tr}} \Phi_m B = \mathsf{C}^{\mathsf{tr}} \Phi_m^{-\mathsf{tr}} \mathsf{C}. \end{aligned}$$

Now, for m even and δ a primitive Hodge cycle we know that $\tilde{\delta}^{\mathsf{pd}} \in H_0^{\frac{m}{2},\frac{m}{2}}$ and the affirmation follows from the second Hodge-Riemann bilinear relations, see [Voi02, Theorem 6.32, p. 152].

Now, we prove the second statement. Let $\lambda : \mathsf{G} \to \mathbb{G}_m$ be a group morphism such that for all $\mathsf{g} \in \mathsf{G}$ we have $\mathsf{g}_m^{\mathsf{tr}} \mathsf{C} = \lambda(\mathsf{g})\mathsf{C}$. From another side we know that $\mathsf{g}_m^{\mathsf{tr}} \Phi_m \mathsf{g}_m = \Phi_m$. Combining these equalities and the fact that $\mathsf{C}^{\mathsf{tr}} \Phi_m^{-\mathsf{tr}} \mathsf{C}$ is non-zero we get $\lambda(\mathsf{g})^2 = \pm 1$. □

The argument used in the proof of Proposition 7.4 fails for non-primitive Hodge cycles. Let us write the Lefschetz decomposition of $\tilde{\delta}^{\mathsf{pd}}$:

$$\tilde{\delta}^{\mathsf{pd}} = \tilde{\delta}_m + \tilde{\delta}_{m-2} + \cdots, \quad \tilde{\delta}_{m-2q} \in H^{m-2q}(X,\mathbb{Q})_0,$$

where each piece $\tilde{\delta}_{m-2q}$ is a primitive Hodge class and they are orthogonal to each other, and so,

$$\langle \tilde{\delta}, \tilde{\delta} \rangle = \langle \tilde{\delta}_m, \tilde{\delta}_m \rangle + \langle \tilde{\delta}_{m-2}, \tilde{\delta}_{m-2} \rangle + \cdots.$$

By the second Hodge–Riemann bilinear relations $\langle \tilde{\delta}_{m-2q}, \tilde{\delta}_{m-2q} \rangle$ is non-zero, however, its sign is $(-1)^{\frac{m}{2}-q}$, which at the end may result in $\langle \tilde{\delta}, \tilde{\delta} \rangle = 0$. This means that the corresponding modular foliation $\mathscr{F}(\mathsf{C})$ might have non-trivial character.

7.4 Isolated Hodge cycles

Let $V \subset \mathrm{Hilb}_P(\mathbb{P}^N_{\mathfrak{k}})$ be a Zariski open subset of a Hilbert scheme parameterizing deformations of a smooth projective variety $Y_0 \subset \mathbb{P}^N_{\mathfrak{k}}$ and and let $Y \to V$ be the corresponding family of smooth projective varieties. Recall from §2.11 that the reductive group $\mathbf{G}$ acts from the left on V. We have mainly in our mind the case in which $\mathbf{G}\backslash V$ is the moduli of the projective variety Y_0, however, we do not assume this for the discussion below. Since for $t \in V$ and $\mathbf{g} \in \mathbf{G}$ the varieties Y_t and $Y_{\mathbf{g} \bullet t}$ are isomorphic, the Hodge locus in V is invariant under the action of $\mathbf{G}$.

Definition 7.6. A Hodge cycle $\delta_0 \in H_m(Y_0, \mathbb{Z})$, $0 \in V$ is called isolated if the local Hodge locus through 0 is a neighborhood of the orbit of $\mathbf{G}$ passing through 0. In other words, the Hodge locus in $\mathbf{G}\backslash V$ crossing 0 is the isolated point 0 itself.

We could also formulate a weaker version of the above definition without talking about Hilbert schemes and action of reductive groups, in the following way. Let $Y \to V$ be a family of smooth projective varieties. A Hodge cycle $\delta_0 \in H_m(Y_0, \mathbb{Z})$, $0 \in V$ is called isolated if for all s in the Hodge locus corresponding to δ_0, we have an isomorphism $Y_s \cong Y_0$ of projective varieties over $\mathbb{C}$. Note that we do not claim that such an isomorphism is given by some automorphism of the ambient projective space. Recall the notation in §6.4 and in particular Proposition 6.4 which says that the orbits of $\mathrm{Stab}(\mathsf{G}, \mathsf{C})_0$ are contained in the leaves of the modular foliation $\mathscr{F}(\mathsf{C})$.

Definition 7.7. Let X/T be a full enhanced family. A Hodge cycle $\delta_0 \in H_m(\mathsf{X}_0, \mathbb{Z})$, $0 \in \mathsf{T}$ is called isolated with constant periods C if the local Hodge locus L_δ with constant periods C through 0 is a neighborhood of 0 in the orbit of $\mathrm{Stab}(\mathsf{G}, \mathsf{C})_0$ through 0. In case X/T is equipped with an action of a reductive group $\mathbf{G}$ as in §2.11, we say that δ_0 is isolated with constant periods C if L_δ is a neighborhood of $\mathbf{G} \cdot 0 \bullet \mathrm{Stab}(\mathsf{G}, \mathsf{C})_0$.

Note that $0 \bullet \mathrm{Stab}(\mathsf{G}, \mathsf{C})_0$ and $\mathbf{G} \cdot 0 \bullet \mathrm{Stab}(\mathsf{G}, \mathsf{C})_0$ in both cases above are contained in the global Hodge locus L_δ and in the above definition we say that near to 0, L_δ does not contain more points. The following proposition tells us that modular foliations arising from isolated Hodge cycles might not be so interesting.

Proposition 7.5. *Let X/T be a full family of enhanced projective varieties, $\delta_0 \in H_m(\mathsf{X}_0, \mathbb{Z})$ be an isolated Hodge cycle with the period vector C and $\mathscr{F} = \mathscr{F}(\mathsf{C})$ be the corresponding modular foliation. Further, assume* *that the stabilizer of all points in* T *with respect to the action of* G *are finite. Then* $\mathscr{F}$ *is trivial in some Zariski open subset of* T *in the sense of Definition 6.4.*

Proof. By our hypothesis the fiber of the local first integral $f = \mathsf{P}^{-\mathrm{tr}}\mathsf{C} : (\mathsf{T}, 0) \to \mathbb{C}^{\mathsf{b}}$ of $\mathscr{F}$ over $\tilde{\mathsf{C}}$ is just the orbit of $\mathrm{Stab}(\mathsf{G}, \mathsf{C})_0$ passing 0. Since the function of dimension of fibers is upper semi-continuous, we conclude that the fiber of f over points near to $\tilde{\mathsf{C}}$ is either empty or the orbit of $\mathrm{Stab}(\mathsf{G}, \mathsf{C})_0$. However, by our hypothesis the dimension of $t \bullet \mathrm{Stab}(\mathsf{G}, \mathsf{C})_0$ is the same as the dimension of $0 \bullet \mathrm{Stab}(\mathsf{G}, \mathsf{C})_0$. All these together imply that the foliation $\mathscr{F}(\mathsf{C})$ in a neighborhood (usual topology) of 0 is the same as the foliation $\mathscr{F}(\mathrm{Stab}(\mathsf{G}, \mathsf{C})_0)$ given by the action of $\mathrm{Stab}(\mathsf{G}, \mathsf{C})_0$. Since both foliations are given by algebraic differential forms, we have the same statement in a Zariski open neighborhood. □

The theory of modular foliations developed in this book is mainly for non-isolated Hodge cycles. This is the case for instance for all Hodge cycles of abelian varieties, see P. Deligne's lecture notes in [DMOS82]. More strongly, he proves that a Hodge cycle of an Abelian variety can be deformed into a Hodge cycle of a CM Abelian variety.

In [MN18] we have constructed the universal family X/T of enhanced mirror quintic Calabi–Yau n-folds, and for n even, modular foliations are trivial in the sense of Definition 6.4, and hence, Hodge cycles are expected to be isolated. More generally, it is reasonable to expect that a general Hodge cycle in the middle cohomology of a projective Calabi–Yau variety of even dimension is isolated. As an example, we can take a smooth hypersurface of degree 6 in the five-dimensional projective space. Its fourth cohomology has the Hodge numbers $1, 426, 1751, 426, 1$. For many examples of (n, d) one can prove that a Fermat hypersurface of degree d and dimension n has isolated Hodge cycles, see [Mov12b, Chapter 16].

Let X_1 and X_2 be two mirror quintic Calabi–Yau threefolds, see [Mov17b], or any two Calabi–Yau threefolds which appear in the list of Almkvist–Enckevort–Straten–Zudilin in [AvEvSZ10]. We take the tensor $H^3_{\mathrm{dR}}(X_1) \otimes_{\mathbb{C}} H^3_{\mathrm{dR}}(X_2)$ which has the Hodge numbers $1, 2, 3, 4, 3, 2, 1$. A leaf L of $\mathscr{F}(2)$ in this case is two-dimensional and restricted to L, a modular foliation $\mathscr{F}(\mathsf{C})$ with C a period vector of Hodge type, is given by three 1-forms which we might expect that they are linearly independent, and hence, $\mathscr{F}(\mathsf{C})$ is trivial. A similar procedure in the case of elliptic curves produces non-isolated Hodge cycles, see Chapter 10.

In order to construct trivial foliations we might first construct isolated Hodge cycles without using the parameter space T of an enhanced family. One example is as follows. We take the n-dimensional Dwork family with n even, see §12.7 and [MN18]. A zero of multiplicity $\frac{n}{2}$ of the holomorphic solution of the corresponding Picard–Fuchs equation is a Hodge cycle. Let us take $n = 4$ and so we have a four-dimensional projective variety X_z with Hodge numbers $1, 1, 1, 1, 1$. The Picard–Fuchs equation L of the holomorphic 4-form in X_z is well-known and we can take three solutions $x_{i1} = \int_{\delta_i} \eta$, $\delta_i \in H_4(X_z, \mathbb{Z})$ $i = 1, 2, 3$ of L represented with explicit formulas. We also define $x_{ij} = (z\frac{\partial}{\partial z})^{j-1} x_{i1}$. We are looking for a Hodge cycle $\delta = a_1\delta_1 + a_2\delta_2 - \delta_3$, $a_1, a_2 \in \mathbb{Q}$. This is equivalent to say

$$a_1 x_{11} + a_2 x_{21} - x_{31} = 0, \quad a_1 x_{12} + a_2 x_{22} - x_{32} = 0.$$

We get

$$\frac{x_{22}x_{31} - x_{21}x_{32}}{x_{11}x_{22} - x_{12}x_{21}} \in \mathbb{Q}, \quad \frac{-x_{12}x_{31} + x_{11}x_{32}}{x_{11}x_{22} - x_{12}x_{21}} \in \mathbb{Q}.$$

Similar to computations in [Mov17b], we may take the inverse of one of the above functions and substitute in the other one, let us call it f. We are looking for a rational point in its domain such that its image is also rational.

7.5 The converse of Cattani–Deligne–Kaplan theorem

We are interested to see whether the converse of the Cattani–Deligne–Kaplan theorem is true. We have formulated this in the following statement. For the definition of a period vector of Hodge type see (6.14).

Property 7.1. For an enhanced family $\mathsf{X} \to \mathsf{T}$ of projective varieties and a period vector C of Hodge type, all defined over $\mathbb{C}$, the irreducible components of the Hodge locus in T are the only algebraic leaves of $\mathscr{F}(\mathsf{C})$.

Property 7.1 can be considered as the converse of Cattani–Deligne–Kaplan theorem in the following way. If a locus of Hodge cycles with arbitrary coefficients is algebraic then such coefficients, up to multiplication by a constant, must be necessarily rational numbers. This is the same as to say that if the set L_δ in (7.1) is algebraic for a continuous family of cycles $\delta_t \in H_m(\mathsf{X}_t, \mathbb{C})$ then up to multiplication by a constant $\delta_t \in H_m(\mathsf{X}_t, \mathbb{Q})$.

Property 7.1 implies that the modular foliation $\mathscr{F}(\mathsf{C})$ is not trivial in the sense of Definition 6.4 in §6.4. For a particular format of C this is essentially the same as to consider non-isolated Hodge classes, that is, Hodge classes that can be transported along a one-dimensional analytic curve inside the moduli space of the underlying variety. It is natural to expect that Property 7.1 is satisfied in many cases such as abelian varieties, as P. Deligne in [DMOS82] has proved that Hodge classes for principally polarized abelian varieties are non-isolated, and isolated Hodge classes may give us counterexamples to the Hodge conjecture. For all consequences of Property 7.1 we need that $\tilde{\mathsf{C}}$ belongs to a subset U of $\mathbb{P}^{\mathsf{b}-1}(\mathbb{C})$ without any local, non-discrete analytic subset inside. In particular, a non-constant holomorphic map from some analytic variety to $\mathbb{P}^{\mathsf{b}-1}(\mathbb{C})$ has its image in U if and only if it is constant.

Property 7.1 must be considered as a variant to Voisin's hypothesis on the non-existence of a constant subvariation of Hodge structures in Theorem 0.6 in [Voi07]. This is as follows. Let $V \subset \mathsf{T}$ be an algebraic leaf of $\mathscr{F}(\mathsf{C})$ and hence by definition it is irreducible. We have a holomorphic flat section δ of $H^m_{\mathrm{dR}}(\mathsf{X}/\mathsf{T})$ in a neighborhood of V such that restricted to V, δ is a linear combination of $\alpha_{m,i}$'s with constant coefficients (see Remark 6.1). Let B be the smooth part of V and Y be any smooth projective compactification of the inverse image of B under $\mathsf{X} \to \mathsf{T}$. For a fixed point $0 \in B$, we have the monodromy representation

$$\rho : \pi_1(B, 0) \to \mathrm{Aut}(H^m(\mathsf{X}_0, \mathbb{Q})). \tag{7.3}$$

The cycle $\delta_0 \in H^m(\mathsf{X}_0, \mathbb{C})$ is invariant under the monodromy. This implies that we have a subspace $H \subset H^m(\mathsf{X}_0, \mathbb{Q})$ whose elements are invariant under $\rho(\pi_1(B, 0))$ and $\delta \in H \otimes_{\mathbb{Q}} \mathbb{C}$. From another side, Deligne's global invariant cycle theorem or "théorèm de la partie fixe", see [Del71a, Theorem 4.1.1], or [Del68], tells us that the space of invariant cycles

$$H^m(\mathsf{X}_0, \mathbb{Q})^\rho := \{\delta \in H^m(\mathsf{X}_0, \mathbb{Q}) \mid \rho(\gamma)(\delta) = \delta, \ \forall \gamma \in \pi_1(B, 0)\}$$

is equal to the image of the restriction map $i^* : H^m(Y, \mathbb{Q}) \to H^m(\mathsf{X}_0, \mathbb{Q})$, where $i : \mathsf{X}_0 \hookrightarrow Y$ is the inclusion map. Moreover, this is a morphism of Hodge structures. Therefore, H lies in the image of i^*. If we assume that $\dim_{\mathbb{Q}} i^*(H^m(Y, \mathbb{Q})) = 1$ then $\dim_{\mathbb{Q}}(H) = 1$ and so δ_0 up to multiplication with a constant is in $H^m(\mathsf{X}_0, \mathbb{Q})$ which is the affirmation of Property 7.1. It is also reasonable to make a weaker assumption that the Hodge structure of $H^m(Y, \mathbb{C})$ (or actually its image under i^*) is trivial, that is it has only the middle piece $H^{\frac{m}{2}, \frac{m}{2}}$. Then all the cycles in $H^m(Y, \mathbb{Q})$ are Hodge. This implies that δ_0 is a $\mathbb{C}$-linear combination of Hodge cycles in $H^m(\mathsf{X}_0, \mathbb{Q})$ all of them invariant under $\rho(\pi_1(B,0))$. Let us write this $\delta_t = \sum_{i=1}^{k} c_i\delta_{i,t}$, $t \in V$. The function $\mathsf{C}_i(t) := \langle \alpha_m, \delta_{i,t}\rangle$, $t \in V$ is regular in V and $\mathsf{C} = \sum_{i=1}^{k} c_i \mathsf{C}_i(t)$ is a constant vector. It turns out that the foliation $\mathscr{F}(\mathsf{C}_i(0))$ restricted to V (a leaf of $\mathscr{F}(\mathsf{C})$) has only algebraic leaves. We do not get the affirmation in Property 7.1.

7.6 Weak absolute cycles

The notion of an absolute Hodge class is introduced by Deligne in [DMOS82]. Voisin in [Voi07] introduced the notion of a weak absolute Hodge class and observed that this notion is more natural when one studies the Hodge locus. In this section, we translate both notions into homological cycles and describe a consequence of Property 7.1.

Let $\bar{\mathfrak{k}} \subset \mathbb{C}$ be an algebraically closed field and let X a smooth projective variety over $\bar{\mathfrak{k}}$. For $\sigma \in \mathrm{Gal}(\bar{\mathfrak{k}}/\mathbb{Q})$, we denote by X_σ the underlying complex manifold after the action of σ on the defining coefficients of X. We have also the map induced in algebraic de Rham cohomologies:

$$\sigma : H^m_{\mathrm{dR}}(X/\bar{\mathfrak{k}}) \to H^m_{\mathrm{dR}}(X/\bar{\mathfrak{k}}), \quad \omega \mapsto \omega_\sigma.$$

Definition 7.8. Let X be a smooth projective variety defined over $\bar{\mathfrak{k}} \subset \mathbb{C}$. The complex numbers

$$\mathsf{P}(\omega,\delta) := (2\pi i)^{-\frac{m}{2}} \int_\delta \omega, \qquad \omega \in H^m_{\mathrm{dR}}(X/\bar{\mathfrak{k}})$$

are called the periods of $\delta \in H_m(X,\mathbb{Q})$. Such a cycle δ is called weak absolute if it has periods in $\bar{\mathfrak{k}}$ and for any $\sigma \in \mathrm{Gal}(\bar{\mathfrak{k}}/\mathbb{Q})$, there is a cycle $\delta_\sigma \in H_m(X_\sigma,\mathbb{Q})$ and $a_\sigma \in \mathbb{C}$ such that

$$\sigma\left(\mathsf{P}(\omega,\delta)\right) = a_\sigma \cdot \mathsf{P}(\omega_\sigma,\delta_\sigma), \quad \forall \omega \in H^m_{\mathrm{dR}}(X/\tilde{\mathfrak{k}}).$$

It is called absolute if moreover for all σ, $a_\sigma = 1$.

Note that in general if X is defined over $\bar{\mathfrak{k}}$ and $\delta \in H_m(X,\mathbb{Q})$ then the periods of δ are defined in an extension $\tilde{\mathfrak{k}}$ of $\bar{\mathfrak{k}}$ with transcendental numbers. Hodge conjecture implies the following.

Conjecture 7.2. *Let X be a smooth projective variety defined over $\bar{\mathfrak{k}} \subset \mathbb{C}$ and $\delta \in H_m(X,\mathbb{Q})$ be a Hodge cycle. We have the following conditions:*

(1) *The periods of δ are in $\bar{\mathfrak{k}}$.*
(2) *The cycle δ is absolute.*

(3) *If* $\tilde{\mathsf{k}} \subset \mathbb{C}$ *is a field extension of* $\bar{\mathsf{k}}$ *and* δ *as a cycle for* $X/\tilde{\mathsf{k}}$ *is absolute then it is also absolute for* $X/\bar{\mathsf{k}}$.

For algebraic cycles Conjecture 7.2 is trivially true, see [DMOS82, Proposition 1.5]. Note that the third item is a consequence of the first and second items. We have reproduced it because for absolute cycles which are not Hodge it seems to be a highly non-trivial statement. The first item can be used in order to study special values of many functions which can be written as periods, see for instance [MR06]. An interesting observation due to Voisin is the following:

Proposition 7.6 (Voisin [Voi07, p. 948]). *For a weak absolute Hodge cycle we have*

$$a_\sigma^2 \in \mathbb{Q}_{>0}.$$

Proof. It is easier to present the proof in cohomologies rather than homologies and using integrals. Let $\tilde{\delta} \in H^m(X, \mathbb{Q})$ be a weak absolute Hodge class. The Lefschetz (or primitive) decomposition of $\tilde{\delta}$ in both $H^m(X, \mathbb{Q})$ and $H^m_{\mathrm{dR}}(X/\mathsf{k})$ is unique. We conclude that the primitive pieces of $\tilde{\delta}$ are weak absolute with the same a_σ as of $\tilde{\delta}$. Now the affirmation follows from the second Hodge–Riemann bilinear relations. □

We might expect that some isolated Hodge cycles are not absolute. This statement must be easier to prove than the statement on the existence of algebraic cycles. For this, we would need only to prove that a bunch of integrals are zero and at least one integral with a proper factor of $2\pi i$ is a transcendental number.

7.7 Consequences of Property 7.1

In this section, we combine Property 7.1 and Theorem 5.18 and we prove the following theorem. It says that Property 7.1 and Hodge conjecture have few common consequences.

Theorem 7.7. *Let* X/T *be an enhanced family defined over the field* $\mathsf{k} \subset \mathbb{C}$ *and assume that Property 7.1 is true for all modular foliation* $\mathscr{F}(\mathsf{C})$ *attached to* X/T *and all period vector* C *of Hodge type and defined over* $\bar{\mathsf{k}}$. *Consider a modular foliation* $\mathscr{F}(\mathsf{C})$ *attached to* X/T *and with a period vector* C *of Hodge type and with an algebraic leaf* L*, both* C *and* L *defined over* $\mathbb{C}$. *If* L *is homologically defined over* $\mathbb{Q}$ *(see Definition 6.3) and it contain a* $\bar{\mathsf{k}}$*-rational* *point t_0 of T then primitive parts of C, up to multiplication with a constant, and L are defined over $\bar{\mathfrak{k}}$.*

Proof. The leaf L is homologically defined over $\mathbb{Q}$, and so, we have a continuous family of cycles $\delta = \{\delta_t\}_{t\in U}, \delta_t \in H_m(X_t, \mathbb{Q})$ such that $L = L_\delta$ and δ_t for $t \in L$ is a Hodge cycle. Here $U = (\mathsf{T}, t_0)$ is a small neighborhood of t_0 in T. Let us assume that the theorem with δ primitive and C a primitive period vector of Hodge type, is true. This implies the theorem for arbitrary δ as follows.

We use Proposition 6.8, and the notations used in its proof. Let

$$\mathsf{C} = \mathsf{C}_m + \mathsf{C}_{m-2} + \cdots + \mathsf{C}_{m-2q} + \cdots, \quad \delta_t = \delta_{t,m} + \delta_{t,m-2} + \cdots + \delta_{t,m-2q} + \cdots$$

be the primitive decomposition of C and δ_t, respectively. The foliation $\mathscr{F}(\mathsf{C}_{m-2q})$ has the leaf $L_{\delta_{t,m-2q}}$ and each $\delta_{t,m-2q}$ is a Hodge cycle. By the Cattani–Deligne–Kaplan theorem $L_{\delta_{t,m-2q}}$ is algebraic. We also know that it contains $t_0 \in \mathsf{T}(\bar{\mathfrak{k}})$. We apply the theorem in the case of primitive cycles and conclude that for all q, $L_{\delta_{t,m-2q}}$ and C_{m-2q} up to multiplication with a constant $c_{m-2q} \in \mathbb{C}$ are defined over $\bar{\mathfrak{k}}$. This implies the theorem for arbitrary δ_t. Note that we do not claim that c_{m-2q}'s are the same for all q. From now on assume that δ is primitive and C is a primitive period vector of Hodge type.

Next, we prove that if C is defined over $\bar{\mathfrak{k}}$, and the algebraic leaf L of $\mathscr{F}(\mathsf{C})$ is weakly homologically defined over $\mathbb{Q}$, see Definition 6.3, then L is defined over $\bar{\mathfrak{k}}$. Assume that this is not the case. By our assumption L is given by $c \cdot \int_\delta \alpha_m = \mathsf{C}$ for some constant $c \in \mathbb{C}$. By the second part of Theorem 5.18, we have varieties Z and $\tilde{L} \subset \mathsf{T} \times Z$ and the projection $\pi : \tilde{L} \to Z$, all defined over $\bar{\mathfrak{k}}$ such that $L_x = \pi^{-1}(x)$, $x \in Z$ is an algebraic leaf of $\mathscr{F}(\mathsf{C})$. This means that we have families of algebraic leaves for $\mathscr{F}(\mathsf{C})$. Further, for some $a \in Z$, $L_a = L$ is the original leaf. From now on we only work with $L_x \cap U$ with $x \in (Z, a)$.

Recall from Proposition 6.2 that $\mathscr{F}(\mathsf{C})$ has the local first integral $f := \mathsf{P}^{-\mathrm{tr}}\mathsf{C} : U \to \mathbb{C}$ such that for the leaf $f^{-1}(\tilde{\mathsf{C}})$ of $\mathscr{F}(\mathsf{C})$ the corresponding continuous family of cycles has coefficients in $\mathbb{Q}$ if and only if $\tilde{\mathsf{C}}$ has rational entries. Since $L_x \cap U$ is a leaf of $\mathscr{F}(\mathsf{C})$ in the sense of Definition 5.6, the restriction of f to $L_x \cap U$ is constant, and hence, we have finitely many constant vectors $\tilde{\mathsf{C}}_{x,i}$, $i = 1, 2, \ldots$ such that f restricted to connected components $L_{x,i}$, $i = 1, 2, \ldots$ of $L_x \cap U$ is $\tilde{\mathsf{C}}_{x,i}$. For this affirmation we may take U smaller such that all the irreducible components of $L_a \cap U$ crosses $0 \in \mathsf{T}$, however, for $x \in (Z, a)$ and $x \neq a$, $L_x \cap U$ might have still finitely many connected components. The conclusion is that the vector $\tilde{\mathsf{C}}_x$ is a holomorphic multi-valued function in $x \in (Z, a)$ and according to Property 7.1, up to multiplication with a constant it has rational entries. This means that $\tilde{\mathsf{C}}_x = c_x \tilde{\mathsf{C}}_1$, where $\tilde{\mathsf{C}}_1$ is a constant vector with rational entries and c_x is a multi-valued holomorphic function in $x \in (Z, a)$. We conclude that L_x is given by $c_x \cdot \int_\delta \alpha_m = \mathsf{C}$. We use the fact that δ_t is primitive and so by Proposition 7.4

$$\delta_t \cdot \delta_t = c_x^{-2} \mathsf{C}^{\mathsf{tr}} \Phi_m^{-\mathsf{tr}} \mathsf{C} > 0$$

and so c_x must be a non-zero constant in $\mathbb{C}$. This implies that Z consists of just the point a, and hence, L is defined over $\bar{\mathfrak{k}}$.

Let $\tilde{\mathfrak{k}}$ be the algebraically closed field over $\bar{\mathfrak{k}}$ generated by the entries of C and the definition field of L (which may contain some transcendental numbers over $\bar{\mathfrak{k}}$). We use the first part of Theorem 5.18 for the triple $(\mathsf{T}, \mathscr{F}(\mathsf{C}), L)$ and obtain an affine variety Z and a triple $(\mathsf{T} \times Z,\ \tilde{\mathscr{F}}, \tilde{L})$. Since the Gauss–Manin connection matrix is defined over $\mathfrak{k}$, in the definition of a modular foliation the only transcendental numbers over $\bar{\mathfrak{k}}$ may occur in C. We can regard C as a vector with entries which are rational functions on Z. In particular, we can evaluate over a Zariski open subset $\tilde{Z}$ of Z. From now on we replace Z with $\tilde{Z}$. Let C_x be the evaluation of C over the point $x \in Z$. We have

$$\tilde{\mathscr{F}} \mid_{\mathsf{T} \times \{x\}} = \mathscr{F}(\mathsf{C}_x), \quad x \in Z.$$

Now we use Property 7.1 for the foliation $\mathscr{F}(\mathsf{C}_x)$ and its algebraic leaf $L_x := \pi^{-1}(x)$ and conclude that there is a continuous family if cycles $\delta_t \in H_m(\mathsf{X}_t, \mathbb{Q})$ and a multi-valued holomorphic function c_x in an open subset of (Z, a) such that the algebraic leaf L_x of $\mathscr{F}(\mathsf{C}_x)$ in U is given by

$$c_x \cdot \int_{\delta_t} \alpha_m = \mathsf{C}_x.$$

Since $H_m(X_t, \mathbb{Q})$ does not contain any analytic subset of $H_m(X_t, \mathbb{C})$, the continuous family of cycles δ is the same for all x. This follows again by considering holomorphic family of local first integrals $f_x := \mathsf{P}^{-\mathsf{tr}} \mathsf{C}_x : U \to \mathbb{C}$, $x \in (Z, a)$. We conclude that the entries of $c_x \int_{\delta_t} \alpha_m - \mathsf{C}_x$, which are multivalued holomorphic functions, vanishes on the algebraic variety $\tilde{L}$.

Let $\Pi : \tilde{L} \to \mathsf{T}$ be the projection map in T coordinate. The pull-back of the holomorphic function $\int_{\delta_t} \alpha_m$ by the projection Π is $c_x^{-1} \mathsf{C}_x$, whose quotient of entries are rational function on $\tilde{L}$. We conclude that a quotient

g of entries of $\int_{\delta_t} \alpha_m$ restricted to the image of Π are algebraic functions, all of them defined over $\bar{\mathfrak{k}}$, see Proposition 7.8. But by our hypothesis the image of Π contains the $\bar{\mathfrak{k}}$-rational point $0 \in \mathsf{T}$. Therefore, all such g's evaluated at t_0 are in $\bar{\mathfrak{k}}$. Performing pull-back by Π, we get the fact that $\mathsf{C}_a = c \cdot \check{\mathsf{C}}$ with $c \in \mathbb{C}$ and $\check{\mathsf{C}} \in \bar{\mathfrak{k}}^{\mathsf{b}}$. This reduces the problem to the case in which C is defined over $\bar{\mathfrak{k}}$ and L is weakly homologically defined over $\mathbb{Q}$. □

In the last step of the proof of Theorem 7.1, we have used the following simple proposition.

Proposition 7.8. *Let $f : X \to Y$ be a surjective morphism of algebraic varieties over an algebraically closed field $\bar{\mathfrak{k}} \subset \mathbb{C}$. If g is a holomorphic function in an small open set U of Y such that $g \circ f$ is a restriction of a regular function of $X/\bar{\mathfrak{k}}$ to $f^{-1}(U)$ then g is also a restriction of a regular function of $Y/\bar{\mathfrak{k}}$ to U.*

Let us consider an enhanced family X/T over the field $\mathfrak{k} \subset \mathbb{C}$ and a Hodge cycle $\delta_0 \in H_m(X_0, \mathbb{Q})$ and define

$$\mathsf{C} := \int_{\delta_0} \alpha_m \tag{7.4}$$

and consider the modular foliation $\mathscr{F}(\mathsf{C})$. One of the consequences of the Hodge conjecture is that if X_0 is defined over $\mathfrak{k}$, that is, if 0 is a $\mathfrak{k}$-rational point of T, then $\mathsf{C} \in (2\pi i)^{\frac{m}{2}} \bar{\mathfrak{k}}^{\mathsf{b}}$, see the first item in Conjecture 7.2.

Proposition 7.9. *Property 7.1 implies that the period vector C of a Hodge cycle of $X/\mathfrak{k}$ up to multiplication by a constant is defined over $\bar{\mathfrak{k}}$, that is $\mathsf{C} \in \mathbb{P}^{\mathsf{b}-1}(\bar{\mathfrak{k}})$.*

Proof. We consider the modular foliation $\mathscr{F}(\mathsf{C})$ and apply Theorem 7.7. □

Note that the Hodge conjecture implies that the transcendental factor of C is $(2\pi i)^{\frac{m}{2}}$. This does not follow from Property 7.1. Note that in the proof of Theorem 7.7 we strongly use the fact that $\langle \alpha_{m,i}, \alpha_{m,j} \rangle$'s are constant. This forces us to work with the finer parameter space T and indicates that a similar statement for a weakly enhanced family as in §3.5 might be false.

Theorem 7.10. *Assume that Property 7.1 is true for all modular foliation $\mathscr{F}(\mathsf{C})$ attached to X/T and all period vector C of Hodge type and defined over $\mathbb{C}$. A Hodge cycle of any variety X_t, $t \in \mathsf{T}(\mathbb{C})$ is weak absolute.*

Proof. Let $\delta_t \in H_m(X_t, \mathbb{Q})$ be a Hodge cycle and $\mathsf{C} := \int_{\delta_t} \alpha_m$. Let L_δ be the leaf of $\mathscr{F}(\mathsf{C})$ crossing the point $t \in \mathsf{T}$. Theorem 7.2 implies that L_δ is algebraic. Let $\tilde{\mathfrak{k}}$ be the algebraically closed subfield of $\mathbb{C}$ generated by the coordinates of t and C. For $\sigma \in \mathrm{Gal}(\tilde{\mathfrak{k}}/\mathbb{Q})$, we have the foliation $\mathscr{F}(\mathsf{C}_\sigma)$ tangent to the algebraic variety $\sigma(L_\delta)$. We take a local analytic irreducible branch L of $\sigma(L_\delta)$ crossing the point $\sigma(t)$ and we have a continuous family of cycles $\tilde{\delta}_s \in H_m(X_s, \mathbb{C})$, $s \in (\mathsf{T}, \sigma(t))$ such that $L \subset L_{\tilde{\delta}}$. By Property 7.1, up to some constant c_σ, $\tilde{\delta}_s$ is in $H_m(X_s, \mathbb{Q})$ and so

$$\sigma\left(\int_{\delta_t} \alpha_m\right) = \sigma(\mathsf{C}) = c_\sigma \int_{\delta_{\sigma(t)}} \alpha_m. \qquad \square$$

Let us be given a family of smooth projective varieties Y/V over $\mathfrak{k}$. We can use the recipe in §3.6 and construct an enhanced family X/T. If Property 7.1 is valid for all constant period vector C of Hodge type then Theorems 7.7 and 7.10 imply that the Hodge loci in V are defined over $\bar{\mathfrak{k}}$, Hodge cycles of the fibers of Y/V are weak absolute and they have periods in the algebraic closure of the field of definition of the fiber. However, Property 7.1 seems to be stronger than these consequences together. For instance, consider an example of $\mathscr{F}(\mathsf{C})$ which is trivial in the sense of Definition 6.4. All the leaves of $\mathscr{F}(\mathsf{C})$ in $\mathsf{T}\backslash\mathrm{Sing}(\mathscr{F}(\mathsf{C}))$ are algebraic, and they are the orbits of G_C. Note that the notion of a trivial foliation is related to the notion of an isolated Hodge cycle.

Chapter 8

Generalized Period Domain

You forget, he said, that all your curses are of limited duration; one hundred and fifty years from today, their force will be spent (A. Weil in "Mathematische Werke, by Gotthold Eisenstein" [Wei79, p. 398]).

8.1 Introduction

We introduce the generalized period domain U which is the target space of period maps from the parameter space of enhanced families. We also study modular foliations, vector fields and the loci of Hodge cycles in U. Some of the material presented in this chapter are taken from [Mov13]. In the previous chapters, we have explained that the moduli of enhanced varieties is a nice object in which modular vector fields and foliations live, whereas classical moduli of varieties do not carry such rich structures. Going to the Hodge structure or periods of varieties, Griffiths introduced a period domain which is mainly responsible for the variation of Hodge structures on classical moduli spaces and hence it is not adapted to our case. The generalized period domain U is responsible for the variation of Hodge structures arising from enhanced varieties. We have slightly modified the same notion in [Mov13] by considering the whole cohomology ring of a variety. The main reason for this is inspired from the case of abelian varieties. In order to study Hodge classes in the middle cohomology of an abelian variety one considers the period domain of the first cohomology (which generates the whole cohomology ring) and not the period domain attached to the middle cohomology. Therefore, we will reproduce many arguments of [Mov13] and, in particular, we will explain our approach in comparison to Griffiths period domain, see for instance [Gri70, Mov08a].

Another advantage of U is that it does contain the full data of periods of a variety, whereas Griffiths period domain contains the data of certain quotient of periods. In simple words, the space U is the transcendental incarnation of the moduli space T of enhanced varieties introduced in §3.11.

Something which would perfectly fit into this chapter is the construction of a similar period domain U starting from a Hermitian symmetric domain and an action of an arithmetic group. This would give the differential equations of the corresponding automorphic forms in a geometric context. We have not done this and the interested reader may think on this starting from Deligne's description of Hermitian symmetric domains as the parameter space for certain special Hodge structures, see [DMOS82, Del79, Milb].

8.2 Polarized Hodge structures

Recall from §§2.7 and 3.2 that for a projective variety $X_0/\mathbb{C}$ of dimension n we have the following algebraic data: the algebraic de Rham cohomology $H^m_{\mathrm{dR}}(X_0)$, $m = 0, 2, \ldots, 2n$, its Hodge filtration $F^* H^*_{\mathrm{dR}}(X_0)$, cup product $\cup$, polarization θ_0 and the trace map Tr. All these satisfy many properties, such as Lefschetz decomposition, Hard Lefschetz theorem and so on. Now consider an embedding $\mathfrak{k} \subset \mathbb{C}$. From this we get two additional structures. First, we get the complex conjugation in the de Rham cohomology. This together with Hodge filtration give us the Hodge decomposition. It satisfies the so-called Hodge–Riemann bilinear relations, see for instance [Voi02, Mov19]. Second, we can look at X_0 as a complex manifold and hence we have the embedding

$$H_m(X_0, \mathbb{Z}) \hookrightarrow H^m_{\mathrm{dR}}(X_0)^\vee, \quad \delta \mapsto \int_\delta,$$

which is an isomorphism after tensoring with $\mathbb{C}$.

Definition 8.1. We call all these data a polarized Hodge structure.

Note that the same notion in the literature usually refers to a part of the above data with fixed m.

8.3 Generalized period domain

Recall the projective variety $X_0/\mathbb{C}$ which we have fixed in §3.10. We consider it as a complex manifold and we define V_0 to be the de Rham cohomology ring of X_0 equipped with cup product, Hodge filtration and polarization:

$$V_0 := (H^*_{\mathrm{dR}}(X_0),\ F_0^*, \cup, \theta_0).$$

We also define $V_{0,\mathbb{Z}}$ to be the ring of homology groups of X_0 together with the intersection of cycles and the polarization element $[Y_0] \in H_{2n-2}(X_0, \mathbb{Z})$:

$$V_{0,\mathbb{Z}} := (H_*(X_0, \mathbb{Z}),\ \cdot\ , [Y_0]).$$

Let us be given an arbitrary embedding

$$\mathsf{u} : V_{0,\mathbb{Z}} \hookrightarrow V_0^\vee \tag{8.1}$$

which sends $H_m(X_0, \mathbb{Z})$ to $H^m_{\mathrm{dR}}(X_0)^\vee$. We define the abstract integral sign through the equality

$$\mathsf{u}(\delta)(\omega) = \int_\delta \omega, \quad \omega \in H^m_{\mathrm{dR}}(X_0), \quad \delta \in H_m(X_0, \mathbb{Z}).$$

Note that this is just the definition of $\int$ and no integration is taking place. The Poincaré dual $\delta^{\mathsf{pd}} \in H^m_{\mathrm{dR}}(X_0)$ of $\delta \in H_m(X_0, \mathbb{Z})$ is defined uniquely through the equality

$$\int_\delta \omega = (2\pi i)^n \cdot \mathrm{Tr}(\omega \cup \delta^{\mathsf{pd}}), \quad \forall \omega \in H^m_{\mathrm{dR}}(X_0).$$

Definition 8.2. The generalized period domain Π, respectively U, is the set of all embeddings (8.1), respectively image of such embeddings, such that

(1) the intersection of cycles in $V_{0,\mathbb{Z}}$ is Poincaré dual to cup product in de Rham cohomology, that is,

$$\delta_1^{\mathsf{pd}} \cup \delta_2^{\mathsf{pd}} = (\delta_1 \cdot \delta_2)^{\mathsf{pd}}, \quad \delta_1, \delta_2 \in H_*(X_0, \mathbb{Z});$$

(2) we get a polarized Hodge structure in the sense of Definition 8.1.

Clearly, we have at least one element of U obtained by the fact that both $V_{0,\mathbb{Z}}$ and V_0 comes from the homologies and cohomologies of the fixed variety X_0.

From now on we also denote an element of U by u. Attached to u we have a pair $(\mathsf{X_u}, \alpha_{\mathsf{u}})$. Here, $\mathsf{X_u}$ is some ghost projective variety (which does not exists), however, we define its algebraic de Rham cohomology,

Hodge filtration, etc., to be the same as of X_0 and so we define α_{u} to be the identity map

$$\alpha : (H^*_{\rm dR}(X_{\mathsf{u}}),\ F, \cup, \theta) \to V_0.$$

We also define $H_*(X_{\mathsf{u}}, \mathbb{Z})$ to be the image of the embedding (8.1). Note that there is no variety X_{u} and we have introduced it in order to produce the same notation as in the algebraic context. We are just imitating the algebraic context in the level of periods. We may call X_{u} a motif. In this way, the generalized period domain is the moduli of enhanced motives, and we can treat U as it was the moduli T of enhanced projective varieties. For the moment we do not feel the necessity of using classical theory of motives in our context. Therefore, we will avoid the motive terminology. The algebraic group G and the discrete group $\Gamma_{\mathbb{Z}}$ acts from the right and left, respectively, on Π:

$$\Gamma_{\mathbb{Z}} \curvearrowright \Pi \curvearrowleft \mathsf{G}.$$

The action of $\Gamma_{\mathbb{Z}}$ in Π is given by

$$A(\mathsf{u}) := \mathsf{u} \circ A^{-1}, \quad A \in \Gamma_{\mathbb{Z}}, \quad \mathsf{u} \in \Pi,$$

and the action of G on Π is given by

$$(\mathsf{u}, \mathsf{g}) \mapsto \mathsf{u} \bullet \mathsf{g} := \mathsf{u} \circ (\mathsf{g}^\vee)^{-1}, \quad \mathsf{u} \in \Pi, \quad \mathsf{g} \in \mathsf{G}.$$

By our construction we have the following proposition.

Proposition 8.1. *The actions of G and $\Gamma_{\mathbb{Z}}$ on Π commutes and are free.*

By definition we have

$$\mathsf{U} = \Gamma_{\mathbb{Z}} \backslash \Pi$$

and we also use $\bullet$ for the action of G on U. We might have $A \in \Gamma_{\mathbb{Z}}$, $\mathsf{u} \in \Pi$ and the non-identity element $\mathsf{g} \in \mathsf{G}$ such that

$$A(\mathsf{u}) = \mathsf{u} \bullet \mathsf{g}.$$

This means that the action of G on U might not be free. This is the case for instance for elliptic curves, see §9.2. However, it is expected that the stablizer of a generic point of U is the trivial identity group.

Definition 8.3. The moduli of polarized Hodge structures is defined to be

$$\Gamma_{\mathbb{Z}} \backslash \Pi / \mathsf{G}.$$

Definition 8.4. The Griffiths period domain is defined to be

$$D := \Pi/\mathsf{G}.$$

This is a slight modification of Griffiths period domain. For varieties such that the cohomology algebra is generated by the elements of a fixed cohomology, our notion of period domain and Griffiths period domain are the same. However, for other cases our period domain is finer. We will still call D the Griffiths period domain. The set $\Gamma_{\mathbb{Z}}\backslash D$ is the moduli of polarized Hodge structures of fixed topological type and it has a canonical structure of a complex analytic space.

8.4 Period maps

Let X/T be an enhanced family of projective varieties with X_0 as a fiber over $0 \in \mathsf{T}$.

Definition 8.5. We have the generalized period map

$$\mathsf{P} : \mathsf{T} \mapsto \mathsf{U}$$

which is defined in the following way. By definition a fiber X of X/T comes with an isomorphism $\alpha : H^*_{\mathrm{dR}}(X) \to H^*_{\mathrm{dR}}(X_0)$. The inverse of the dual of this isomorphism, call it $(\alpha^\vee)^{-1}$ sends $H_*(X, \mathbb{Z})$ to a lattice in $H^*_{\mathrm{dR}}(X_0)^\vee$ and this gives a unique point of U.

Note that this definition is in a complete harmony with our ghost notation $(X_{\mathsf{u}}, \alpha_{\mathsf{u}})$. Moreover, note that if $\mathsf{P}(X_1, \alpha_1) = \mathsf{P}(X_2, \alpha_2)$ then $(\alpha_2^\vee) \circ (\alpha_1^\vee)^{-1}$: $H^*_{\mathrm{dR}}(X_1)^\vee \to H^*_{\mathrm{dR}}(X_2)^\vee$ maps $H_*(X_1, \mathbb{Z})$ isomorphically to $H_*(X_2, \mathbb{Z})$, and hence, it is an isomorphism of Hodge structures. The following definition will be needed in Proposition 8.4.

Definition 8.6. For $m \in \mathbb{N} \cup \{0\}$ we say that the generalized period map P is m-injective if all the components of

$$\{(t_1, t_2) \in \mathsf{T} \times \mathsf{T} | \mathsf{P}(t_1) = \mathsf{P}(t_2)\}$$

except for the diagonal, are of dimension $< m$.

Let $\tilde{\mathsf{T}}$ be the monodromy covering of T, that is, $\tilde{\mathsf{T}}$ consists of (t, δ), where $t \in \mathsf{T}$ and $\delta : (H_*(X, \mathbb{Z}), \cdot, [Y]) \cong (H_*(X_0, \mathbb{Z}), \cdot, [Y_0])$ as in §4.3 for $X := X_t$.

Definition 8.7. The generalized period map

$$\mathsf{P} : \tilde{\mathsf{T}} \mapsto \mathsf{U},$$

is defined in the following way: for $\tilde{t} = (t, \delta) \in \tilde{\mathsf{T}}$, $X := X_t$ with $\delta : (H_*(X, \mathbb{Z}), \cdot, [Y]) \cong (H_*(X_0, \mathbb{Z}), \cdot, [Y_0])$ as in Definition 4.2, and $\alpha : (H^*_{\mathrm{dR}}(X), F^*, \cup, \theta) \cong (H^*_{\mathrm{dR}}(X_0), F^*_0, \cup, \theta_0)$ which comes from the definition of the enhanced family X/T. We get the the following diagram which is not necessarily commutative:

$$\begin{array}{ccc} H_*(X, \mathbb{Z}) & \overset{\delta}{\to} & H_*(X_0, \mathbb{Z}) \\ i_X \downarrow & & \downarrow i_{X_0} \\ H^*_{\mathrm{dR}}(X)^\vee & \overset{\alpha^\vee}{\leftarrow} & H^*_{\mathrm{dR}}(X_0)^\vee \end{array}, \tag{8.2}$$

where i_X and i_{X_0} are usual integration maps. The image $\mathsf{P}(\tilde{t})$ of $\tilde{t} \in \tilde{\mathsf{T}}$ under P is $(\alpha^\vee)^{-1} i_X \circ \delta^{-1}$.

Recall the monodromy covering $\mathbb{H}$ in §4.3.

Definition 8.8. The classical period map

$$\mathsf{P} : \mathbb{H} \to D$$

is defined in the following way: for $w = (X, \delta) \in \mathbb{H}$ with $\delta : (H_*(X, \mathbb{Z}), \cdot, [Y]) \cong (H_*(X_0, \mathbb{Z}), \cdot, [Y_0])$ as in Definition 4.2, we first consider an arbitrary enhancement (X, α) with $\alpha : (H^*_{\mathrm{dR}}(X), F^*, \cup, \theta) \cong (H^*_{\mathrm{dR}}(X_0), F^*_0, \cup, \theta_0)$. We get the diagram (8.2) which is not necessarily commutative. The image $\mathsf{P}(w)$ of w under the classical period map is $(\alpha^\vee)^{-1} i_X \circ \delta^{-1}$. The choices of different enhancements α will result in the orbit of G in Π, and hence, a well-defined element $\mathsf{P}(w)$ in D.

Note that if T is the universal family of smooth enhanced varieties then $\mathbb{H} = \mathsf{T}/\mathsf{G}$ and the classical period map is induced by the generalized period map after taking quotient by G.

8.5 τ and t maps

The generalizations of automorphic forms in the present text are done in both algebraic and holomorphic context. The discussion in this section provides a precise translation from one to another. It gives us a bridge between the algebraic and holomorphic worlds.

Definition 8.9. A meromorphic map

$$\tau : D \dashrightarrow \mathsf{\Pi} \tag{8.3}$$

such that the composition $D \dashrightarrow \mathsf{\Pi} \to \mathsf{\Pi}/\mathsf{G} = D$ is the identity map, and it is an embedding outside the set of its indeterminacy points, is called the τ map and its image is called the τ loci.

There is no general recipe to define the τ map. It has been constructed in the case of abelian varieties, see §11.11 and mirror quintic Calabi–Yau threefolds, see [Mov17b, Chapter 4]. The discussion in §13.9 gives us the description of such a map for arbitrary Calabi–Yau threefolds. The following proposition is a direct consequence of the above definition.

Proposition 8.2. *For any* $\mathsf{u} \in \mathsf{\Pi}$ *there is a unique* $\mathsf{g} \in \mathsf{G}$ *such that*

$$\tau(x) = \mathsf{u} \bullet \mathsf{g},$$

where x *is the projection of* u *in* $D = \mathsf{\Pi}/\mathsf{G}$.

Proof. Since $\tau(x)$ and u induce the same element in D, there is a unique element $\mathsf{g} \in \mathsf{G}$ such that $\tau(x) = \mathsf{u} \bullet \mathsf{g}$ holds. □

Let us consider a τ map and recall the monodromy covering $\mathbb{H}$ in §4.3. Let also T be the moduli of enhanced varieties as in §3.11.

Proposition 8.3. *For any* τ*-map, there is a meromorphic map*

$$\mathsf{t} : \mathbb{H} \dashrightarrow \mathsf{T},$$

which is characterized by the fact that the following diagram is commutative:

$$\begin{array}{ccc} \mathbb{H} & \overset{\mathsf{t}}{\dashrightarrow} & \mathsf{T} \\ \downarrow & & \downarrow, \\ D & \overset{\tau}{\to} & \mathsf{U} \end{array}$$

and the composition $\mathbb{H} \dashrightarrow \mathsf{T} \to \mathsf{T}/\mathsf{G} = \mathsf{M}$ *is the canonical map* $(X, \delta) \mapsto X$. *Here, the down arrows are respectively the classical and generalized period maps.*

Proof. Let $\tilde{\mathsf{T}}$ be the moduli of (X, δ, α). It is enough to construct the map $\tilde{\mathsf{t}} : \mathbb{H} \to \tilde{\mathsf{T}}$ such that

$$\begin{array}{ccc} \mathbb{H} & \overset{\tilde{\mathsf{t}}}{\dashrightarrow} & \tilde{\mathsf{T}} \\ \downarrow & & \downarrow, \\ D & \overset{\tau}{\to} & \sqcap \end{array}$$

commutes and the composition $\mathbb{H} \dashrightarrow \tilde{\mathsf{T}} \to \tilde{\mathsf{T}}/\mathsf{G} = \mathbb{H}$ is the identity map. This is because $\mathsf{T} := \Gamma_{\mathbb{Z}} \backslash \tilde{\mathsf{T}}$ and $\mathsf{U} = \Gamma_{\mathbb{Z}} \backslash \sqcap$. For a point $w = (X, \delta) \in \mathbb{H}$ we first choose an arbitrary enhancement (X, α). By Proposition 8.2, there is a unique element $\mathsf{g} \in \mathsf{G}$ and $x \in D$ such that $\tau(x) = \mathsf{P}(X, \delta, \alpha) \bullet \mathsf{g}$. The point x is just the projection of $\mathsf{P}(X, \delta, \alpha)$ under $\sqcap \to D$, and so, $x = \mathsf{P}(w)$. We replace (X, α) with $(X, \alpha) \bullet \mathsf{g}$ and get $\tau(\mathsf{P}(w)) = \mathsf{P}(X, \delta, \alpha)$. Therefore, we must define $\tilde{\mathsf{t}}(w) := (X, \delta, \alpha)$ and with this definition the above diagram is commutative. □

Definition 8.10. The map t in Proposition 8.3 is called the t map.

Remark 8.1. Since the action of G on $\sqcap$ is free, the map $\tilde{\mathsf{t}} : \mathbb{H} \to \tilde{\mathsf{T}}$ is unique. If we have two such maps $\tilde{\mathsf{t}}_i$, $i = 1, 2$ then $\mathsf{P}(\tilde{\mathsf{t}}_1(w)) = \mathsf{P}(\tilde{\mathsf{t}}_2(w))$, $\forall w \in \mathbb{H}$. We have a unique $\mathsf{g} \in \mathsf{G}$ such that $\tilde{\mathsf{t}}_2(w) = \tilde{\mathsf{t}}_1(w) \bullet \mathsf{g}$ and so $\mathsf{P}(\tilde{\mathsf{t}}_1(w)) = \mathsf{P}(\tilde{\mathsf{t}}_1(w)) \bullet \mathsf{g}$, and so, $\mathsf{g} = 1$. If the action of G on $\sqcap$ is generically free, then we might have a similar statement for t.

Remark 8.2. Note that if the τ map is holomorphic then the t map is holomorphic too. This is the case, for instance, for abelian varieties. For mirror quintic both τ and t maps are meromorphic, see [Mov17b, Chapter 4].

In §4.3, we have introduced few meromorphic functions in $\mathbb{H}$. These might be used in order to give a local coordinate system in $\mathbb{H}$. The map $\mathsf{t} : \mathbb{H} \dashrightarrow \mathsf{T}$ with such a coordinate system in $\mathbb{H}$ gives us solutions to modular vector fields in T. Moreover, the image of t is going to be a leaf of the foliation $\mathscr{F}(2)$ in §6.10.

8.6 Action of the monodromy group

The monodromy group $\Gamma_{\mathbb{Z}}$ acts on $\mathbb{H}$ from the left by composition of maps and it is natural to ask for the functional equation of the t map with respect to this action. Recall that $\dim(\mathsf{M}) = \dim(\mathbb{H})$ is the dimension of the classical moduli of X_0.

Proposition 8.4. *Let m be the dimension of the classical moduli of X_0. If the generalized period map $\mathsf{P} : \mathsf{T} \to \mathsf{U}$ is m-injective (see Definition 8.6), then*

$$\mathsf{t}(w) = \mathsf{t}(A(w)) \bullet \mathsf{g}(A, w), \quad \forall A \in \Gamma_{\mathbb{Z}}, \quad w \in \mathbb{H},$$

where $\mathsf{g}(A, w) \in \mathsf{G}$ is defined using the equality

$$A(\tilde{\tau}(w)) = \tilde{\tau}(A(w)) \bullet \mathsf{g}(A, w) \tag{8.4}$$

obtained from Proposition 8.2, $\tilde{\tau} = \tau \circ \mathsf{P} : \mathbb{H} \to \Pi$ and P is the classical period map.

Proof. Let $w = (X, \delta) \in \mathbb{H}$ and $\mathsf{t}(w) = (X, \alpha)$. Since the action of $\Gamma_{\mathbb{Z}}$ and G on Π commutes, we can re write (8.4) as

$$A\left(\tilde{\tau}(w) \bullet \mathsf{g}(A, w)^{-1}\right) = \tilde{\tau}(A(w)),$$

which is $\tilde{\tau}(w) \bullet \mathsf{g}(A, w)^{-1} = \tilde{\tau}(A(w))$ in U. Since the diagram in Proposition 8.3 commutes, we have $\mathsf{P}(\mathsf{t}(w)) \bullet \mathsf{g}(A, w)^{-1} = \mathsf{P}(\mathsf{t}(A(w))$. Now, we use the m-injectivity of the period map and we get $\mathsf{t}(w) \bullet \mathsf{g}(A, w)^{-1} = \mathsf{t}(A(w))$. □

8.7 Period matrix

Let α be a $\mathbb{C}$-basis of V_0 and δ a basis $V_{0,\mathbb{Z}}$ as in Chapter 4. Let $\mathsf{u} \in \Pi$. Since u comes with an embedding in (8.1) and α_{u} is the identity map, we have automatically a basis of $H_*(X_{\mathsf{u}}, \mathbb{Z})$ and $H^*_{\mathrm{dR}}(X_{\mathsf{u}})$ which we denote it again by δ and α, respectively. We define the mth period matrix of u in the following way:

$$\begin{aligned}
\mathsf{P}_m(\mathsf{u}) &:= (2\pi i)^{-\frac{m}{2}} \cdot \int_{\delta_m} \alpha_m^{\mathsf{tr}} \\
&= (2\pi i)^{-\frac{m}{2}} \begin{pmatrix}
\int_{\delta_{m,1}} \alpha_{m,1} & \int_{\delta_{m,1}} \alpha_{m,2} & \cdots & \int_{\delta_{m,1}} \alpha_{m,\mathsf{b}_m} \\
\int_{\delta_{m,2}} \alpha_{m,1} & \int_{\delta_{m,2}} \alpha_{m,2} & \cdots & \int_{\delta_{m,2}} \alpha_{m,\mathsf{b}_m} \\
\vdots & \vdots & \vdots & \vdots \\
\int_{\delta_{m,\mathsf{b}_m}} \alpha_{1,m} & \int_{\delta_{m,\mathsf{b}_m}} \alpha_{2,m} & \cdots & \int_{\delta_{m,\mathsf{b}_m}} \alpha_{m,\mathsf{b}_m}
\end{pmatrix}.
\end{aligned}$$

If we have an enhanced family then we can replace t with u and the integrations in the above matrix are the usual ones. Instead of the period matrix it is useful to use the matrix q_m defined by

$$\alpha_m = \mathsf{q}_m \delta^{\mathsf{pd}}_{2n-m}.$$

Then we have

$$\mathsf{P}_m = \Psi^{\mathsf{tr}}_{2n-m} \cdot \mathsf{q}^{\mathsf{tr}}_m.$$

Combining these two equalities we have

$$\alpha_m = \mathsf{P}^{\mathsf{tr}}_m \Psi^{-1}_{2n-m} \delta^{\mathsf{pd}}_{2n-m}. \tag{8.5}$$

Remark 8.3. The entries of period matrices P_m, $m = 0, 1, 2, \ldots, 2n$ satisfy many polynomial equations which have been described in Proposition 4.1. Considering such entries as variables, this gives us an affine variety $V \subset \prod_{m=0}^{2n} \mathrm{Mat}(\mathsf{b}_m, \mathbb{C})$ defined over $\mathbb{Q}$. The generalized period domain $\sqcap$ can be considered as an open subset of V (using the usual topology of V).

8.8 Griffiths period domain as a quotient of real Lie groups

In this section, we first recall from [Del71b, CK78] some classical construction related to Griffiths period domain and then reformulate them in terms of the generalized period domain. Let

$$\Gamma_k := \mathrm{Aut}(H_*(X_0, \mathbb{K}), \cdot, [Y_0]), \quad \mathbb{K} = \mathbb{R}, \mathbb{C},$$

which is defined in a similar way as with $\Gamma_{\mathbb{Z}}$ in (4.9). The group $\Gamma_{\mathbb{R}}$ acts from the left on $\sqcap$, U and D in a canonical way. Let

$$\begin{aligned} \mathrm{Lie}(\Gamma_{\mathbb{R}}) := \{ \mathfrak{g} \in \mathrm{End}_{\mathbb{R}}(H_*(X_0, \mathbb{R})) | \\ &\mathbb{R}\text{-linear, respects the homology grading,} \\ &\forall x, y \in H_*(X_0, \mathbb{R}), \quad \mathfrak{g}x \cdot y + x \cdot \mathfrak{g}y = \mathfrak{g}(x \cdot y), \ \mathfrak{g}([Y_0]) = 0 \}. \end{aligned} \tag{8.6}$$

Note that for a fixed m this is just

$$\mathrm{Lie}(\Gamma_{\mathbb{R}}) := \{ \mathfrak{g} \in \mathrm{End}_{\mathbb{R}}(H_m(X_0, \mathbb{R})) | \forall x, y \in H_m(X_0, \mathbb{R}), \quad \mathfrak{g}x \cdot y + x \cdot \mathfrak{g}y = 0 \}. \tag{8.7}$$

For any point $\alpha \in D$, there is a natural filtration in $\mathrm{Lie}(\Gamma_{\mathbb{C}})$

$$F^i \mathrm{Lie}(\Gamma_{\mathbb{C}}) = \{ \mathfrak{g} \in \mathrm{Lie}(\Gamma_{\mathbb{C}}) \mid \mathfrak{g}^{\mathsf{pd}}(F^p) \subset F^{p+i}, \ \forall p \in \mathbb{Z} \}, \ i = 0, -1, -2, \ldots,$$

where $F^\bullet$ is the Hodge filtration associated to α and $\mathfrak{g}^{\mathsf{pd}}$ is the Poincaré dual of the linear map $\mathfrak{g}$. We get a natural filtration of the tangent bundle of D at α:

$$T^h_\alpha D := \frac{F^{-1}(\mathrm{Lie}(\Gamma_{\mathbb{C}}))}{F^0(\mathrm{Lie}(\Gamma_{\mathbb{C}}))} \subset \frac{F^{-2}(\mathrm{Lie}(\Gamma_{\mathbb{C}}))}{F^0(\mathrm{Lie}(\Gamma_{\mathbb{C}}))} \subset \cdots \subset \frac{\mathrm{Lie}(\Gamma_{\mathbb{C}})}{F^0(\mathrm{Lie}(\Gamma_{\mathbb{C}}))} = T_\alpha D.$$

One usually calls $T^h_\alpha D$ the horizontal tangent bundle.

Recall that the marked projective variety X_0 gives us a point in D. By abuse of notation, we also denote it by X_0. We define

$$\mathrm{Stab}(\Gamma_{\mathbb{R}}, X_0) := \{A \in \Gamma_{\mathbb{R}} | A \cdot X_0 = X_0\},$$

where we have considered $X_0 \in D$.

Proposition 8.5. *We have*

$$\mathrm{Lie}(\mathsf{G}) = F^0(\mathrm{Lie}(\Gamma_{\mathbb{C}})) = \mathrm{Lie}\left(\mathrm{Stab}(\Gamma_{\mathbb{C}}, X_0)\right),$$

where G *is the algebraic group introduccd in §3.3.*

Proof. The first equality is just the definition of G via Poincaré duality. The second equality follows from the definition of the Lie algebra of a Lie group. □

Proposition 8.6. *The subgroup* $\mathrm{Stab}(\Gamma_{\mathbb{R}}, X_0)$ *of* $\Gamma_{\mathbb{R}}$ *is compact.*

Proof. See [Del71b, CK78] or [Mov08a, Proposition 2]. □

The map

$$\alpha : \Gamma_{\mathbb{R}}/\mathrm{Stab}(\Gamma_{\mathbb{R}}, X_0) \to D,\ \alpha(A) = A \cdot X_0 \tag{8.8}$$

is an isomorphism and so we may identify D with the left-hand side of (8.8). In general, $\mathrm{Stab}(\Gamma_{\mathbb{R}}, X_0)$ may not be maximal. It is connected and is contained in a unique maximal compact subgroup K of $\Gamma_{\mathbb{R}}$. When $K \neq \mathrm{Stab}(\Gamma_{\mathbb{R}}, X_0)$, then there is a fibration of $D = \Gamma_{\mathbb{R}}/\mathrm{Stab}(\Gamma_{\mathbb{R}}, X_0) \to \Gamma_{\mathbb{R}}/K$ with compact fibers isomorphic to $K/\mathrm{Stab}(\Gamma_{\mathbb{R}}, X_0)$ which are complex subvarieties of D. In this case, we have $T_\alpha(D) = T^h_\alpha(D) \oplus T^v_\alpha(D)$, where $T^v(D)$ restricted to a fibre of π coincides with the tangent bundle of that fiber.

Example 8.1. For mirror quintic Calabi–Yau threefolds discussed in [Mov17b] the period domain D is of dimension four. In this case, $\mathrm{Lie}(\mathsf{G})$

(respectively, $F^{-1}\mathrm{Lie}(\Gamma_{\mathbb{C}})$) is of dimension 6 (respectively, 8) generated by the matrices:

$$A_1 := \begin{bmatrix} 0 & 0 & 0 & 0 \\ 0 & -1 & 0 & 0 \\ 0 & 0 & 1 & 0 \\ 0 & 0 & 0 & 0 \end{bmatrix}, \quad A_2 = \begin{bmatrix} -1 & 0 & 0 & 0 \\ 0 & 0 & 0 & 0 \\ 0 & 0 & 0 & 0 \\ 0 & 0 & 0 & 1 \end{bmatrix}, \quad A_3 = \begin{bmatrix} 0 & 0 & 0 & 0 \\ -1 & 0 & 0 & 0 \\ 0 & 0 & 0 & 0 \\ 0 & 0 & 1 & 0 \end{bmatrix}$$

$$A_4 = \begin{bmatrix} 0 & 0 & 0 & 0 \\ 0 & 0 & 0 & 0 \\ 0 & 1 & 0 & 0 \\ 0 & 0 & 0 & 0 \end{bmatrix}, \quad A_5 = \begin{bmatrix} 0 & 0 & 0 & 0 \\ 0 & 0 & 0 & 0 \\ 1 & 0 & 0 & 0 \\ 0 & 1 & 0 & 0 \end{bmatrix}, \quad A_6 = \begin{bmatrix} 0 & 0 & 0 & 0 \\ 0 & 0 & 0 & 0 \\ 0 & 0 & 0 & 0 \\ 1 & 0 & 0 & 0 \end{bmatrix}$$

$$\left(\text{respectively, those above and } A_7 = \begin{bmatrix} 0 & 1 & 0 & 0 \\ 0 & 0 & 0 & 0 \\ 0 & 0 & 0 & -1 \\ 0 & 0 & 0 & 0 \end{bmatrix}, \quad A_8 = \begin{bmatrix} 0 & 0 & 0 & 0 \\ 0 & 0 & 1 & 0 \\ 0 & 0 & 0 & 0 \\ 0 & 0 & 0 & 0 \end{bmatrix}\right).$$

In this example $\dim \mathbf{T}^h_\alpha D = 2$, whereas the dimension of the moduli of mirror quintics is 1. Note that $F^{-1}\mathrm{Lie}(\Gamma_{\mathbb{C}})$ is not a sub-Lie algebra of $\mathrm{Lie}(\Gamma_{\mathbb{C}})$ as $[A_7, A_8]$ is not inside $\mathrm{Lie}(\Gamma_{\mathbb{C}})$.

8.9 Gauss–Manin connection matrix

We consider the trivial bundle $\mathscr{H} = \mathsf{U} \times V_0$ on U and call it the ghost cohomology bundle. On $\mathscr{H}$ we have a well-defined integrable connection

$$\nabla : \mathscr{H} \to \Omega^1_{\mathsf{U}} \otimes_{\mathcal{O}_{\mathsf{U}}} \mathscr{H}$$

such that a section s of $\mathscr{H}$ in a small open set $V \subset \mathsf{U}$ with the property

$$s(\mathsf{u}) \in \{\mathsf{u}\} \times H_m(\mathsf{X}_{\mathsf{u}}, \mathbb{Z}), \ \mathsf{u} \in \mathsf{U}$$

is flat. We will call this the ghost Gauss–Manin connection. Let α be a basis of V_0 as before. We can consider $\alpha_{m,i}$ as a global section of $\mathscr{H}$ and so we have

$$\nabla \alpha_m = \mathsf{A}_m \otimes \alpha_m, \ \mathsf{A}_m = \begin{pmatrix} \mathsf{A}_{m,11} & \mathsf{A}_{m,12} & \cdots & \mathsf{A}_{m,1\mathsf{b}_m} \\ \mathsf{A}_{m,21} & \mathsf{A}_{m,22} & \cdots & \mathsf{A}_{m,2\mathsf{b}_m} \\ \vdots & \vdots & \ddots & \vdots \\ \mathsf{A}_{m,\mathsf{b}_m 1} & \mathsf{A}_{m,\mathsf{b}_m 2} & \cdots & \mathsf{A}_{m,\mathsf{b}_m \mathsf{b}_m} \end{pmatrix}, \ \mathsf{A}_{m,ij} \in H^0(\mathsf{U}, \Omega^1_{\mathsf{U}}). \tag{8.9}$$

A_m is called the (mth) ghost connection matrix of ∇ in the basis α_m. The connection ∇ is integrable and so

$$d\mathsf{A}_m = \mathsf{A}_m \wedge \mathsf{A}_m$$

which is

$$d\mathsf{A}_{m,ij} = \sum_{k=1}^{\mathsf{b}_m} \mathsf{A}_{m,ik} \wedge \mathsf{A}_{m,kj}, \ i,j = 1,2,\ldots,\mathsf{b}_m. \tag{8.10}$$

A fundamental system for the linear differential equation $dY = \mathsf{A}_m \cdot Y$ in U is given by $Y = \mathsf{P}_m^{\mathsf{tr}}$. It is sometimes useful to write

$$\mathsf{A}_m = d\mathsf{P}_m^{\mathsf{tr}} \cdot \mathsf{P}_m^{-\mathsf{tr}}. \tag{8.11}$$

In the next discussion, for simplicity, we drop the subindex m from our notations.

Proposition 8.7. *We have the equality*

$$\mathsf{A}_{i1} \wedge \mathsf{A}_{i2} \wedge \cdots \wedge \mathsf{A}_{i\mathsf{b}} = \frac{1}{\det(\mathsf{P})} d\mathsf{P}_{1i} \wedge d\mathsf{P}_{2i} \wedge \cdots \wedge d\mathsf{P}_{\mathsf{b}i}, \ i = 1,2,\ldots,\mathsf{b},$$

and hence

$$\bigwedge_{i=1}^{\mathsf{b}} \bigwedge_{j=1}^{\mathsf{b}} \mathsf{A}_{ij} = \frac{1}{\det(\mathsf{P})^{\mathsf{b}}} \bigwedge_{i=1}^{\mathsf{b}} \bigwedge_{j=1}^{\mathsf{b}} d\mathsf{P}_{ji}.$$

Proof. The proposition follows from the identity $\mathsf{A}^{\mathsf{tr}} = \mathsf{P}^{-1} \cdot d\mathsf{P}$ derived from (4.16). □

Note that we have

$$\Phi = \mathsf{P}^{\mathsf{tr}} \Psi^{-\mathsf{tr}} \mathsf{P}. \tag{8.12}$$

In particular, up to a minus sign we have

$$\det(\mathsf{P}) = \pm\sqrt{\frac{\det(\Phi)}{\det(\Psi)}},$$

which is a constant.

8.10 Griffiths transversality distribution

Motivated by the Griffiths's transversality theorem we have the following definition.

Definition 8.11. The Griffiths transversality distribution $\mathscr{F}_{\rm gr}$ in U is the $\mathscr{O}_{\mathsf{U}}$ submodule of Ω^1_{U} generated by the differential 1-forms in the entries of

$$\mathsf{A}^{ij}_m,\ j-i\geq 2,\quad m=1,2,\ldots,2n-1. \tag{8.13}$$

These are %-entries of the Gauss–Manin connection matrices:

$$\mathsf{A}_m=\begin{pmatrix} * & * & \% & \% & \cdots & \% & \% \\ * & * & * & \% & \cdots & \% & \% \\ \vdots & \vdots & \vdots & \ddots & \ddots & \vdots & \vdots \\ * & * & * & \cdots & * & \% & \% \\ * & * & * & \cdots & * & * & \% \\ * & * & * & \cdots & * & * & * \\ * & * & * & \cdots & * & * & * \end{pmatrix},\quad m=0,1,2,\ldots,2n.$$

In general, the distribution $\mathscr{F}_{gr}$ on U is not integrable, see §5.2. This is, for instance, the case of mirror quintic Calabi–Yau threefolds, see [Mov11a] and the section on the τ-locus.

Definition 8.12. A holomorphic map $f: V\to \mathsf{U}$, where V is an analytic variety, is called a ghost period map if it is tangent to the Griffiths transversality distribution, that is, for all 1-form ω in (8.13) we have $f^{-1}\omega=0$.

Proposition 8.8. *We have the following $\mathbb{C}$-linear relations between the differential forms (8.13):*

$$0=\sum_{k=m-j}^{m}\mathsf{A}^{ik}\Phi^{k,j}+\sum_{k=m-i}^{m}\Phi^{i,k}(\mathsf{A}^{jk})^{\mathsf{tr}},i+j\leq m-2.$$

We have also

$$0=\sum_{k=m-j}^{m}\mathsf{A}^{ik}\Phi^{k,j}+\sum_{k=m-i}^{m}\Phi^{i,k}(\mathsf{A}^{jk})^{\mathsf{tr}},i+j=m-1,$$

which includes the entries of $A^{ij}, j-i=1$ and modulo the differential forms (8.13) is:

$$0=\mathsf{A}^{i,i+1}\Phi^{i+1,j}+\Phi^{i,m-i}(\mathsf{A}^{m-i-1,m-i})^{\mathsf{tr}}.$$

Proof. We have just opened the equality

$$0 = d\Phi = \mathsf{A}\Phi + \Phi\mathsf{A}^{\mathsf{tr}}.$$

This equality looks like $(m = 4)$

$$0 = \begin{pmatrix} * & * & \# & \# & \# \\ * & * & * & \# & \# \\ * & * & * & * & \# \\ * & * & * & * & * \\ * & * & * & * & * \end{pmatrix} \begin{pmatrix} 0 & 0 & 0 & 0 & * \\ 0 & 0 & 0 & * & * \\ 0 & 0 & * & * & * \\ 0 & * & * & * & * \\ * & * & * & * & * \end{pmatrix} + \begin{pmatrix} 0 & 0 & 0 & 0 & * \\ 0 & 0 & 0 & * & * \\ 0 & 0 & * & * & * \\ 0 & * & * & * & * \\ * & * & * & * & * \end{pmatrix} \begin{pmatrix} * & * & * & * & * \\ * & * & * & * & * \\ \# & * & * & * & * \\ \# & \# & * & * & * \\ \# & \# & \# & * & * \end{pmatrix}.$$

Here, $*$ indicates the entries which do not interest us and $\#$ indicates the entries where the differential forms (8.13) appear. Now, we write the above equality for the entries $\%$ described below

$$\begin{pmatrix} \% & \% & \% & * & * \\ \% & \% & * & * & * \\ \% & * & * & * & * \\ * & * & * & * & * \\ * & * & * & * & * \end{pmatrix}$$

and get the desired linear relations between the differential forms (8.13). We look for the identities obtained from the entries $\%$ in:

$$\begin{pmatrix} * & * & * & \% & * \\ * & * & \% & * & * \\ * & \% & * & * & * \\ \% & * & * & * & * \\ * & * & * & * & * \end{pmatrix}$$

and obtain the desired linear relations for the entries of $A^{i,i+1}$, $i = 0, 1, \ldots, m-1$ and the differential forms (8.13). □

Proposition 8.9. *The distribution $\mathscr{F}_{gr}$ is invariant under the action of* G.

Proof. An element $g \in \mathsf{G}$ induces a biholomorphism on U which we denote it for simplicity by g again. From $\mathsf{A} = d\mathsf{P}^{\mathsf{tr}} \cdot \mathsf{P}^{-\mathsf{tr}}$ it follows that

$$g^*\mathsf{A} = g^{\mathsf{tr}}\mathsf{A}g^{-\mathsf{tr}}.$$

This implies that g^* sends the vector space generated by the 1-forms (8.13) to itself. To see this one may draw $g^{\mathsf{tr}}\mathsf{A}g^{-\mathsf{tr}}$:

$$g^{\mathsf{tr}}Ag^{-\mathsf{tr}} = \begin{pmatrix} * & 0 & 0 & 0 & 0 \\ * & * & 0 & 0 & 0 \\ * & * & * & 0 & 0 \\ * & * & * & * & 0 \\ * & * & * & * & * \end{pmatrix} \begin{pmatrix} * & * & \# & \# & \# \\ * & * & * & \# & \# \\ * & * & * & * & \# \\ * & * & * & * & * \\ * & * & * & * & * \end{pmatrix} \begin{pmatrix} * & 0 & 0 & 0 & 0 \\ * & * & 0 & 0 & 0 \\ * & * & * & 0 & 0 \\ * & * & * & * & 0 \\ * & * & * & * & * \end{pmatrix}.$$

□

Remark 8.4. Proposition 8.8 implies that in general the distribution $\mathscr{F}_{gr}$ has not the expected codimension $\mathsf{h}^{0,m}\mathsf{h}^2+\mathsf{h}^{1,m-1}\mathsf{h}^3+\cdots+\mathsf{h}^{m-2,2}\mathsf{h}^m$ which is the number of ω_{ij} in (8.13).

Remark 8.5. The equality $d\mathsf{A} = \mathsf{A}\wedge\mathsf{A}$ modulo the differential forms (8.13) has the form:

$$d\mathsf{A} = \begin{pmatrix} * & * & 0 & 0 & 0 \\ * & * & * & 0 & 0 \\ * & * & * & * & 0 \\ * & * & * & * & * \\ * & * & * & * & * \end{pmatrix} \wedge \begin{pmatrix} * & * & 0 & 0 & 0 \\ * & * & * & 0 & 0 \\ * & * & * & * & 0 \\ * & * & * & * & * \\ * & * & * & * & * \end{pmatrix} = \begin{pmatrix} * & * & * & 0 & 0 \\ * & * & * & * & 0 \\ * & * & * & * & * \\ * & * & * & * & * \\ * & * & * & * & * \end{pmatrix}.$$

Therefore,

$$d\mathsf{A}_{ij},\ i\leq h^{m-x},\ j>h^{m-x-2},\ x=0,1,\ldots,m-3$$

are in the Ω^1_{U}-module generated by the differential forms (8.13). However,

$$d\mathsf{A}^{i,i+2} = \mathsf{A}^{i,i+1}\wedge\mathsf{A}^{i+1,i+2},\ \text{ modulo } \mathscr{F}_{gr}, \tag{8.14}$$

which violates the integrability condition.

Remark 8.6. In the case of mirror quintic, the Griffiths transversality distribution $\mathscr{F}_{gr} := \mathscr{F}(\mathsf{A}^{02},\mathsf{A}^{03},\mathsf{A}^{13})$ is of codimension 2 and it is not integrable. In order to see this we proceed as follows. The matrix equality $0 = d\Phi = \mathsf{A}\Phi + \Phi\mathsf{A}^{\mathsf{tr}}$ gives us six independent equalities:

$$\begin{aligned} &\mathsf{A}^{13}-\mathsf{A}^{02}=0,\ \mathsf{A}^{23}+\mathsf{A}^{01}=0,\ \ \mathsf{A}^{00}+\mathsf{A}^{33}=0,\\ &\mathsf{A}^{22}+\mathsf{A}^{11}=0,\ \mathsf{A}^{32}+\mathsf{A}^{10}=0,\\ &\mathsf{A}^{31}-\mathsf{A}^{20}=0, \end{aligned} \tag{8.15}$$

derived from its $(1,2),(1,3),(1,4),(2,3),(2,4),(3,4)$ entries, respectively. Using $d\mathsf{A}=\mathsf{A}\wedge\mathsf{A}$ we have

$$\begin{aligned}
&\mathsf{A}^{13}\wedge\mathsf{A}^{14}\wedge\mathsf{A}^{24}\wedge d\mathsf{A}^{13}=\mathsf{A}^{13}\wedge\mathsf{A}^{14}\wedge\mathsf{A}^{14}\wedge\mathsf{A}^{12}\wedge\mathsf{A}^{23},\\
&\mathsf{A}^{13}\wedge\mathsf{A}^{14}\wedge\mathsf{A}^{24}\wedge d\mathsf{A}^{14}=0,\\
&\mathsf{A}^{13}\wedge\mathsf{A}^{14}\wedge\mathsf{A}^{24}\wedge d\mathsf{A}^{24}=\mathsf{A}^{13}\wedge\mathsf{A}^{14}\wedge\mathsf{A}^{14}\wedge\mathsf{A}^{23}\wedge\mathsf{A}^{34}.
\end{aligned}$$

The right-hand side of the above equalities is not zero.

Remark 8.7. If all the linear relations in Proposition 8.8 were independent from each other we could conclude that the Griffiths transversality is a consequence of the definition if an enhanced family. For instance, this is the case for the Hodge numbers $h^{20}=h^{02}=1$, in which the collection of differential forms (8.13) is just $\omega_{1,\mathsf{b}}$ and the equality in the $(1,1)$ entries is $\mathsf{A}_{1\mathsf{b}}\Phi_{\mathsf{b}1}+\Phi_{\mathsf{b}1}\mathsf{A}_{1\mathsf{b}}=0$. Since $\Phi_{\mathsf{b}1}=\Phi_{1\mathsf{b}}$ is not identically zero, we conclude that $\mathsf{A}_{1\mathsf{b}}$ is identically zero.

Remark 8.8. Is U an Stein variety? The answer to this question can be the first step toward the algebraization of U. To investigate this question, one may start with the article [LN99] of A. Lins Neto in which a theorem of A. Takeuchi in 1967 and G. Elencwajg in 1975 is used.

8.11 Modular foliations and vector fields

From now on we can use all the discussion in Chapters 3 and 6 replacing T with U. Instead of algebraic objects like X_t we use their ghost versions like X_u. The only difference lies in the format of Gauss–Manin connections. In T the Gauss–Manin connection due to Griffiths transversality has a special format with bunch of zeros, whereas in U we have the Griffiths transversality distribution introduced in §8.10. We can now define modular foliations and Hodge cycles, etc., in the same style that we did it for T in Chapter 6. It is too premature to claim a kind of Cattani–Deligne–Kaplan theorem for U, that is, to say that a locus of Hodge cycles with constant periods is a part of a global analytic subvariety of U. However, it seems to the author that the following statement is true: Let $p:\mathsf{T}\to\mathsf{U}$ be an analytic map from an algebraic variety T to the generalized period domain U which is tangent to the Griffiths transversality distribution. Then the pull-back of a locus of Hodge cycles with constant periods by p is an algebraic subvariety of T. The possible proof must be reconstructed from the arguments in [CDK95].

Recall that the group $\Gamma_{\mathbb{Z}}$ acts from the left on the generalized period domain $\sqcap$ and $\mathsf{U} = \Gamma_{\mathbb{Z}}\backslash\sqcap$. We are interested to find vector fields in $\sqcap$ that are $\Gamma_{\mathbb{Z}}$-invariants and hence can be lifted to vector fields in U. For simplicity, we sometimes drop the subindex m from our notations. We denote a vector field v on the matrix space $\mathrm{Mat}(\mathsf{b}, \mathbb{C})$ with $[\mathsf{v}_{ij}(\mathsf{P})]$. In usual notations, this is

$$\mathsf{v} := \sum_{i,j} \mathsf{v}_{ij}(\mathsf{P}) \frac{\partial}{\partial \mathsf{P}_{ij}}.$$

For $A \in \Gamma_{\mathbb{Z}}$, let us consider the map

$$f_A : \mathrm{Mat}(\mathsf{b}, \mathbb{C}) \to \mathrm{Mat}(\mathsf{b}, \mathbb{C}),\ \mathsf{P} \mapsto A\mathsf{P}.$$

Since f_A is linear in P-coordinates, the vector field v in $\sqcap$ is $\Gamma_{\mathbb{Z}}$-invariant if

$$A[\mathsf{v}_{ij}(\mathsf{P})] = [\mathsf{v}_{ij}(A\mathsf{P})].$$

In particular, this is the case for vector fields of the form $[\mathsf{v}_{ij}(\mathsf{P})] = \mathsf{P} \cdot B$, where B is a constant matrix. We use Remark 8.3 and conclude that the mentioned vector field is tangent to $\sqcap$ if B is in $\mathrm{Lie}(\Gamma_{\mathbb{C}})$ defined in (8.6).

Proposition 8.10. *For a matrix $B \in \mathrm{Lie}(\Gamma_{\mathbb{C}})$, the vector field $[\mathsf{v}_{ij}(\mathsf{P})] = \mathsf{P} \cdot B$ gives us a vector field in the generalized period domain U.*

Recall that G acts from the right on U and so we have a canonical embedding

$$i : \mathrm{Lie}(\mathsf{G}) \hookrightarrow H^0(\mathsf{U}, \Theta_{\mathsf{U}}). \tag{8.16}$$

Recall also from §8.8 that $\mathrm{Lie}(\mathsf{G}) \subset \mathrm{Lie}(\Gamma_{\mathbb{C}})$.

Proposition 8.11. *For $\mathfrak{g} \in \mathrm{Lie}(\mathsf{G})$ the vector field $i(\mathfrak{g})$ in the mth period coordinates of U is given by*

$$\mathsf{v}_{\mathfrak{g}}(\mathsf{P}) := \mathsf{P}_m \cdot \mathfrak{g}_m.$$

Proof. This is just the reformulation of Proposition 8.10. □

Proposition 8.12. *The Gauss–Manin connection matrix A_m composed with the vector field $\mathsf{v}_{\mathfrak{g}}$, $\mathfrak{g} \in \mathrm{Lie}(\mathsf{G})$ is $\mathfrak{g}_m^{\mathsf{tr}}$, that is,*

$$(\mathsf{A}_m)_{\mathsf{v}_{\mathfrak{g}}} = \mathfrak{g}_m^{\mathsf{tr}}.$$

Proof. We have $\mathsf{A}_m = d\mathsf{P}_m^{\mathsf{tr}} \cdot \mathsf{P}_m^{-\mathsf{tr}}$ and so

$$(\mathsf{A}_m)_{\mathsf{v}_{\mathfrak{g}}} = (\mathsf{P}_m \mathfrak{g}_m)^{\mathsf{tr}} \mathsf{P}_m^{-\mathsf{tr}} = \mathfrak{g}_m^{\mathsf{tr}}.$$ □

Definition 8.13. A modular vector field v in U is a vector field such that $(\mathsf{A}_m)_\mathsf{v}$ for all $m = 0, 1, 2, \ldots, 2n$ is upper Hodge block triangular and all its Hodge blocks in the diagonal are also zero.

8.12 Space of leaves

The foliation $\mathscr{F}(\mathsf{C})$ in U has local first integral $\mathsf{P}^{-\mathrm{tr}}\mathsf{C}$, that is, its leaves are the inverse image of points by $\mathsf{P}^{-\mathrm{tr}}\mathsf{C}$. We consider the global $L_{\check{\mathsf{C}}}$, that is, we consider P as a multivalued function in U (the multivaluedness arises from the choice of the basis in homology), and

$$L_{\check{\mathsf{C}}} := \{\mathsf{u} \in \mathsf{U} | \mathsf{P}^{-\mathrm{tr}}\mathsf{C} = \check{\mathsf{C}}\}. \tag{8.17}$$

The adjectives local or global will distinguish both sets from each other.

Proposition 8.13. *The global set $L_{\check{\mathsf{C}}}$ consists of finitely many connected components.*

Note that Proposition 8.13 is valid also in the algebraic context of enhanced families X/T if we assume that the period map $\mathsf{P} : \mathsf{T} \to \mathsf{U}$ is a biholomorphism. This hypothesis can be verified for elliptic curves, K3 surfaces, cubic fourfolds and abelian varieties. In general, partial compactifications of T might be used to prove statement similar to Proposition 8.13.

Proof of Proposition 8.13. The generalized period domain Π is an open subset (using the usual topology) of the affine variety V of $\mathbb{C}^{\mathsf{b}}$ given by the quadratic polynomials in (8.12). Its boundary in V is given by zeros of polynomials in the entries of P and its complex conjugate $\bar{\mathsf{P}}$. These are points for which the Hodge–Riemann bilinear relations fails. Any algebraic subset of V in holomorphic variables P intersects Π in a finite number of connected components. In particular, for fixed $\mathsf{C}, \check{\mathsf{C}} \in \mathbb{C}^{\mathsf{b}}$ the set

$$\check{L}_{\check{\mathsf{C}}} := \{\mathsf{P} \in \Pi | \mathsf{P}^{\mathrm{tr}}\check{\mathsf{C}} = \mathsf{C}\} \tag{8.18}$$

has finitely many connected components. Its image under the canonical map $\Pi \to \mathsf{U}$ is the set $L_{\check{\mathsf{C}}}$ defined in U. □

For $A \in \Gamma_{\mathbb{Z}}$, the biholomorphism $\Pi \to \Pi$, $\mathsf{P} \mapsto A\mathsf{P}$ induces a biholomorphism $\check{L}_{A^{\mathrm{tr}}\check{\mathsf{C}}} \to \check{L}_{\check{\mathsf{C}}}$ and so in the quotient $\mathsf{U} := \Gamma_{\mathbb{Z}} \backslash \Pi$ they induce the same leaf of $\mathscr{F}(\mathsf{C})$. Therefore, $\Gamma_{\mathbb{Z}}$ acts on the set A of connected components of $L_{\check{\mathsf{C}}}$. Note that such components might have different dimensions. Let

$\{[L_1], [L_2], \ldots, [L_s]\}$ be the decomposition of A into equivalency classes and

$$\Gamma_{\mathbb{Z},i} := \{A \in \Gamma_{\mathbb{Z}} | A(L_i) = L_i\}, \quad i = 1, 2, \ldots, s.$$

Proposition 8.14. *The sum of indices of $\Gamma_{\mathbb{Z},i}$, $i = 1, 2, \ldots, s$ inside $\Gamma_{\mathbb{Z}}$ is equal to the number of connected components of $L_{\tilde{\mathsf{C}}}$ in U.*

Proof. Using Proposition 8.13, we know that the map from $\Gamma_{\mathbb{Z}}/\Gamma_{\mathbb{Z},i}$ to the set of connected components of $L_{\tilde{\mathsf{C}}}$ given by $A \mapsto A(\tilde{L})$ is injective. Considering all $i = 1, 2, \ldots, s$ we get the desired statement. □

8.13 Transcendental degree of automorphic forms

Let $D \subset \mathbb{C}^N$ be an open Hermitian symetric domain and let Γ be an arithmetic group acting on D, see [Milb]. Let also $j : D \times \Gamma \to \mathbb{C}^*$ be an automorphy factor. It follows from the Baily–Borel theorem in [BJB66] that the transcendental degree of the field of automorphic forms for (D, Γ, j) is the dimension of D. One may also ask what is the transcendental degree of the algebra generated by automorphic forms and their derivatives. When $\Gamma \backslash D$ is a moduli space of Hodge structures of weight w, we can answer this question. The answer in this case is the dimension of the moduli space of enhanced Hodge structures. We denote by $z = (z_1, z_2, \ldots, z_N)$ the coordinate system in $D \subset \mathbb{C}^N$ and by $\partial_i := \frac{\partial}{\partial z_i}$ the derivation with respect to z_i. The field generated by automorphic forms for $\Gamma \curvearrowright D$ and their derivatives under ∂_i is expected to be of transcendence degree $\dim(\mathsf{T})$ over $\mathbb{C}$. For $D = \mathbb{H}_g$, the Siegel upper half-plane and $\Gamma = \mathrm{Sp}(2g, \mathbb{Z})$, this has been proved in [BZ01, BZ03].

Chapter 9

Elliptic Curves

Though his [D. Northcott's] thesis was in analysis under G. H. Hardy, he attended Artin's seminar, and when one of the first speakers mentioned the characteristic of a field, Northcott raised his hand and asked what that meant. His question begot laughter from several students, whereupon Artin delivered a short lecture on the fact that one could be a fine mathematician without knowing what the characteristic of a field was. And, indeed, it turned out that Northcott was the most gifted student in that seminar (J. Tate in [RS11, p. 446]).

9.1 Introduction

The case of elliptic curves is the founding stone of the present text, and therefore, the content of this chapter must be read alongside any other chapter in this text. Unsatisfied with P. Griffiths' formulation of period domain [Gri70], K. Saito's formulation of primitive forms [Sai01] and N. Katz's description of the relation between the Eisenstein series E_2 and the Gauss–Manin connection of a family of elliptic curves [Kat73], the author had to rewrite the case of elliptic curves in [Mov08b, Mov08c]. Later, more details were gathered in the lecture notes [Mov12b]. The main objective in this lecture notes is to derive the theory of quasi-modular forms in the purely geometric context of elliptic curves. Therefore, in this chapter I will omit many discussions regarding quasi-modular forms. Even the reader who is interested in Hodge cycles and Hodge loci will find the content of this chapter relevant. The only geometric phenomena which could happen for a single elliptic curve are either getting singular or having a complex multiplication. Both phenomena are isolated points in the moduli of elliptic curves. However, the dynamics of modular foliations and the non-existence of algebraic leaves in this case become non-trivial topics. This will prepare the reader to the content of Chapter 10 in which we will encounter the first non-trivial Hodge locus arising from the isogeny of elliptic curves and the corresponding modular curves.

9.2 Enhanced elliptic curves

It is a well-known fact that any elliptic curve over a field $\mathfrak{k}$ can be embedded in the projective space of dimension 2. Therefore, we are going to consider degree 3 smooth curves X in $\mathbb{P}^2$. Note that an elliptic curve over $\mathfrak{k}$ by definition is marked with a $\mathfrak{k}$-rational point $0 \in X(\mathfrak{k})$. We can neglect the polarization θ in all of our discussions, because it is uniquely determined by $\mathrm{Tr}(\theta) = 3$. We only need to discuss the middle cohomology $H^1_{\mathrm{dR}}(X)$ and so we drop the subindex m from our notations. For many missing details the reader is referred to [Mov12b].

An enhanced elliptic curve turns out to be a pair $(X, \{\alpha, \omega\})$, where α is a regular differential form in X and $\omega \in H^1_{\mathrm{dR}}(X)$ such that $\langle \alpha, \omega \rangle = 1$. In this case we have a universal family of enhanced elliptic curves $\mathsf{X} \to \mathsf{T}$, where

$$\mathsf{X} : y^2 - 4(x - t_1)^3 + t_2(x - t_1) + t_3 = 0, \quad \alpha = \left[\frac{dx}{y}\right], \quad \omega = \left[\frac{xdx}{y}\right] \tag{9.1}$$

which is written in the affine coordinate (x, y), and

$$\mathsf{T} := \mathrm{Spec}\left(\mathfrak{k}\left[t_1, t_2, t_3, \frac{1}{27t_3^2 - t_2^3}\right]\right).$$

We have

$$\Phi = \Psi = \begin{bmatrix} 0 & 1 \\ -1 & 0 \end{bmatrix}.$$

The algebraic group G is

$$\mathsf{G} = \left\{ \begin{bmatrix} k & k' \\ 0 & k^{-1} \end{bmatrix} \middle| \ k' \in \mathfrak{k}, k \in \mathfrak{k} - \{0\} \right\} = \mathrm{Spec}\left(\mathfrak{k}\left[k, k', \frac{1}{k}\right]\right) \tag{9.2}$$

and its action on T is given by

$$t \bullet \mathsf{g} := (t_1 k^{-2} + k' k^{-1}, t_2 k^{-4}, t_3 k^{-6}),$$

$$t = (t_1, t_2, t_3) \in \mathsf{T}, \quad \mathsf{g} = \begin{bmatrix} k & k' \\ 0 & k^{-1} \end{bmatrix} \in \mathsf{G}.$$

The Gauss–Manin connection matrix of the family X/T is

$$\mathsf{A} = \frac{1}{\Delta} \begin{bmatrix} -\frac{3}{2}t_1\alpha - \frac{1}{12}d\Delta & \frac{3}{2}\alpha \\ \Delta dt_1 - \frac{1}{6}t_1 d\Delta - \left(\frac{3}{2}t_1^2 + \frac{1}{8}t_2\right)\alpha & \frac{3}{2}t_1\alpha + \frac{1}{12}d\Delta \end{bmatrix}, \tag{9.3}$$
$$\Delta = 27t_3^2 - t_2^3, \ \alpha = 3t_3dt_2 - 2t_2dt_3.$$

This computation with $t_1 = 0$ is well-known, see [Sas74, p. 304] or [Sai01] (in the first reference the formula has a sign mistake). From a historical point of view another family of elliptic curves turns out to be important as well. From this family we are going to derive Darboux's differential equation. This can be also served as a motivation to enlarge our notion of moduli of enhanced projective varieties by adding more ingredients in the definition of an enhanced variety. Let

$$\mathsf{X} : y^2 - 4(x - t_1)(x - t_2)(x - t_3) = 0, \ \alpha = \left[\frac{dx}{y}\right], \quad \omega = \left[\frac{xdx}{y}\right],$$
$$\mathsf{T} = \mathrm{Spec}\left(\mathfrak{k}\left[t_1, t_2, t_3, \frac{1}{(t_1 - t_2)(t_2 - t_3)(t_3 - t_1)}\right]\right). \tag{9.4}$$

The family X/T is the universal family for the moduli of 3-tuple $(X, (P, Q), \{\alpha, \omega\})$, where $(X, \{\alpha, \omega\})$ is an enhanced elliptic curve and P and Q are points of $X(\bar{\mathfrak{k}})$ that generate the 2-torsion subgroup of X with Weil pairing $e(P, Q) = -1$. The points P and Q are given by $(t_1, 0)$ and $(t_2, 0)$. Its Gauss–Manin connection matrix is given by

$$\begin{aligned} \mathsf{A} = &\frac{dt_1}{2(t_1 - t_2)(t_1 - t_3)} \begin{bmatrix} -t_1 & 1 \\ t_2t_3 - t_1(t_2 + t_3) & t_1 \end{bmatrix} \\ &+ \frac{dt_2}{2(t_2 - t_1)(t_2 - t_3)} \begin{bmatrix} -t_2 & 1 \\ t_1t_3 - t_2(t_1 + t_3) & t_2 \end{bmatrix} \\ &+ \frac{dt_3}{2(t_3 - t_1)(t_3 - t_2)} \begin{bmatrix} -t_3 & 1 \\ t_1t_2 - t_3(t_1 + t_2) & t_3 \end{bmatrix}. \end{aligned} \tag{9.5}$$

9.3 Modular vector fields

In this section, we briefly recall the material of §6.11 in the case of elliptic curves. We first consider the family (9.1). There are unique vector fields

e, h, f in T such that the Gauss–Manin connection matrix A of the family X/T along these vector fields has the form:

$$\mathsf{A}_h = \begin{bmatrix} 1 & 0 \\ 0 & -1 \end{bmatrix}, \quad \mathsf{A}_f = \begin{bmatrix} 0 & 1 \\ 0 & 0 \end{bmatrix}, \quad \mathsf{A}_e = \begin{bmatrix} 0 & 0 \\ 1 & 0 \end{bmatrix}.$$

Since we know the explicit formula of the Gauss–Manin connection in (9.3), this affirmation can be checked easily, and in fact, we can compute explicit expressions for e, f, h:

$$\begin{aligned} f &= -\left(t_1^2 - \frac{1}{12}t_2\right)\frac{\partial}{\partial t_1} - (4t_1t_2 - 6t_3)\frac{\partial}{\partial t_2} - \left(6t_1t_3 - \frac{1}{3}t_2^2\right)\frac{\partial}{\partial t_3}, \\ h &= -6t_3\frac{\partial}{\partial t_3} - 4t_2\frac{\partial}{\partial t_2} - 2t_1\frac{\partial}{\partial t_1}, \quad e = \frac{\partial}{\partial t_1}. \end{aligned} \tag{9.6}$$

The $\mathfrak{k}$-vector space generated by these vector fields equipped with the classical bracket of vector fields is isomorphic to the Lie Algebra $\mathfrak{sl}_2$, and hence, it gives a representation of $\mathfrak{sl}_2$ in the polynomial ring $\mathbb{Q}[t_1, t_2, t_3]$ which is infinite dimensional:

$$[h, e] = 2e, \quad [h, f] = -2f, \quad [e, f] = h. \tag{9.7}$$

Definition 9.1. We call $\mathsf{R} := -f$ in (9.6) the Ramanujan vector field.

One can check the relations (9.7) directly, however, there is a geometric way to see this by using the composition of vector fields with the Gauss–Manin connection matrix. For this we use Proposition 6.17 and we have for instance $\mathsf{A}_{[f,e]} = [\mathsf{A}_e, \mathsf{A}_f]$. Note that after composing with A the role of h and e is exchanged. A similar discussion can be done for the family (9.4). In this case we have the following definition.

Definition 9.2. There is a unique vector field D in T such that

$$\mathsf{A}_{\mathsf{D}} = \begin{bmatrix} 0 & -1 \\ 0 & 0 \end{bmatrix}.$$

This is

$$\mathsf{D} = (t_1(t_2+t_3) - t_2t_3)\frac{\partial}{\partial t_1} + (t_2(t_1+t_3) - t_1t_3)\frac{\partial}{\partial t_2} + (t_3(t_1+t_2) - t_1t_2)\frac{\partial}{\partial t_3}$$

and it is called the Darboux–Halphen vector field.

The algebraic morphism $\alpha : \mathbb{A}^3_{\mathfrak{k}} \to \mathbb{A}^3_{\mathfrak{k}}$ defined by

$$\alpha : (t_1, t_2, t_3) \mapsto \left(T, 4 \sum_{1 \le i < j \le 3} (T - t_i)(T - t_j), 4(T - t_1)(T - t_2)(T - t_3) \right), \tag{9.8}$$

where $T := \frac{1}{3}(t_1 + t_2 + t_3)$ connects the families (9.1) and (9.4), that is, if in (9.1) we replace t with $\alpha(t)$ we obtain the family (9.4). The Gauss–Manin connection matrix associated to (9.4) is just the pull-back of the Gauss–Manin connection associated to (9.1), and so, the Darboux–Halphen vector field is mapped to Ramanujan vector field under α.

9.4 Quasi-modular forms

The Eisenstein series are defined as follows:

$$E_{2k}(\tau) = 1 + (-1)^k \frac{4k}{B_k} \sum_{n \ge 1} \sigma_{2k-1}(n) q^n, \quad k = 1, 2, 3, \ \tau \in \mathbb{H}, \tag{9.9}$$

where $q = e^{2\pi i \tau}$ and $\sigma_i(n) := \sum_{d|n} d^i$ and B_i's are the Bernoulli numbers:

$$B_1 = \frac{1}{6}, \ B_2 = \frac{1}{30}, \ B_3 = \frac{1}{42}, \ \ldots .$$

Ramanujan in [Ram16] proved that

$$g = (g_1, g_2, g_3) = (a_1 E_2, a_2 E_4, a_3 E_6), \tag{9.10}$$

with

$$(a_1, a_2, a_3) = \left(\frac{2\pi i}{12}, 12 \left(\frac{2\pi i}{12} \right)^2, 8 \left(\frac{2\pi i}{12} \right)^3 \right),$$

satisfies the ODE's

$$\mathsf{R} : \begin{cases} \dot{t}_1 = t_1^2 - \frac{1}{12} t_2, \\ \dot{t}_2 = 4 t_1 t_2 - 6 t_3, \\ \dot{t}_3 = 6 t_1 t_3 - \frac{1}{3} t_2^2, \end{cases} \tag{9.11}$$

where the derivation is with respect to τ. The constants a_i's have appeared in our geometric treatment of Ramanujan differential equation R, and up to these constants, (9.11) is the same as (1.1).

In 1881, G. Halphen considered the nonlinear differential system

$$\begin{cases} \dot{t}_1 + \dot{t}_2 = 2t_1t_2, \\ \dot{t}_2 + \dot{t}_3 = 2t_2t_3, \\ \dot{t}_1 + \dot{t}_3 = 2t_1t_3, \end{cases}$$

(see [Hal81b, Hal81c, Hal81a]) which originally appeared in G. Darboux's work in 1878 on triply orthogonal surfaces in $\mathbb{R}^3$ (see [Dar78b]). We write the above equations in the ordinary differential equation form:

$$\mathsf{D}: \begin{cases} \dot{t}_1 = t_1(t_2 + t_3) - t_2t_3, \\ \dot{t}_2 = t_2(t_1 + t_3) - t_1t_3, \\ \dot{t}_3 = t_3(t_1 + t_2) - t_1t_2. \end{cases} \tag{9.12}$$

Halphen expressed a solution of the system (9.12) in terms of the logarithmic derivatives of the theta series:

$$\begin{aligned} u_1 &= 2(\ln\theta_4(0|\tau))', \\ u_2 &= 2(\ln\theta_2(0|\tau))', \qquad ' = \frac{\partial}{\partial\tau}, \\ u_3 &= 2(\ln\theta_3(0|\tau))' \end{aligned}$$

where

$$\begin{cases} \theta_2(0|\tau) := \sum_{n=-\infty}^{\infty} q^{\frac{1}{2}(n+\frac{1}{2})^2}, \\ \theta_3(0|\tau) := \sum_{n=-\infty}^{\infty} q^{\frac{1}{2}n^2}, \\ \theta_4(0|\tau) := \sum_{n=-\infty}^{\infty} (-1)^n q^{\frac{1}{2}n^2}, \end{cases} \qquad q = e^{2\pi i\tau}, \ \tau \in \mathbb{H}.$$

9.5 Halphen property

For $\mathsf{R} = \mathbb{Z}, \mathbb{R}, \mathbb{C}$, etc., let us define

$$\mathrm{SL}(2,\mathsf{R}) := \left\{ \begin{bmatrix} a & b \\ c & d \end{bmatrix} \middle| a, b, c, d \in \mathsf{R},\ ad - bc = 1 \right\}$$

and

$$\Gamma(d) := \left\{ A \in \mathrm{SL}(2,\mathbb{Z}) \middle| A \equiv \begin{bmatrix} 1 & 0 \\ 0 & 1 \end{bmatrix} \bmod d \right\}, \quad d \in \mathbb{N}.$$

For a holomorphic function defined in $\mathbb{H}$ and $A = \begin{bmatrix} a & b \\ c & d \end{bmatrix} \in \mathrm{SL}(2,\mathbb{C})$, $m \in \mathbb{N}$, let also

$$(f \,|_m^0\, A)(\tau) := (c\tau + d)^{-m} f(A\tau),$$
$$(f \,|_m^1\, A)(\tau) := (c\tau + d)^{-m} f(A\tau) - c(c\tau + d)^{-1}.$$

The following property of the differential equations R and D is known as the Halphen property.

Proposition 9.1. *If ϕ_i, $i = 1, 2, 3$ are the coordinates of a solution of the Ramanujan differential equation R (respectively, the Darboux–Halphen differential equation D) then*

$$\phi_1 \,|_2^1\, A,\ \phi_2 \,|_4^0\, A,\ \phi \,|_6^0\, A,$$
$$(\textit{respectively, } \phi_i \,|_2^1\, A,\ i = 1, 2, 3),$$

are also coordinates of a solution of R (respectively, D) for all $A \in \mathrm{SL}(2,\mathbb{C})$. The subgroup of $\mathrm{SL}(2,\mathbb{C})$ which fixes the solution given by Eisenstein series (respectively, theta series) is $\mathrm{SL}(2,\mathbb{Z})$ (respectively, $\Gamma(2)$).

Proof. The first part of the proposition is a mere calculation and it is in fact true for a general Halphen equation in §9.12. The second part is easy and it is left to the reader. □

It is useful to consider the special case $A = \begin{bmatrix} \sqrt{d} & 0 \\ 0 & \sqrt{d}^{-1} \end{bmatrix}$. Recall the Eisenstein series (g_1, g_2, g_3) in §9.4. The following

$$(d \cdot g_1(d \cdot \tau), \quad d^2 \cdot g_2(d \cdot \tau), \quad d^3 \cdot g_3(d \cdot \tau))$$

is also a solution of the Ramanujan differential equation. For more discussion on Halphen property, see [Mov12a].

9.6 Period domain and map

For the rest of our discussion in this chapter, we will only consider the family (9.1). In this section, we take $\mathsf{k} = \mathbb{C}$. The generalized period domain in the case of elliptic curves is

$$\mathsf{\Pi} := \left\{ \begin{bmatrix} x_1 & x_2 \\ x_3 & x_4 \end{bmatrix} \middle| x_i \in \mathbb{C},\ x_1x_4 - x_2x_3 = 1,\ \mathrm{Im}(x_1\overline{x_3}) > 0 \right\}, \tag{9.13}$$

which seems to appear in the literature for the first time in [Mov08b]. In the case of elliptic curves, the discrete group $\Gamma_{\mathbb{Z}}$ turns out to be $\mathrm{SL}(2,\mathbb{Z})$. This group (respectively, G in (9.2)) acts from the left (respectively, right) on $\mathsf{\Pi}$ by usual multiplication of matrices. We also call

$$\mathsf{U} := \mathrm{SL}(2,\mathbb{Z})\backslash\mathsf{\Pi}$$

the period domain.

Theorem 9.2. *The period map*

$$\mathsf{P} : \mathsf{T} \to \mathsf{U},\ t \mapsto \left[\frac{1}{\sqrt{2\pi i}} \begin{bmatrix} \int_\delta \frac{dx}{y} & \int_\delta \frac{xdx}{y} \\ \int_\gamma \frac{dx}{y} & \int_\gamma \frac{xdx}{y} \end{bmatrix} \right] \tag{9.14}$$

is a biholomorphism of complex manifolds.

The first bracket $[\cdot]$ in (9.14) means the equivalence class in the quotient $\mathrm{SL}(2,\mathbb{Z})\backslash\mathsf{\Pi}$, and the other refers to a 2×2 matrix. We have chosen a basis $\delta, \gamma \in H_1(X,\mathbb{Z})$ with $\langle\delta,\gamma\rangle = -1$. It is well-defined because different choices of δ,γ lead to the action of $\mathrm{SL}(2,\mathbb{Z})$ from the left on $\mathsf{\Pi}$ which is already absorbed in the quotient U. Theorem 9.2 follows from the fact that the classical period map $\mathbb{C} \to \mathrm{SL}(2,\mathbb{Z})\backslash\mathbb{H}$, which for $j \in \mathbb{C}$ it associates the quotient of two elliptic integrals attached to the elliptic curve with the j-invariant j, is a biholomorphism and its inverse is given by the classical j-function. For details see [Mov08c] or [Mov08b].

The period map satisfies

$$\mathsf{P}(t \bullet \mathsf{g}) = \mathsf{P}(t)\cdot\mathsf{g},\ t \in \mathsf{T},\ \mathsf{g} \in \mathsf{G}. \tag{9.15}$$

The push-forward of the vector fields f, e, h by the period map are, respectively, given by

$$\begin{aligned} f &= x_2\frac{\partial}{\partial x_1} + x_4\frac{\partial}{\partial x_3}, \\ e &= x_1\frac{\partial}{\partial x_2} + x_3\frac{\partial}{\partial x_4}, \\ h &= x_1\frac{\partial}{\partial x_1} - x_2\frac{\partial}{\partial x_2} + x_3\frac{\partial}{\partial x_3} - x_4\frac{\partial}{\partial x_4}. \end{aligned}$$

Here, the vector fields act as derivations on the ring $\mathscr{O}_{\mathsf{U}}$ of holomorphic functions in U. This follows from $d\mathsf{P} = \mathsf{P}\mathsf{A}^{\mathsf{tr}}$ and

$$d\mathsf{P}(f) = \begin{bmatrix} x_2 & 0 \\ x_4 & 0 \end{bmatrix},$$

$$d\mathsf{P}(e) = \begin{bmatrix} 0 & x_1 \\ 0 & x_3 \end{bmatrix},$$

$$d\mathsf{P}(h) = \begin{bmatrix} x_1 & -x_2 \\ x_3 & -x_4 \end{bmatrix}.$$

9.7 Modular foliations

We analyze the modular foliations in the case of elliptic curves. The action of G on $\mathbb{A}^2_{\mathfrak{k}} - \{0\}$ has two orbits represented by

$$\mathsf{C} = \begin{bmatrix} 1 \\ 0 \end{bmatrix}, \begin{bmatrix} 0 \\ 1 \end{bmatrix}.$$

The modular foliation $\mathscr{F}(\mathsf{C})$ corresponding to the first one is given by the Ramanujan vector field R in (9.6) and the modular foliation $\mathscr{F}(\mathsf{C})$ corresponding to the second C is given by the vector field e. Therefore, the moduli of modular foliations consists of two points. The Ramanujan vector field gives us also the foliation $\mathscr{F}(2)$ and so

$$\mathscr{F}(2) = \mathscr{F}(\mathsf{R}) = \mathscr{F}(\mathsf{C}), \quad \mathsf{C} := [1, 0]^{\mathsf{tr}}.$$

We call this the Ramanujan foliation. The singularities of $\mathscr{F}(2)$ are given by

$$\mathrm{Sing}(\mathscr{F}(2)) := \{(a, 12a^2, 8a^3) | a \in \mathfrak{k}\}. \tag{9.16}$$

We consider the foliation $\mathscr{F}$ in $\mathbb{A}^3_{\mathfrak{k}}$ given by the $\mathfrak{k}$-vector space generated by e, f, h (Fig. 9.1). The foliation $\mathscr{F}$ is three-dimensional and its unique general leaf is given by the whole space $\mathbb{A}^3_{\mathfrak{k}}$. The vector fields e, h, f are tangent to the discriminant locus $\{\Delta = 0\} \subset \mathbb{A}^3_{\mathfrak{k}}$ and

$$(6t_3)f - \left(6t_3t_1 - \frac{1}{3}t_2^2\right)h + \left(\frac{2}{3}t_1t_2^2 - 6t_1^2t_3 - \frac{1}{2}t_2t_3\right)e = \frac{4}{3}\Delta\frac{\partial}{\partial t_2}. \tag{9.17}$$

Therefore, $\mathrm{Sing}(\mathscr{F}) = \{\Delta = 0\}$. There is natural stratification of $\mathbb{A}^3_{\mathfrak{k}}$ according to linear independence of e, f, h which is given by

$$\{(t_1, 0, 0) \mid t_1 \in \mathfrak{k}\} \subset \{\Delta = 0\} \subset \mathbb{A}^3_{\mathfrak{k}}.$$

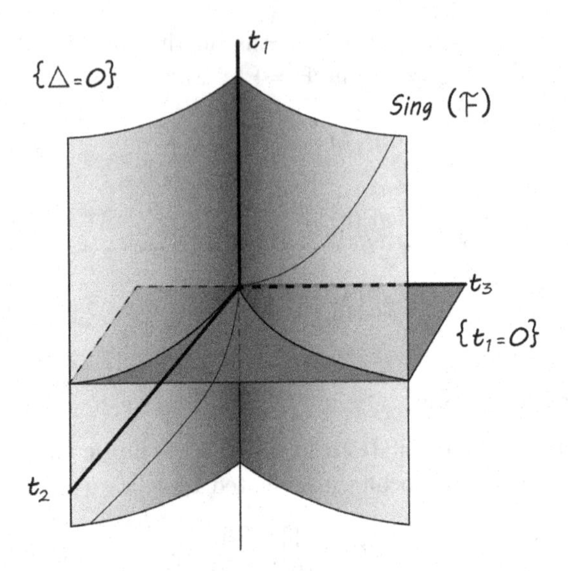

Fig. 9.1. Singularities of Ramanujan vector field.

9.8 Moduli of Hodge decompositions

The moduli of Hodge decompositions T_1 in the case of elliptic curves is the set of pairs $(E, H^{0,1})$, where E is an elliptic curve over $\mathfrak{k}$ and $H^{0,1} \neq F^1 H^1_{\mathrm{dR}}(E)$ is a one-dimensional subspace of $H^1_{\mathrm{dR}}(E)$. It turns out that this moduli is the quotient of T under the action

$$\mathsf{T} \times \mathbb{G}_m \to \mathsf{T}, \quad (t_1, t_2, t_3), k \mapsto (k \cdot t_1, k^2 \cdot t_2, k^3 \cdot t_3),$$

and so, such a moduli space is

$$\mathsf{T}_1 = \mathbb{P}^{1,2,3}_{\mathfrak{k}} - \mathbb{P}\{\Delta = 0\}.$$

This has a natural compactification in the weighted projective space $\mathbb{P}^{1,2,3}$ which is the projectivization of $\mathbb{A}^3_{\mathfrak{k}}$ with the coordinate system (t, t_2, t_3) and weights $\deg(t_i) = i$, $i = 1, 2, 3$. The foliation $\mathscr{F}(2)$ in T_1 induced by the Griffiths transversality can be written explicitly in the affine coordinate system $[1 : x : y]$ for $\mathbb{P}^{1,2,3}$ with

$$(x, y) := \left(\frac{t_2}{t_1^2}, \frac{t_3}{t_1^3} \right).$$

This is as follows. The Ramanujan vector field R as an ordinary differential equation gives us

$$\begin{cases} \dot{x} = t_1 \cdot (2x - 6y + \frac{1}{6}x^2), \\ \dot{y} = t_1 \cdot (3y - \frac{1}{3}x^2 + \frac{1}{4}xy). \end{cases}$$

Therefore,

$$\mathscr{F}(2): \quad \left(2x - 6y + \frac{1}{6}x^2\right) dy - \left(3y - \frac{1}{3}x^2 + \frac{1}{4}xy\right) dx = 0. \tag{9.18}$$

Proposition 9.3. *The only algebraic leaf of the foliation* (9.18) *in* $\mathbb{C}^2$ *is the curve*

$$27y^2 - x^3 = 0.$$

The reader who is familiar with foliations in two-dimensional surfaces, might try to prove this theorem using the Camacho–Sad index theorem, see for instance [LNS]. This can be done in the same way as one proves that the Jouanolou foliation has no algebraic leaf. Our proof of Proposition 9.3 does not use this and it follows from Theorem 9.7 in which we use the moduli interpretation of the coordinate system (t_1, t_2, t_3). Note that the natural projectivization of the foliation (9.18) is in $\mathbb{P}^{1,2,3}$ for which the line at infinity is not invariant. However, if we look at (9.18) in the usual projective space $\mathbb{P}^2$ then the line at infinity is invariant. Note also that singular points of the foliation (9.18) in $\mathbb{C}^2$ are $(x, y) = (0, 0), (12, 8)$.

9.9 Algebraic group acting on modular vector fields

Let us consider the left action of the Borel group G on the moduli space T. It induces an action of G from the left on the space of vector fields on T in a canonical way.

Proposition 9.4. *We have*

$$\begin{bmatrix} f \bullet \mathsf{g} \\ h \bullet \mathsf{g} \\ e \bullet \mathsf{g} \end{bmatrix} = \begin{bmatrix} 1 & -k' & -k'^2 \\ 0 & 1 & 2k' \\ 0 & 0 & 1 \end{bmatrix} \begin{bmatrix} f \\ h \\ e \end{bmatrix}, \quad \mathsf{g} = \begin{bmatrix} 1 & k' \\ 0 & 1 \end{bmatrix} \in \mathsf{G},$$

$$\begin{bmatrix} f \bullet \mathsf{g} \\ h \bullet \mathsf{g} \\ e \bullet \mathsf{g} \end{bmatrix} = \begin{bmatrix} k^2 & 0 & 0 \\ 0 & 1 & 0 \\ 0 & 0 & k^{-2} \end{bmatrix} \begin{bmatrix} f \\ h \\ e \end{bmatrix}, \quad \mathsf{g} = \begin{bmatrix} k & 0 \\ 0 & k^{-1} \end{bmatrix} \in \mathsf{G}.$$

Proof. We just need to compute the derivation of the action

$$\mathbb{A}^3_{\mathfrak{k}} \to \mathbb{A}^3_{\mathfrak{k}},\ (t_1,t_2,t_3) \mapsto (t_1k^{-2} + k'k^{-1}, t_2k^{-4}, t_3k^{-6}),\quad \mathsf{g} = \begin{bmatrix} k & k' \\ 0 & k^{-1} \end{bmatrix} \in \mathsf{G}.$$

□

The group G_C with $\mathsf{C} = [1,0]^{\mathsf{tr}}$ in (6.11) turns out to be

$$\mathsf{G}_\mathsf{C} := \left\{ \begin{bmatrix} k & 0 \\ 0 & k^{-1} \end{bmatrix} \mid k \in \mathfrak{k}^* \right\},$$

and the character $\lambda : \mathsf{G}_\mathsf{C} \to \mathbb{G}_m$ is an isomorphism of groups given by $\begin{bmatrix} k & 0 \\ 0 & k^{-1} \end{bmatrix} \mapsto k$. We conclude that

Proposition 9.5. *The foliation $\mathscr{F}(2)$ is invariant under the action of $\mathsf{g} = \begin{bmatrix} k & 0 \\ 0 & k^{-1} \end{bmatrix}$ on T, that is, g maps a leaf of $\mathscr{F}(\mathsf{C})$ to another leaf.*

Proof. The statement follows from $f \bullet \mathsf{g} = k^2 f$. In fact, g maps the leaf $L_{\tilde{\mathsf{C}}} : \mathsf{P}^{-\mathsf{tr}}\mathsf{C} = \tilde{\mathsf{C}}$ to $L_{k^{-1}\tilde{\mathsf{C}}}$. Note that we have used the particular format of $\mathsf{C} = [1,0]^{\mathsf{tr}}$. □

Note that $\mathsf{T}_1 = \mathsf{T}/\mathsf{G}_\mathsf{C}$ turns out to be the moduli of Hodge decompositions discussed in §9.8. The foliation $\mathscr{F}(2)$ is our first example of a modular foliation with a non-trivial character. We conclude that there is a foliation $\mathscr{F}(\mathsf{C},\lambda)$ containing the leaves of $\mathscr{F}(2)$ which is introduced in §6.4. This turns out to be the foliation $\mathscr{F}(f,h)$.

Proposition 9.6. *If the leaf of $\mathscr{F}(2)$ passing through $p \in \mathbb{C}^3$ is algebraic then the leaf of $\mathscr{F}(f,h)$ passing through p is also algebraic.*

Proof. Note that $[h,f] = -2f$ and so the distribution given by f and h in the tangent space of T is integrable. The assertion follows from Proposition 9.5 and the fact that the $\mathbb{G}_m$-action used in this proposition is given by h. □

9.10 Leaves

The period map in the case of elliptic curves is a biholomorphism between T and U and under this biholomorphism the notations in Chapters 6 and 8 are translated to each other. For instance, $\mathscr{F}(2)$ in T is mapped to $\mathscr{F}(2)$

in U. A leaf of the Ramanujan foliation $\mathscr{F}(2)$ in T is given by the equality

$$L_{\tilde{\mathsf{C}}}: \quad \begin{bmatrix} x_1 & x_2 \\ x_3 & x_4 \end{bmatrix}^{-\mathsf{tr}} \begin{bmatrix} 1 \\ 0 \end{bmatrix} = \begin{bmatrix} x_4 \\ -x_2 \end{bmatrix} = \tilde{\mathsf{C}} \tag{9.19}$$

for some constant vector $\tilde{\mathsf{C}} \in \mathbb{C}^2 \backslash \{0\}$. In particular, the set $\mathscr{L}(\mathsf{C})$ in (6.12) is

$$\mathscr{L}(\mathsf{C}) = \left\{ \tilde{\mathsf{C}} \in \mathbb{C}^2, \ \tilde{\mathsf{C}} \neq 0 \right\}. \tag{9.20}$$

Two such leaves $L_{\tilde{\mathsf{C}}_i}$, $i = 1, 2$ are the same if and only if $\tilde{\mathsf{C}}_1 = A\tilde{\mathsf{C}}_2$ for some $A \in \mathrm{SL}(2, \mathbb{Z})$. This means that the converse of Proposition 6.6 is true in the case of elliptic curves. The space of leaves $\mathrm{SL}(2, \mathbb{Z}) \backslash \mathscr{L}(\mathsf{C})$ for points $\tilde{\mathsf{C}}$ with $\tilde{\mathsf{C}}_1 / \tilde{\mathsf{C}}_2 \in \mathbb{R}$ does not enjoy any reasonable structure. For further study of the dynamics of $\mathscr{F}(2)$ see [Mov12a].

A leaf of $\mathscr{F}(f, h)$ is given by

$$L_c: \qquad \frac{x_2}{x_4} = c, \quad \text{for some constant } c \in \mathbb{P}^1. \tag{9.21}$$

Two such leaves L_{c_i}, $i = 1, 2$ are the same if and only if $c_1 = Ac_2$ for some $A \in \mathrm{SL}(2, \mathbb{Z})$. Here, we have used the Möbius action of $\mathrm{SL}(2, \mathbb{Z})$ on $\mathbb{P}^1$. We expect that all the algebraic leaves of modular foliations are again moduli spaces of certain structures in Algebraic Geometry. Viewed in this way, the following can be considered as a first evidence to such a statement.

Theorem 9.7. *No leaf of the Ramanujan foliation $\mathscr{F}(2)$ in T : $\mathbb{C}^3 \backslash \{27t_3^2 - t_2^3 = 0\}$ is algebraic.*

Proof. Using Proposition 9.6 it is enough to prove the same statement for $\mathscr{F}(f, h)$. Two leaves L_{c_1}, L_{c_2} of $\mathscr{F}(f, h)$ given in (9.21) are the same if and only if $A(c_1) = c_2$ for some $A \in \mathrm{SL}(2, \mathbb{Z})$. If a leaf L_c of $\mathscr{F}(f, h)$ is algebraic then the orbit of c under $\mathrm{SL}(2, \mathbb{Z})$-action is a closed subset of $\mathbb{P}^1$. Otherwise, such an orbit have an accumulation point b. This implies that the leaf L_c accumulates on the leaf L_b which is a contradiction. It turns out that such a closed orbit does not exist and the theorem follows. □

9.11 An alternative proof of Theorem 9.7

There are two essential steps in the proof of Theorem 9.7. The first one is the fact that the action of $\mathrm{SL}(2, \mathbb{Z})$ in $\mathbb{P}^1$ has no finite orbit. The second one is the usage of Theorem 9.2. From this theorem what we need is the following. The loci of parameters $t \in \mathsf{T}$ such that the quotient $\int_\delta \frac{xdx}{y} / \int_\gamma \frac{xdx}{y}$ is constant for some choice of δ and γ with $\langle \delta, \gamma \rangle = -1$, is a "connected" leaf of $\mathscr{F}(f,h)$. The connectedness may fail in general and this set might be a union of infinite number of leaves. This must be considered the most critical part of the proof which makes it hard for generalizations. Another way to look at this issue is that for any family of elliptic curves, which is not isotrivial, that is, all the smooth fibers are not isomorphic to each other, the image of the corresponding monodromy group in $\mathrm{SL}(2,\mathbb{Z})$ has finite index. For an algebraic leaf of R in T the monodromy group contains only $A \in \mathrm{SL}(2,\mathbb{Z})$ such that $Ac = c$. Up to the action of $\mathrm{SL}(2,\mathbb{Z})$, there are four possibilities for c:

$$c = \pm i, \pm \frac{-1 + i\sqrt{3}}{2}. \tag{9.22}$$

For the first (respectively, the second) two c's the group of such matrices A is a cyclic group of order 4 (respectively, 6). Therefore, such a group is far from having finite index in $\mathrm{SL}(2,\mathbb{Z})$.

There are several other ideas which lead us to new proofs for Theorem 9.7. Since they might be useful for further generalizations we mention them here.

The first idea is as follows. We know that the Eisenstein series E_2, E_4, E_6 up to multiplication with a constant form a solution of R. From another side we know that they are algebraically independent over $\mathbb{C}$. We can actually derive this from Theorem 9.2 and the fact that the inverse of the period map can be expressed in terms of Eisenstein series, see [Mov08b]. For an elementary proof see Theorem A.9 and [MR05]. We conclude that at least one solution of R is Zariski sense. Now, we use the Halphen property of R (see §9.5) and conclude the statement for other solutions, for more details see [Mov08c].

The idea of the second proof is as follows. Take any one parameter family of elliptic curves which is not isotrivial, that is, all the smooth fibers are not isomorphic to each other. Leaves of R in T can be parameterized by elliptic integrals of this family. The differential Galois group of the corresponding Picard–Fuchs equation is $\mathrm{SL}(2,\mathbb{C})$, and from this, we can deduce that the transcendental degree of the field generated by such elliptic integrals is exactly three, and in particular, the parameterization of leaves of R by elliptic integrals has three algebraically independent coordinates over $\mathbb{C}$, for more details see [Mov12b]. A similar statement in the case of mirror-quintic is proved using this method, see [Mov11a].

The third proof is the following. Using the abelian subvariety theorem, one can show that all the leaves of R in T intersects $\bar{\mathbb{Q}}^3$ at most at one point. For more details see [Mov08c]. The following is slightly stronger than Theorem 9.7.

Theorem 9.8. *The leaves of the Ramanujan foliation $\mathscr{F}(2)$ in $\mathsf{T} := \mathbb{C}^3 \backslash \{27t_3^2 - t_2^3 = 0\}$ are Zariski dense.*

Proof. By contradiction let L be a leaf of $\mathscr{F}(2)$ in T contained in $P(t_1, t_2, t_3) = 0$ for some polynomial P with coefficients in $\mathbb{C}$ for which we can further assume that P is irreducible over $\mathbb{C}$. We have $L \subset \{dP(\mathsf{R}) = 0\}$, and since by Theorem 9.7 L is a transcendental leaf, P divides $dP(\mathsf{R})$. This means that $P = 0$ is tangent to R. From another side using the equality $dP(\mathsf{R}) = P \cdot Q$, for some polynomial Q, we can take the last homogeneous piece of P which satisfies a similar equality, and so, we can assume that P is homogeneous. This is equivalent to say that $dP(h) = \deg(P) \cdot P$ and so $P = 0$ is a leaf of $\mathscr{F}(f, h)$. In Theorem 9.7 we have shown that this foliation has no algebraic leaf in T. □

9.12 Halphen differential equation

In a series of article [Hal81b, Hal81c, Hal81a] Halphen studied the following system of ODE's:

$$\mathrm{H}(\alpha) : \begin{cases} \dot{t}_1 = (1 - \alpha_1)(t_1 t_2 + t_1 t_3 - t_2 t_3) + \alpha_1 t_1^2, \\ \dot{t}_2 = (1 - \alpha_2)(t_2 t_1 + t_2 t_3 - t_1 t_3) + \alpha_2 t_2^2, \\ \dot{t}_3 = (1 - \alpha_3)(t_3 t_1 + t_3 t_2 - t_1 t_2) + \alpha_3 t_3^2, \end{cases} \tag{9.23}$$

with $\alpha_i \in \mathbb{C} \cup \{\infty\}$ (if for instance $\alpha_1 = \infty$ then the first row is replaced with $-t_1 t_2 - t_1 t_3 + t_2 t_3 + t_1^2$). He showed the so called Halphen property for $\mathrm{H}(\alpha)$.

Proposition 9.9. *If ϕ_i, $i = 1, 2, 3$ are the coordinates of a solution of $\mathrm{H}(\alpha)$ then*

$$\frac{1}{(cz+d)^2} \phi_i \left(\frac{az+b}{cz+d} \right) - \frac{c}{cz+d}, \ i = 1, 2, 3, \ \begin{bmatrix} a & b \\ c & d \end{bmatrix} \in \mathrm{SL}(2, \mathbb{C})$$

are also coordinates of a solution of $\mathrm{H}(\alpha)$.

Proof. The proof is based on explicit calculations and is left to the reader. □

Halphen concluded that it is enough to find one solution of $\mathrm{H}(\alpha)$ and then used Proposition 9.9 to obtain the general solution. He then constructed a particular solution of $\mathrm{H}(\alpha)$ using the Gauss hypergeometric function.

Let a, b, c be defined by the equations:

$$1-\alpha_1 = \frac{a-1}{a+b+c-2},$$
$$1-\alpha_2 = \frac{b-1}{a+b+c-2},$$
$$1-\alpha_3 = \frac{c-1}{a+b+c-2}.$$

We consider the following matrix of differential 1-forms in $(t_1, t_2, t_3) \in \mathbb{C}^3$:

$$\mathsf{A} = \sum_{i=1}^{3} \mathsf{A}_i dt_i, \tag{9.24}$$

where

$$\mathsf{A}_1 = \frac{1}{(t_1-t_2)(t_1-t_3)} \cdot \begin{pmatrix} \frac{1}{2}((a+c-1)t_2+(a+b-1)t_3+(b+c-2)t_1) & at_2t_3+(b-1)t_1t_3+(c-1)t_1t_2 \\ -a-b-c+2 & -\frac{1}{2}((a+c-1)t_2+(a+b-1)t_3+(b+c-2)t_1) \end{pmatrix} \tag{9.25}$$

and A_2 (respectively, A_3) is obtained from A_1 by changing the role of t_1 and t_2 (respectively, t_1 and t_3). It is a mere computation to see that A gives us an integrable connection on $\mathbb{C}^3$, that is, $d\mathsf{A} = -\mathsf{A} \wedge \mathsf{A}$. From Lie theoretic point of view, Halphen's differential equation should not be considered on its own but together with the attached $\mathfrak{sl}_2$ structure, see for instance Guillot's article [Gui07]. Let $\mathrm{H} = \mathrm{H}(\alpha)$ be the vector field in $\mathbb{C}^3$ corresponding to the Halphen's differential equation (9.23) and

$$f := -\mathrm{H}, \quad h = -\sum_{i=1}^{3} 2t_i \frac{\partial}{\partial t_i}, \quad e = \sum_{i=1}^{3} \frac{\partial}{\partial t_i}.$$

The following is a mere computation.

Proposition 9.10. *We have*

$$\mathsf{A}_f = \begin{bmatrix} 0 & 1 \\ 0 & 0 \end{bmatrix}, \ \mathsf{A}_h = \begin{bmatrix} 1 & 0 \\ 0 & -1 \end{bmatrix}, \ \mathsf{A}_e = \begin{bmatrix} 0 & 0 \\ 1 & 0 \end{bmatrix}.$$

It follows that the $\mathbb{C}$-vector space generated by e, f, h forms the classical Lie algebra $\mathfrak{sl}_2$: $[h, e] = 2e,\ [h, f] = -2f,\ [e, f] = h$. The reader is referred to [Mov12a] for the geometric interpretation of the connection matrix A. In particular, one can find the proof of the following proposition.

Proposition 9.11. *The integrals*

$$p \int_\delta \frac{x dx}{(x - t_1)^a (x - t_2)^b (x - t_3)^c}, \tag{9.26}$$

where

$$p := (t_1 - t_3)^{-\frac{1}{2}(1-a-c)} (t_1 - t_2)^{-\frac{1}{2}(1-a-b)} (t_2 - t_3)^{-\frac{1}{2}(1-b-c)}$$

and δ is path in $\mathbb{C} \backslash \{t_1, t_2, t_3\}$ connecting two points in t_1, t_2, t_3, ∞ or it is a Pochhammer cycle, as local multivalued functions in t_1, t_2, t_3 are constant along the solutions of the Halphen differential equation $\mathrm{H}(\alpha)$.

The following slight modification of the above discussion might be useful. Consider the connection matrix

$$\mathsf{A} = \sum_{i=1}^{3} \mathsf{A}_i dt_i, \tag{9.27}$$

where

$$\mathsf{A}_1 := \frac{1}{(t_1 - t_2)(t_1 - t_3)} \cdot \begin{bmatrix} -at_1 + (a + c - 1)t_2 + (a + b - 1)t_3 & -a - b - c + 2 \\ at_2 t_3 + (b - 1)t_1 t_3 + (c - 1)t_1 t_2 & (-a - b - c + 2)t_1 \end{bmatrix} \tag{9.28}$$

and A_2 (respectively, A_3) is obtained from A_1 by changing the role of t_1 and t_2 (respectively, t_1 and t_3). It is obtained from the connection matrix in (9.24) by subtracting it from $\frac{dp}{p} I_{2\times 2}$, where p is defined in Proposition 9.11

and I_2 is the identity 2×2 matrix. The vector field H such that for $f := (a+b+c-2)^{-1} \cdot \mathrm{H}$ we have $\mathsf{A}_f = \begin{bmatrix} 0 & 1 \\ 0 & 0 \end{bmatrix}$ is given by

$$\begin{cases} \dot{t}_1 = (a-1)t_2t_3 + bt_1t_3 + ct_1t_2, \\ \dot{t}_2 = at_2t_3 + (b-1)t_1t_3 + ct_1t_2, \\ \dot{t}_3 = at_2t_3 + bt_1t_3 + (c-1)t_1t_2. \end{cases} \tag{9.29}$$

The integral

$$\int_\delta \frac{xdx}{(x-t_1)^a(x-t_2)^b(x-t_3)^c},$$

where δ is as in Proposition 9.11, is constant along the trajectories of (9.29). For the computer codes used in this section, see the author's webpage.

Finally, let us state and prove the generalization of Theorem 9.7 for the Halphen differential equation.

Theorem 9.12. *If one of $a+c$, $a+b$ or $b+c$ is equal to one and $a, b, c \notin \mathbb{Z}$ then the leaves of the Halphen differential equation in $\mathbb{C}^3 \setminus \cup_{i,j=1,2,3,\ i<j} \{t_i - t_j = 0\}$ are Zariski dense.*

Proof. We only sketch a proof. For more details see [Mov12a; DGMS13, Theorem 3]. The integral (9.26) for $t_1 = 0$, $t_2 = 1$, $t_3 = z$ satisfies a second-order linear differential equation L whose monodromy group is inside $\mathrm{SL}(2,\mathbb{C})$. This is essentially the Gauss hypergeometric equation. Moreover, since one of $a+c$, $a+b$ or $b+c$ is equal to one and a, b, c are non-integral it contains a nilpotent element $\begin{bmatrix} 1 & \alpha \\ 0 & 1 \end{bmatrix}$, $\alpha \neq 0$. This implies that the differential Galois group of L is $\mathrm{SL}(2,\mathbb{C})$. From another side, the leaves of the Halphen differential equation in the Zariski open set $t_i \neq t_j$ can be parameterized by three algebraically independent elements in the differential field generated by solutions of L. This completes the proof. □

In the proof of Theorem 9.12 what we actually need is that the Galois group of the corresponding Gauss hypergeometric equation is $\mathrm{SL}(2,\mathbb{C})$. One might use the result of Beukers and Heckman in [BH89] and give a complete classification of Halphen differential equations with Zariski dense general leaves.

Chapter 10

Product of Two Elliptic Curves

The applications of modular curves and modular functions to number theory are especially exciting: you use GL_2 to study GL_1, so to speak! There is clearly a lot more to come from that direction ... may be even a proof of the Riemann Hypothesis some day! (J. P. Serre in [CTC01]).

10.1 Introduction

This chapter deals with enhanced pairs of elliptic curves, their moduli space T and a modular foliation $\mathscr{F}$ on T. The reader interested in Hodge loci will encounter its first non-trivial example, this is namely, the modular curve $X_0(d)$ which parameterizes pairs of elliptic curves with an isogeny of degree d. In our context, we replace $X_0(d)$ with a three-dimensional affine subvariety V_d of T, which itself is six-dimensional. We prove that V_d's are the only algebraic leaves of $\mathscr{F}$. This will be the first verification of Property 7.1. This chapter is the continuation of the previous work [Mov15b] and §9, in which we have studied V_d using Eisenstein series.

10.2 Enhanced elliptic curves

We set $X = X_1 \times X_2$, where X_1 and X_2 are two elliptic curves over $\mathfrak{k}$. The subindex $i = 1, 2$ will attach an object to the first and second elliptic curves respectively. This example does not fit exactly to the notion of enhanced varieties and their moduli introduced in Chapter 3. It shows that we have to enlarge our moduli spaces T by adding product structures or fixed algebraic cycles in X. In this example, such fixed algebraic cycles are given by $\{p_1\} \times X_2$ and $X_1 \times \{p_2\}$ for some $p_i \in X_i(\mathfrak{k})$. For the cohomology

$$H^2_{\mathrm{dR}}(X_1 \times X_2)_j := H^1_{\mathrm{dR}}(X_1) \otimes_{\mathfrak{k}} H^1_{\mathrm{dR}}(X_2)$$

we use the basis

$$[\alpha_1 \otimes \alpha_2, \alpha_1 \otimes \omega_2, \omega_1 \otimes \alpha_2, \omega_1 \otimes \omega_2]^{\mathsf{tr}}. \tag{10.1}$$

This is compatible with the Hodge filtration

$$0 = F^3 \subset F^2 \subset F^1 \subset F^0 = H^2_{\mathrm{dR}}(X_1 \times X_2)_j,$$

where

$$F^1 H^2_{\mathrm{dR}}(X_1 \times X_2)_j = F^1 H^1_{\mathrm{dR}}(X_1) \otimes_{\mathfrak{k}} H^1_{\mathrm{dR}}(X_2) + H^1_{\mathrm{dR}}(X_1) \otimes_{\mathfrak{k}} F^1 H^1_{\mathrm{dR}}(X_2),$$
$$F^2 H^2_{\mathrm{dR}}(X_1 \times X_2)_j = F^1 H^1_{\mathrm{dR}}(X_1) \otimes_{\mathfrak{k}} F^1 H^1_{\mathrm{dR}}(X_2).$$

Therefore, the Hodge numbers are $(\mathsf{h}^{20}, \mathsf{h}^{11}, \mathsf{h}^{02}) = (1, 2, 1)$. For the homology group

$$H_2(X_1 \times X_2, \mathbb{Z})_j := H_1(X_1, \mathbb{Z}) \otimes_{\mathbb{Z}} H_1(X_2, \mathbb{Z}),$$

we use the basis

$$[\delta_1 \otimes \delta_2, \quad \delta_1 \otimes \gamma_2, \gamma_1 \otimes \delta_2, \quad \gamma_1 \otimes \gamma_2]^{\mathsf{tr}}.$$

The intersection matrices both in cohomology and homology are equal:

$$\Phi = \Psi = \begin{bmatrix} 0 & 0 & 0 & -1 \\ 0 & 0 & 1 & 0 \\ 0 & 1 & 0 & 0 \\ -1 & 0 & 0 & 0 \end{bmatrix}. \tag{10.2}$$

Note that both in cohomology and homology we have

$$\langle a_1 \otimes a_2, b_1 \otimes b_2 \rangle = -\langle a_1, b_1 \rangle \cdot \langle a_2, b_2 \rangle. \tag{10.3}$$

This follows from the relation between the cup product in cohomology (respectively, intersection bilinear map in homology) and Künneth formula in cohomology (respectively, homology). For now, the reader may consider (10.3) as a wedge product with four pieces, and interchanging a_2 with b_1 contributes a minus sign to the product. Note that the duality between homology and cohomology is given by the integration formula:

$$\int_{\delta_1 \otimes \delta_2} \omega_1 \otimes \omega_2 := \int_{\delta_1} \omega_1 \cdot \int_{\delta_2} \omega_2, \quad \delta_i \in H_1(X_i, \mathbb{Z}), \quad \omega_i \in H^1_{\mathrm{dR}}(X_i), \quad i = 1, 2. \tag{10.4}$$

10.3 The Gauss–Manin connection and modular vector fields

Let us now consider two copies $\mathsf{X}_i/\mathsf{T}_i$, $i = 1, 2$ of the universal family of enhanced elliptic curves (9.1). Let A_i, $i = 1, 2$ be the Gauss–Manin connection matrix of $\mathsf{X}_i/\mathsf{T}_i$ in the basis $[\alpha_i, \omega_i]^{\mathsf{tr}}$, see (9.3). In the basis (10.1), the Gauss–Manin connection matrix of the family $\mathsf{X}_1 \times \mathsf{X}_2 \to \mathsf{T}_1 \times \mathsf{T}_2$ is given by

$$\mathsf{A} = \begin{bmatrix} (\mathsf{A}_1)_{11} & 0 & (\mathsf{A}_1)_{12} & 0 \\ 0 & (\mathsf{A}_1)_{11} & 0 & (\mathsf{A}_1)_{12} \\ (\mathsf{A}_1)_{21} & 0 & (\mathsf{A}_1)_{22} & 0 \\ 0 & (\mathsf{A}_1)_{21} & 0 & (\mathsf{A}_1)_{22} \end{bmatrix} + \begin{bmatrix} (\mathsf{A}_2)_{11} & (\mathsf{A}_2)_{12} & 0 & 0 \\ (\mathsf{A}_2)_{21} & (\mathsf{A}_2)_{22} & 0 & 0 \\ 0 & 0 & (\mathsf{A}_2)_{21} & (\mathsf{A}_2)_{22} \\ 0 & 0 & (\mathsf{A}_2)_{21} & (\mathsf{A}_2)_{22} \end{bmatrix}.$$

The Gauss–Manin connection matrix composed with the vector fields h_i, e_i, f_i, $i = 1, 2, 3$ is of the form:

$$\mathsf{A}_{h_1} = \begin{bmatrix} 1 & 0 & 0 & 0 \\ 0 & 1 & 0 & 0 \\ 0 & 0 & -1 & 0 \\ 0 & 0 & 0 & -1 \end{bmatrix}, \quad \mathsf{A}_{f_1} = \begin{bmatrix} 0 & 0 & 1 & 0 \\ 0 & 0 & 0 & 1 \\ 0 & 0 & 0 & 0 \\ 0 & 0 & 0 & 0 \end{bmatrix}, \quad \mathsf{A}_{e_1} = \begin{bmatrix} 0 & 0 & 0 & 0 \\ 0 & 0 & 0 & 0 \\ 1 & 0 & 0 & 0 \\ 0 & 1 & 0 & 0 \end{bmatrix},$$

$$\mathsf{A}_{h_2} = \begin{bmatrix} 1 & 0 & 0 & 0 \\ 0 & -1 & 0 & 0 \\ 0 & 0 & 1 & 0 \\ 0 & 0 & 0 & -1 \end{bmatrix}, \quad \mathsf{A}_{f_2} = \begin{bmatrix} 0 & 1 & 0 & 0 \\ 0 & 0 & 0 & 0 \\ 0 & 0 & 0 & 1 \\ 0 & 0 & 0 & 0 \end{bmatrix}, \quad \mathsf{A}_{e_2} = \begin{bmatrix} 0 & 0 & 0 & 0 \\ 1 & 0 & 0 & 0 \\ 0 & 0 & 0 & 0 \\ 0 & 0 & 1 & 0 \end{bmatrix}.$$

From these computations we can easily derive

$$\mathscr{F}(2) = \mathscr{F}(f_1, f_2).$$

10.4 Modular foliations

For the product of two elliptic curves, the algebraic group G is given by

$$\mathsf{G} = \mathsf{G}_1 \otimes \mathsf{G}_2 := \left\{ \begin{bmatrix} k_1 k_2 & k_1 k_2' & k_1' k_2 & k_1' k_2' \\ 0 & k_1 k_2^{-1} & 0 & k_1' k_2^{-1} \\ 0 & 0 & k_1^{-1} k_2 & k_1^{-1} k_2' \\ 0 & 0 & 0 & k_1^{-1} k_2^{-1} \end{bmatrix}, k_1, k_2 \in \mathfrak{k}^*, k_1', k_2' \in \mathfrak{k} \right\}, \tag{10.5}$$

where $\otimes$ is the Kronecker product of matrices. It acts from the right on $\mathbb{A}_{\mathfrak{k}}^4$ by the usual multiplication of matrices. For $\mathsf{C} \in \mathbb{A}_{\mathfrak{k}}^4$ and $\mathsf{g} \in \mathsf{G}$ we have

$$\begin{aligned} \mathsf{C} \bullet \mathsf{g} := \mathsf{g}^{\mathrm{tr}}\mathsf{C}[k_1k_2\mathsf{C}_1, k_1k_2'\mathsf{C}_1 + k_1k_2^{-1}\mathsf{C}_2, k_1'k_2\mathsf{C}_1 \\ + k_1^{-1}k_2\mathsf{C}_3, k_1'k_2'\mathsf{C}_1 + k_1'k_2^{-1}\mathsf{C}_2 + k_1^{-1}k_2'\mathsf{C}_3 + k_1^{-1}k_2^{-1}\mathsf{C}_4]^{\mathrm{tr}}. \end{aligned} \tag{10.6}$$

Proposition 10.1. *The quotient* $(\mathbb{A}_{\mathfrak{k}}^4 - \{0\})/\mathsf{G}$ *is isomorphic to the following set*

$$\begin{bmatrix}1\\0\\0\\a\end{bmatrix}, \begin{bmatrix}0\\a\\1\\0\end{bmatrix}, \quad a \in \mathfrak{k} - \{0\}, \quad \begin{bmatrix}1\\0\\0\\0\end{bmatrix}, \begin{bmatrix}0\\1\\0\\0\end{bmatrix}, \begin{bmatrix}0\\0\\1\\0\end{bmatrix}, \begin{bmatrix}0\\0\\0\\1\end{bmatrix}. \tag{10.7}$$

Proof. We only use the definition (10.6). For $\mathsf{C} \in \mathbb{A}_{\mathfrak{k}}^4$, $\mathsf{C} \neq 0$ if $\mathsf{C}_1 \neq 0$ then in its orbit under G we have a unique element of the form $\mathsf{C} = [1, 0, 0, a]^{\mathrm{tr}}$, $a \in \mathfrak{k}$. The case $a = 0$ is separated and it is the third matrix in (10.7). If $\mathsf{C}_1 = 0$ then $\mathsf{C}_2 \cdot \mathsf{C}_3$ is invariant under the action of G. The four cases $\mathsf{C}_2 \cdot \mathsf{C}_3 \neq 0$, $\mathsf{C}_2 \neq 0 \& \mathsf{C}_3 = 0$, $\mathsf{C}_2 = 0 \& \mathsf{C}_3 \neq 0$ and $\mathsf{C}_2 = 0 \& \mathsf{C}_3 = 0$ give us the full classification. $\square$

Remark 10.1. The map

$$\mathbb{A}_{\mathfrak{k}}^2 \times \mathbb{A}_{\mathfrak{k}}^2 \to \mathbb{A}_{\mathfrak{k}}^4, \quad ([c_1, c_2]^{\mathrm{tr}}, [b_1, b_2]^{\mathrm{tr}}) \mapsto [c_1b_1, c_1b_2, c_2b_1, c_2b_2]^{\mathrm{tr}}$$

is invariant under the action of G on both sides. Therefore, we have four orbits of G in $\mathbb{A}_{\mathfrak{k}}^4$ corresponding to the product of $[1, 0]^{\mathrm{tr}}$ and $[0, 1]^{\mathrm{tr}}$. This gives us the last four C in (10.7).

We are going to discuss only the modular foliation attached to the second C in (10.7). For a brief discussion of other cases see Remark 10.2. For reasons which come from [Mov15b] and will be explained in §10.7, instead of the second matrix in (10.6) we define

$$\mathsf{C}_d := \begin{bmatrix}0\\d\\-1\\0\end{bmatrix}. \tag{10.8}$$

Here, C_d stands for both its entry, if $d = 1, 2, 3$ or 4, and this new notation. I hope this will not produce any confusion. Let $d, \check{d} \in \mathbb{N}$ be two natural numbers. Two modular foliations $\mathscr{F}(\mathsf{C}_d)$, $\mathscr{F}(\mathsf{C}_{\check{d}})$ are isomorphic through the action of G:

$$\mathsf{g}_{\check{d},d}^{\mathsf{tr}} \cdot \mathsf{C}_d = \left(\frac{\check{d}}{d}\right)^{-\frac{1}{2}} \mathsf{C}_{\check{d}}, \quad \mathsf{g}_{\check{d},d} := I_{2\times 2} \otimes \begin{bmatrix} (\frac{\check{d}}{d})^{-\frac{1}{2}} & 0 \\ 0 & (\frac{\check{d}}{d})^{\frac{1}{2}} \end{bmatrix}, \tag{10.9}$$

see Proposition 6.3. Therefore, we only need to study the foliation $\mathscr{F}(\mathsf{C}_d)$ with $d = 1$. For reasons that will be revealed next section, we will keep working with an arbitrary d.

Proposition 10.2. *The modular foliation $\mathscr{F}(\mathsf{C}_d)$ with C_d in (10.8) is given by the vector fields*

$$h_1 + h_2, \; d \cdot f_1 + f_2, \; e_1 + d \cdot e_2. \tag{10.10}$$

Proof. Using the explicit computations in §10.3, we have

$$\mathsf{A}_{h_1} \cdot \mathsf{C}_d = \begin{bmatrix} 0 \\ d \\ 1 \\ 0 \end{bmatrix}, \; \mathsf{A}_{f_1} \cdot \mathsf{C}_d = \begin{bmatrix} -1 \\ 0 \\ 0 \\ 0 \end{bmatrix}, \; \mathsf{A}_{e_1} \cdot \mathsf{C}_d = \begin{bmatrix} 0 \\ 0 \\ 0 \\ d \end{bmatrix},$$

$$\mathsf{A}_{h_2} \cdot \mathsf{C}_d = \begin{bmatrix} 0 \\ -d \\ -1 \\ 0 \end{bmatrix}, \; \mathsf{A}_{f_2} \cdot \mathsf{C}_d = \begin{bmatrix} d \\ 0 \\ 0 \\ 0 \end{bmatrix}, \; \mathsf{A}_{e_2} \cdot \mathsf{C}_d = \begin{bmatrix} 0 \\ 0 \\ 0 \\ -1 \end{bmatrix}.$$

The space of vector fields $\mathsf{v} := a_1 h_1 + a_2 e_1 + a_3 f_1 + b_1 h_2 + b_2 e_2 + b_3 f_2$ with $a_i, b_i \in \mathsf{k}$ such that $\mathsf{A}_\mathsf{v} \mathsf{C}_d = 0$ is generated by (10.10). □

Proposition 10.3. *The singular set of $\mathscr{F}(\mathsf{C}_d)$ in $\mathbb{A}_\mathsf{k}^3 \times \mathbb{A}_\mathsf{k}^3$ consists of two irreducible components of dimensions 2 and 3:*

$$\{(t_1, 0, 0, s_1, 0, 0) | t_1, s_1 \in \mathsf{k}\} \cup \{(t_1, t_2, t_3, t_1, t_2, t_3) | 27t_3^2 - t_2^3 = 27s_3^2 - s_2^3$$
$$= d \cdot (2t_1 - t_3^{\frac{1}{3}}) - (2s_1 - s_3^{\frac{1}{3}}) = 0\}.$$

Proof. Assume that at some point of $\mathbb{A}_\mathsf{k}^3 \times \mathbb{A}_\mathsf{k}^3$ the three vectors (10.10) are linearly dependent. This gives us linear relations for h_i, e_i, f_i, $i = 1, 2$

$$a_1 h_1 + d a_2 f_1 + a_3 e_1 = 0, \;\; a_1 h_2 + a_2 f_2 + d a_3 e_2 = 0$$

and so such a point is in $\Delta \times \Delta$. For the rest of the proof we have used a computer. □

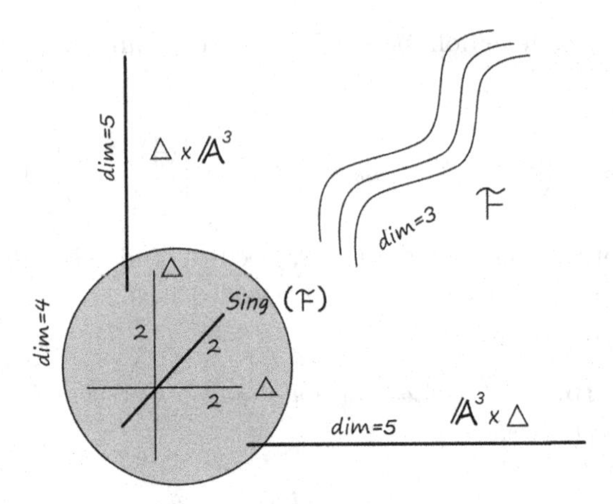

Fig. 10.1. Modular foliation for two elliptic curves.

Figure 10.1 might be useful for analyzing the foliation $\mathscr{F}(\mathsf{C}_d)$. The vector fields e, h, f are tangent to V, where V is one of

$$\{(t_1, 0, 0) \in \mathbb{A}^3_{\mathfrak{k}}\},\ \Delta := \{(t_1, t_2, t_3) \in \mathbb{A}^3_{\mathfrak{k}} | 27t_3^2 - t_2^3 = 0\},\ \text{or}\ \mathbb{A}^3_{\mathfrak{k}}.$$

Therefore, the three vectors (10.10) induce a foliation in $V_1 \times V_2$ for three choices of V_1 and V_2. In particular, $\mathscr{F}(\mathsf{C}_d)$ induces a foliation of dimension 3 in $\Delta \times \Delta$ with the singular set as in Proposition 10.3.

Remark 10.2. The modular foliations $\mathscr{F}(\mathsf{C})$ with C being the last four matrices in (10.7) do not seem to be of interest. The foliation $\mathscr{F} := \mathscr{F}(\mathsf{C})$ with C being the first matrix in (10.7) is given by

$$h_1 - h_2,\ f_1 - a \cdot e_2,\ a \cdot e_1 - f_2.$$

It seems that this foliation has no algebraic leaves in T as this is a particular case of Property 7.1. Note that the phenomenon of an isogeny between elliptic curves is controlled by the modular foliation $\mathscr{F}(\mathsf{C}_d)$, see §10.7. Note also that $\mathscr{F}$ induces a foliation in $\mathbb{P}^{1,2,3,-1,-2,-3}$. This also follows from computing G_C defined in (6.11) and Proposition 6.5.

10.5 Period domain

The period domain in the case of product of two elliptic curves is

$$\mathsf{U} := \mathsf{U}_1 \times \mathsf{U}_2 = \mathrm{SL}(2, \mathbb{Z}) \backslash \Pi_1 \times \mathrm{SL}(2, \mathbb{Z}) \backslash \Pi_2, \tag{10.11}$$

where U_i, $i = 1, 2$ are two copies of the period domain (9.13) in the case of elliptic curves. The period matrix for $(\mathsf{P}_1, \mathsf{P}_2) \in \Pi_1 \times \Pi_2$ is the Kronecker product of P_1 and P_2:

$$\mathsf{P} := \mathsf{P}_1 \otimes \mathsf{P}_2 = \begin{pmatrix} x_1\mathsf{P}_2, \, x_2\mathsf{P}_2 \\ x_3\mathsf{P}_2, \, x_4\mathsf{P}_2 \end{pmatrix}. \tag{10.12}$$

We consider the modular foliation $\mathscr{F} := \mathscr{F}(\mathsf{C})$ with $\mathsf{C} = \mathsf{C}_d, \ d = 1$. The leaves of the modular foliation $\mathscr{F}$ in U is given by $\tilde{\mathsf{C}}\mathsf{P} = \mathsf{C}^{\mathsf{tr}}$ for constant 1×4 matrices $\tilde{\mathsf{C}}$ with coefficients in $\mathbb{C}$. This can be written as an equality of 2×2 matrices:

$$\mathsf{P}_1^{\mathsf{tr}} f \mathsf{P}_2 = \begin{bmatrix} 0 & 1 \\ -1 & 0 \end{bmatrix}, \tag{10.13}$$

where f is a constant 2×2 matrix whose rows written one after another give us $\tilde{\mathsf{C}}$. Therefore, the local first integral of the foliation $\mathscr{F}$ is the function

$$\begin{aligned} &f : \mathsf{U} \to \mathrm{Mat}(2 \times 2, \mathbb{C}), \\ &f(\mathsf{P}_1, \mathsf{P}_2) := \mathsf{P}_1^{-\mathsf{tr}} \begin{bmatrix} 0 & 1 \\ -1 & 0 \end{bmatrix} \mathsf{P}_2^{-1}. \end{aligned}$$

We have actions of $\mathrm{SL}(2, \mathbb{Z})$ on Π_i, $i = 1, 2$. This means that analytic continuations of f will result in the multiplications of f both from the left and right with elements of $\mathrm{SL}(2, \mathbb{Z})$.

10.6 Character

Recall the group G_{C} in (6.11) and let C_d be as in (10.8).

Proposition 10.4. *The algebraic group $\mathsf{G}_{\mathsf{C}_d}$ has two connected components*

$$\mathsf{G}_{\mathsf{C}} = \mathsf{G}_{\mathsf{C}_d,+} \cup \mathsf{G}_{\mathsf{C}_d,-},$$

where

$$\mathsf{G}_{\mathsf{C}_d,\epsilon} := \left\{ \begin{bmatrix} k_1 & k_1' \\ 0 & k_1^{-1} \end{bmatrix} \times \begin{bmatrix} k_2 & k_2' \\ 0 & k_2^{-1} \end{bmatrix} \middle| k_2 = \epsilon k_1, \ k_2' = \epsilon d \cdot k_1' \right\}.$$

The character λ restricted to $\mathsf{G}_{\mathsf{C}_d,+}$ is 1 and restricted to $\mathsf{G}_{\mathsf{C}_d,-}$ is -1.

Proof. The proposition follows directly from the definition of G_{C} and the explicit expression of the algebraic group G in (10.5). □

In the case of product of two elliptic curves, λ is a discrete character, in the sense that its image is a finite subgroup of $\mathbb{G}_m$. For such characters we have $\mathscr{F}(\mathsf{C},\lambda) = \mathscr{F}(\mathsf{C})$. We have computed the algebraic group $\mathsf{G}_{\mathsf{C}_d}$ and we have seen that it is of dimension two, containing one copy of the multiplicative group $\mathbb{G}_m$ and one copy of the additive group $\mathbb{G}_a$. Its subgroup $\mathsf{G}_{\mathsf{C}_d,+}$ containing the identity element acts in each leaf of $\mathscr{F}(\mathsf{C}_d)$. This will be used in §10.10 in order to transport $\mathscr{F}(\mathsf{C}_d)$ into a four-dimensional ambient space.

10.7 Modular curves as three-dimensional affine varieties

We recall from [Mov15b] the following definition.

Definition 10.1. The affine variety $V_d \subset \mathsf{T}_1 \times \mathsf{T}_2$ is the locus of isogenies

$$f : X_1 \to X_2, \quad \deg(f) = d, \quad f^*\omega_2 = d \cdot \omega_1, \quad f^*\alpha_2 = \alpha_1. \tag{10.14}$$

Here, $f^* : H^1_{\mathrm{dR}}(X_2) \to H^1_{\mathrm{dR}}(X_1)$ is the map induced in de Rham cohomologies.

For an isogeny f as above let also $f_* : H_1(X_1,\mathbb{Z}) \to H_1(X_2,\mathbb{Z})$ be the map induced in homologies. Note that the definition in [Mov15b] is done using $f^*\omega_2 = \omega_1$ which is slightly different.

Proposition 10.5. *The graph of an isogeny $f : X_1 \to X_2$ is represented by the homology class*

$$\delta_1 \otimes f_*\gamma_1 - \gamma_1 \otimes f_*\delta_1 \in H_2(X_1 \times X_2,\mathbb{Z})_j, \tag{10.15}$$

(up to $\pm$ sign) and its Poincaré dual is

$$d \cdot \omega_1 \otimes \alpha_2 - \alpha_1 \otimes \omega_2 \in H^2_{\mathrm{dR}}(X_1 \times X_2)_j.$$

Proof. Let $\delta \in H_2(X_1 \times X_2,\mathbb{Z})_j$ be the homology class of the image of the map $X_1 \to X_1 \times X_2$, $x \mapsto (x, f(x))$. We have

$$\frac{1}{2\pi i}\left[\int_\delta \alpha_1 \otimes \alpha_2, \int_\delta \alpha_1 \otimes \omega_2, \int_\delta \omega_1 \otimes \alpha_2, \int_\delta \omega_1 \otimes \omega_2\right] = [0, d, -1, 0],$$

where we have used $\int_{X_1} \alpha_1 \wedge \omega_1 = 2\pi i$ and (10.14). For instance,

$$\int_\delta \alpha_1 \otimes \alpha_2 = \int_{X_1} \alpha_1 \wedge f^*\alpha_2 = \int_{X_1} \alpha_1 \wedge \alpha_1 = 0. \qquad \square$$

Proposition 10.6. *The set V_d is a leaf of $\mathscr{F}(\mathsf{C}_d)$.*

Proof. Note that the projection $\pi_i : V_d \to \mathsf{T}_i$, $i = 1, 2$ is a $\sigma_1(d) := \sum_{i|d} i$ degree covering of T_i and V_d is a subvariety of $\mathsf{T}_1 \times \mathsf{T}_2$ of dimension 3. The statement follows from the period computation in Proposition 10.5 and the definition of a modular foliation in Proposition 6.1. □

10.8 Eisenstein series and modular curves

Consider the Eisenstein series $s_i = E_{2i}$ in (1.2). Recall that the map $s = (s_1, s_2, s_3) : \mathbb{H} \to \mathbb{C}^3$, up to multiplication of s_i's with explicit rational numbers, is a solution of the Ramanujan vector field. We denote by

$$\psi(d) := d \prod_{p|d} \left(1 + \frac{1}{p}\right)$$

the Dedekind ψ function, where p runs through primes p dividing d.

Theorem 10.7. *For $i = 1, 2, 3$ and $d \in \mathbb{N}$, there is a homogeneous polynomial $P_{d,i}$ of degree $i \cdot \psi(d)$ in the weighted ring*

$$\mathbb{Q}[t_i, s_1, s_2, s_3], \text{ weight}(t_i) = i, \text{ weight}(s_j) = j, \ j = 1, 2, 3 \qquad (10.16)$$

and monic in the variable t_i such that $t_i(\tau) := s_i(d \cdot \tau), s_1(\tau), s_2(\tau), s_3(\tau)$ satisfy the algebraic relation:

$$P_{d,i}(t_i, s_1, s_2, s_3) = 0.$$

Moreover, for $i = 2, 3$ the polynomial $P_{d,i}$ does not depend on s_1.

Note that the map $\tau \mapsto d \cdot \tau$ in q variable is $q \mapsto q^d$. The above theorem for $i = 2, 3$ must be considered classical, however, for $i = 1$ it has been mainly ignored in the literature. For a proof see [Mov15b, Theorem 1]. We consider s_i, t_i, $i = 1, 2, 3$ as indeterminate variables and for simplicity we do not introduce new notations in order to distinguish them from the Eisenstein series. We regard $(t, s) = (t_1, t_2, t_3, s_1, s_2, s_3)$ as coordinates of the affine variety $\mathbb{A}^6_{\mathfrak{k}}$, where $\mathfrak{k}$ is any field of characteristic zero and not necessarily algebraically closed. It can be shown that the curve given by $I_{d,2} = I_{d,3} = 0$ in the weighted projective space $\mathbb{P}^{(2,3,2,3)}_{\mathbb{C}}$ with the coordinates (t_2, t_3, s_2, s_3)

is a singular model for the modular curve

$$X_0(d) := \Gamma_0(d)\backslash(\mathbb{H} \cup \mathbb{Q}), \quad \text{where}$$

$$\Gamma_0(d) := \left\{ \begin{bmatrix} a_1 & a_2 \\ a_3 & a_4 \end{bmatrix} \in \mathrm{SL}(2, \mathbb{Z}) \middle| a_3 \equiv 0 \pmod{d} \right\}.$$

Computing explicit equations for $X_0(d)$ in terms of the variables $j_1 = 1728\frac{t_2^3}{t_2^3 - 27t_3^2}$ and $j_2 = 1728\frac{s_2^3}{s_2^3 - 27s_3^2}$ has many applications in number theory and it has been done by many authors, see for instance [Yui78] and the references therein.

Theorem 10.8. *The variety V_d (see Definition 10.1) in $\mathsf{T}_1 \times \mathsf{T}_2$ is given by the zero set of the ideal*

$$\langle I_{d,1}, I_{d,2}, I_{d,3} \rangle \subset \mathsf{k}[s, t],$$

where $I_{d,i}$ is the pull-back of $P_{d,i}$ by the map

$$(t_1, t_2, t_3, s_1, s_2, s_3) \mapsto (d^{-1} \cdot t_1,\ d^{-2} \cdot t_2,\ d^{-3} \cdot t_3,\ s_1, s_2, s_3).$$

The proof can be found in [Mov15b].

10.9 Algebraic invariant sets

The main result of this section is Theorem 10.9. This is the first verification of a general conjecture which is formulated in Property 7.1. It turns out that the verification of this property involves the study of dynamics of modular foliations.

Theorem 10.9. *Any algebraic leaf of the foliation*

$$\mathscr{F}\left(h_1 + h_2,\ d \cdot f_1 + f_2,\ \ e_1 + d \cdot e_2\right), \quad d \in \mathbb{N}$$

in $\mathsf{T}_1 \times \mathsf{T}_2$ is of the form $L_{\check{d}}$, where $L_{\check{d}}$ parameterizes pairs of enhanced elliptic curves $(X_1, \alpha_1, \omega_1)$ and $(X_2, \alpha_2, \omega_2)$ with

$$f : X_1 \to X_2, \quad \deg(f) = \check{d}, \quad f^*\omega_2 = d^{\frac{1}{2}}\check{d}^{\frac{1}{2}}\omega_1, \quad f^*\alpha_2 = d^{-\frac{1}{2}}\check{d}^{\frac{1}{2}}\alpha_1.$$

In the coordinate system (t, s), this is the zero set of the ideal $\langle Q_1, Q_2, Q_3 \rangle \subset \mathsf{k}[s, t]$, where Q_i is the pull-back of $P_{d,i}$ by the map

$$(t_1, t_2, t_3, s_1, s_2, s_3) \mapsto (a_1 \cdot t_1,\ a_2 \cdot t_2,\ a_3 \cdot t_3,\ s_1, s_2, s_3),$$

and $a_i = d^{\frac{i}{2} - 2i}\check{d}^{\frac{i}{2}}$, $i = 1, 2, 3$.

Proof. Using (10.9) and Proposition 6.3 we know that under the map $\mathsf{T}_1 \times \mathsf{T}_2 \to \mathsf{T}_1 \times \mathsf{T}_2$, $(t,s) \mapsto (t,s) \bullet \mathsf{g}_{\check{d},d}^{-1}$, the foliation $\mathscr{F}(\mathsf{C}_{\check{d}})$ is mapped to $\mathscr{F}(\mathsf{C}_d)$. The first foliation has the algebraic leaf $V_{\check{d}}$ which is mapped to

$$L_{\check{d}} := V_{\check{d}} \bullet \mathsf{g}_{\check{d},d}^{-1}, \quad d \in \mathbb{N}. \tag{10.17}$$

Therefore, $L_{\check{d}}$'s are algebraic leaves of $\mathscr{F}(\mathsf{C}_d)$.

Now let us prove that $L_{\check{d}}$'s are the only algebraic leaves of $\mathscr{F}(\mathsf{C}_d)$. We do not have a purely algebraic proof for this theorem, in the sense that it does not work over an arbitrary field $\mathfrak{k}$. We consider the complex context $\mathfrak{k} = \mathbb{C}$ and study the dynamic of $\mathscr{F}(\mathsf{C}_d)$. Since the vector fields h, e, f are tangent to the discriminant locus $\Delta = 0$ in $\mathbb{A}^3_{\mathfrak{k}}$, we know that the modular foliation $\mathscr{F}(\mathsf{C}_d)$ is tangent to the hypersurfaces

$$\{\Delta = 0\} \times \mathbb{A}^3_{\mathfrak{k}}, \quad \mathbb{A}^3_{\mathfrak{k}} \times \{\Delta = 0\}$$

and hence $\Delta \times \mathbb{A}^3_{\mathfrak{k}}$, $\mathbb{A}^3_{\mathfrak{k}} \times \Delta$ are $\mathscr{F}(\mathsf{C}_d)$-invariant. Let L be an algebraic leaf of $\mathscr{F}(\mathsf{C}_d)$ in $\mathsf{T}_1 \times \mathsf{T}_2$. This implies that the closure $\bar{L}$ of L intersects these two hypersurfaces in $\mathrm{Sing}(\mathscr{F}(\mathsf{C}_d))$, which is given in (10.3). Note that a non-algebraic leaf can accumulate in both singularities and other leaves. Recall the notation of leaves of modular foliations in §6.3. Using this notation the set of algebraic leaves $L_{\check{d}}$, $\check{d} \in \mathbb{N}$ is the same as the set of $L_{\tilde{\mathsf{C}}}$, $[\tilde{\mathsf{C}}] \in \mathbb{P}^3(\mathbb{Q})$. Note that in the latter set, there are many repetitions. Our theorem follows from the next proposition. □

Proposition 10.10. *A leaf $L_{\tilde{\mathsf{C}}}$ of $\mathscr{F}(\mathsf{C}_d)$ in $\mathsf{T}_1 \times \mathsf{T}_2$ accumulates on another leaf in $\mathbb{A}^3_{\mathfrak{k}} \times \Delta \cup \Delta \times \mathbb{A}^3_k$ if and only if $[\tilde{\mathsf{C}}] \notin \mathbb{P}^3(\mathbb{Q})$.*

Proof. Let us prove the non-trivial direction $\Leftarrow$ of the proposition. We study the behavior of the leaves of $\mathscr{F}(\mathsf{C}_d)$ under the composition

$$\mathbb{A}^3_{\mathfrak{k}} \times \mathbb{A}^3_{\mathfrak{k}} \to (\mathbb{A}^3_{\mathfrak{k}} \times \mathbb{A}^3_{\mathfrak{k}})/\mathsf{G} \cong \mathbb{P}^1 \times \mathbb{P}^1 \cong (\mathrm{SL}(2,\mathbb{Z})\backslash\mathbb{H}^*) \times (\mathrm{SL}(2,\mathbb{Z})\backslash\mathbb{H}^*),$$

where $\mathbb{H}^* := \mathbb{H} \cup \mathbb{Q}$ and G is the algebraic group (10.5). The last isomorphism is obtained from the isomorphism of the period map in Theorem 9.2. Let us denote by $\infty := \mathrm{SL}(2,\mathbb{Z})\backslash\mathbb{Q}$ the cusp of $\mathrm{SL}(2,\mathbb{Z})\backslash\mathbb{H}^*$. A leaf $L_{\tilde{\mathsf{C}}}$ of the modular foliation $\mathscr{F}(\mathsf{C}_d)$ is mapped to a locally analytic curve X in $(\mathrm{SL}(2,\mathbb{Z})\backslash\mathbb{H}) \times (\mathrm{SL}(2,\mathbb{Z})\backslash\mathbb{H})$. If $L_{\tilde{\mathsf{C}}}$ is algebraic in $\mathbb{A}^3_{\mathfrak{k}} \times \mathbb{A}^3_{\mathfrak{k}}$ then so it is X in the compactified moduli $(\mathrm{SL}(2,\mathbb{Z})\backslash\mathbb{H}^*) \times (\mathrm{SL}(2,\mathbb{Z})\backslash\mathbb{H}^*)$. From this we are going to conclude that $\tilde{\mathsf{C}} \in \mathbb{P}^3(\mathbb{Q})$. Let $\tilde{\mathsf{C}} = [a_1, a_2, a_3, a_4]^{\mathsf{tr}}$. The definition of $L_{\tilde{\mathsf{C}}}$ includes the equality $\int_\delta \alpha_1 \otimes \alpha_2 = 0$, where

$$\delta := a_1\delta_1 \otimes \delta_2 + a_2\delta_1 \otimes \gamma_2 + a_3\gamma_1 \otimes \delta_2 + a_4\gamma_1 \otimes \gamma_2.$$

This implies that for $\tau_k = \frac{\int_{\delta_k} \alpha_k}{\int_{\gamma_k} \alpha_k}$, $k = 1, 2$ we have

$$\tau_2 = A(\tau_1),\ A := \begin{bmatrix} a_1 & a_3 \\ -a_2 & -a_4 \end{bmatrix}, \tag{10.18}$$

where $A(\tau_1)$ is the Möbius transformation of τ_1. By an analytic continuation argument the equality (10.18) can be transformed into $\tau_2 = B_1AB_2\tau_1$ for $B_1, B_2 \in \mathrm{SL}(2, \mathbb{Z})$. We know that X is a closed algebraic curve in the compactification $(\mathrm{SL}(2, \mathbb{Z})\backslash\mathbb{H}^*) \times (\mathrm{SL}(2, \mathbb{Z})\backslash\mathbb{H}^*)$. From this we only use the following: for a fixed $[\tau_1] \in \mathrm{SL}(2, \mathbb{Z})\backslash\mathbb{H}^*$ there are finite number of $[\tau_2] \in \mathrm{SL}(2, \mathbb{Z})\backslash\mathbb{H}^*$ with (10.18). This is translated into the following group theory problem:

$$\#\mathrm{SL}(2, \mathbb{Z})\backslash (\mathrm{SL}(2, \mathbb{Z}) \cdot A \cdot \mathrm{SL}(2, \mathbb{Z})) < \infty. \tag{10.19}$$

This happens if and only if the matrix A, up to multiplication with a constant, has rational entries. □

A better analyzing of the above proof reveals that the curves X for $\tilde{\mathsf{C}} \notin \mathbb{P}^3(\mathbb{Q})$ has a very complicated behavior in the compactification $\mathbb{P}^1 \times \mathbb{P}^1$ of the classical moduli of two elliptic curves. Such curves can be also described by differential equations, see for instance the box equation in [CDLW09, DMWH16], however, attaching a geometry to such differential equations seems to be hopeless.

One of the most important pieces in the proof of Theorem 10.9 is (10.19). For further generalization, the following problem must play an important role. For a discrete subgroup $\Gamma \subset \mathrm{GL}(n, \mathbb{C})$ classify all $n \times n$ matrices with complex entries such that

$$\#\Gamma\backslash (\Gamma \cdot A \cdot \Gamma) < \infty. \tag{10.20}$$

10.10 A vector field in four dimensions

Recall the notations in §10.6 and C_d in (10.8). From now on we take $\mathsf{C} = \mathsf{C}_d$ with $d = 1$. The foliation $\mathscr{F}(\mathsf{C})$ induces a one-dimensional foliation in the quotient space

$$(\mathbb{A}^3_{\mathfrak{k}} \times \mathbb{A}^3_{\mathfrak{k}})/\mathsf{G}_{\mathsf{C}_d,+} \cong \mathbb{P}^{2,3,2,3,1}, \ (t,s) \mapsto [t_2; t_3; s_2; s_3; t_1 - s_1], \qquad (10.21)$$

which is of dimension four. Since the foliation $\mathscr{F}(\mathsf{C})$ is given by the vector fields in (10.10) and from this $e_1 + e_2$ and $h_1 + h_2$ are tangent to the orbits of $\mathsf{G}_{\mathsf{C}_d,+}$, only $f_1 + f_2$ will contribute to the tangent space of $\mathscr{F}(\mathsf{C})$ as a foliation $\mathscr{F}$ in the quotient space (10.21). In order to compute this foliation explicitly we take the affine coordinate system

$$(x_2, x_3, y_2, y_3) := \left(\frac{t_2}{(t_1 - s_1)^2}, \ \frac{t_3}{(t_1 - s_1)^3}, \ \frac{s_2}{(t_1 - s_1)^2}, \ \frac{s_3}{(t_1 - s_1)^3}\right)$$

and consider two Ramanujan differential equations (9.11) in (t_1, t_2, t_3) and (s_1, s_2, s_3) variables. We make derivations of x_i and y_i variables and divide them over $t_1 - s_1$ and conclude that the foliation $\mathscr{F}$ in the quotient space (10.21) and in the affine chart $(x_2, x_3, y_2, y_3) \in \mathbb{C}^4$ is given by the quadratic vector field:

$$\left(2x_2 - 6x_3 + \frac{1}{6}(x_2 - y_2)x_2\right)\frac{\partial}{\partial x_2} + \left(3x_3 - \frac{1}{3}x_2^2 + \frac{1}{4}(x_2 - y_2)x_3\right)\frac{\partial}{\partial x_3}$$
$$- \left(2y_2 - 6y_3 + \frac{1}{6}(y_2 - x_2)y_2\right)\frac{\partial}{\partial y_2} - \left(3y_3 - \frac{1}{3}y_2^2 + \frac{1}{4}(y_2 - x_2)y_3\right)\frac{\partial}{\partial y_3}. \qquad (10.22)$$

There are no x_1, y_1 variables and the indices for x_i and y_i are chosen because of their natural weights. For further details on this vector field and its foliation see [Mov18].

10.11 Elliptic curves with complex multiplication

We discuss the case $X \times X$, where X is an elliptic curve. In this context, for the first time we will have isolated Hodge classes corresponding to isogenies, and trivial modular foliations. We consider the reduced de Rham cohomology and homology as follow:

$$H^2_{\mathrm{dR}}(X \times X)_j = \frac{H^1_{\mathrm{dR}}(X) \otimes_{\mathfrak{k}} H^1_{\mathrm{dR}}(X)}{\mathfrak{k}(\alpha_1 \otimes \omega_1 - \omega_1 \otimes \alpha_1)},$$

$$H_2(X \times X, \mathbb{Z})_j = (\delta_1 \otimes \gamma_1 - \gamma_1 \otimes \delta_1)^{\perp} \text{ in } H_1(X_1, \mathbb{Z}) \otimes_{\mathbb{Z}} H_1(X_2, \mathbb{Z}).$$

Here $\delta_1 \otimes \gamma_1 - \gamma_1 \otimes \delta_1$ is the homological cycle corresponding to the diagonal map $X \hookrightarrow X \times X$ (up to sign) and $\alpha_1 \otimes \omega_1 - \omega_1 \otimes \alpha_1$ is its Poincaré dual. In the definition of the reduced de Rham cohomology we can neglect $\otimes$ and assume that the products are commutative. We consider the basis $\alpha_1\alpha_1, \alpha_1\omega_1, \omega_1\omega_1$ for $H^2_{\mathrm{dR}}(X \times X)_j$. The intersection form in this basis is given by

$$\Phi_0 := \begin{bmatrix} 0 & 0 & 1 \\ 0 & -1 & 0 \\ 1 & 0 & 0 \end{bmatrix}.$$

The enhanced moduli space of $X \times X$ is the same as the moduli of elliptic curves. We get the following representation of the algebraic group G of elliptic curves

$$\begin{bmatrix} k & k' \\ 0 & k^{-1} \end{bmatrix} \mapsto \begin{bmatrix} k^2 & kk' & k'^2 \\ 0 & 1 & 2kk' \\ 0 & 0 & k^{-2} \end{bmatrix}.$$

The orbits of G acting on $\mathbb{A}^3_k$ have representatives

$$\begin{bmatrix} 1 \\ 0 \\ 0 \end{bmatrix}, \begin{bmatrix} 0 \\ 0 \\ 1 \end{bmatrix}, \begin{bmatrix} 0 \\ a \\ 0 \end{bmatrix}, \ a \in \mathfrak{k}. \tag{10.23}$$

We have also

$$\mathsf{A}_e = \begin{bmatrix} 0 & 2 & 0 \\ 0 & 0 & 1 \\ 0 & 0 & 0 \end{bmatrix}, \ \mathsf{A}_f = \begin{bmatrix} 0 & 0 & 0 \\ 1 & 0 & 0 \\ 0 & 2 & 0 \end{bmatrix}, \ \mathsf{A}_h = \begin{bmatrix} 1 & 0 & 0 \\ 0 & -1 & 0 \\ 0 & 0 & 1 \end{bmatrix}.$$

Let C be the last vector in (10.23) with $a \neq 0$. There is no linear relations between $\mathsf{A}_e\mathsf{C}$, $\mathsf{A}_f\mathsf{C}$, $\mathsf{A}_h\mathsf{C}$ and so the modular foliation $\mathscr{F}(\mathsf{C})$ is of dimension zero.

Chapter 11

Abelian Varieties

Le présent Mémoire est consacré à l'edtude des surfaces remarquables, introduites dans la Science par M. Picard, et pour lesquelles les coordonnées non homogènes d'un point quelconque peuvent s'exprimer en fonctione uniformc, quadruplcmcnt pćriodiquc, dc dcux paramètrcs [...] nous désignerons ces surfaces sous le nom de surfaces hyperelliptiques (M. G. Humbert in [Hum93]).

11.1 Introduction

The quotation above from Humbert's article refers to the first appearances of abelian surfaces in mathematics under the name "hyperelliptic surfaces". The term soon was discarded as "after Weil's books, the term 'abelian varieties' had taken precedence", (D. Mumford, personal communication, January 15, 2016). We develop the theory of modular foliations and vector fields on the moduli space T of enhanced principally polarized abelian varieties. If there is no danger of confusion we simply use "abelian variety", being clear that it is polarized. The literature on abelian varieties is huge and the reader may consult [LB92, Mil86, Mum66, Mum08, FC90] for missing details.

The content of present chapter was first formulated in [Mov13, §4.1] and later the author gave the idea of the proof of existence and uniqueness of modular vector fields in a complex geometric context in his personal webpage. T. J. Fonseca in [Fon21] developed these initial ideas in the framework of algebraic stacks and gave applications in transcendental number theory, similar to applications of Ramanujan vector field by P. Nesterenko in [NP01].

11.2 De Rham cohomologies

Let X be an abelian variety of dimension n over a field $\mathfrak{k}$ of characteristic 0. The de Rham cohomology ring $(H^*_{\mathrm{dR}}(X), \cup)$ of X is freely generated by $H^1_{\mathrm{dR}}(X)$:

$$H^m_{\mathrm{dR}}(X) = \bigwedge_{i=1}^{m} H^1_{\mathrm{dR}}(X), \quad m = 1, 2, \ldots, 2n. \tag{11.1}$$

Therefore, the Betti numbers b_m, $m = 0, 1, \ldots, 2n$ are respectively

$$1, \binom{2n}{1}, \ldots, \binom{2n}{m}, \ldots, \binom{2n}{2n-1}, 1.$$

We choose a basis $\alpha_1, \alpha_2, \ldots, \alpha_{2n}$ for $H^1_{\mathrm{dR}}(X)$ such that $\alpha_1, \alpha_2, \ldots, \alpha_n$ form a basis of $F^1 H^1_{\mathrm{dR}}(X)$. A basis of $H^m_{\mathrm{dR}}(X)$ is given by

$$\alpha_i := \alpha_{i_1} \wedge \alpha_{i_2} \wedge \cdots \wedge \alpha_{i_m}, \quad 1 \le i_1 < i_2 < \cdots < i_m \le 2n.$$

The quotient F^p/F^{p+1} in the Hodge filtration of the mth de Rham cohomology $H^m_{\mathrm{dR}}(X)$ has the basis α_i with

$$1 \le i_1 < i_2 < \cdots i_p \le n < i_{p+1} < \cdots < i_m \le 2n.$$

Therefore, the Hodge numbers $\mathsf{h}^{m,0}$, $\mathsf{h}^{m-1,1}, \cdots, \mathsf{h}^{0,m}$ are respectively

$$\binom{n}{m}\binom{n}{0}, \binom{n}{m-1}\binom{n}{1}, \ldots, \binom{n}{0}\binom{n}{m}.$$

In particular, the Hodge numbers in the middle cohomology $m = n$ are

$$\binom{n}{0}^2, \binom{n}{1}^2, \ldots, \binom{n}{n}^2.$$

11.3 Polarization

The embedding $X \subset \mathbb{P}^N$ gives us an element $\theta \in H^2_{\mathrm{dR}}(X)$ which we have called it the polarization. It gives us the bilinear map

$$\langle \cdot, \cdot \rangle : H^1_{\mathrm{dR}}(X) \times H^1_{\mathrm{dR}}(X) \to \mathfrak{k}, \ \langle \alpha, \beta \rangle := \mathrm{Tr}\left(\alpha \cup \beta \cup \theta^{n-1}\right), \tag{11.2}$$

which is non-degenerate.

Proposition 11.1. *In some basis α_i, $i = 1, 2, \ldots, 2n$ of $H^1_{\mathrm{dR}}(X)$ with $\alpha_1, \alpha_2, \ldots, \alpha_n$ a basis of $F^1 H^1_{\mathrm{dR}}(X)$, we have*

$$\theta = \alpha_1 \wedge \alpha_{n+1} + \alpha_2 \wedge \alpha_{n+3} + \cdots + \alpha_n \wedge \alpha_{2n} \tag{11.3}$$

and in particular, the intersection form (11.2) is of the form

$$[\langle\alpha_i,\alpha_j\rangle]=\Phi,\quad \textit{where } \Phi:=\begin{bmatrix}0 & I_n\\ -I_n & 0\end{bmatrix}. \tag{11.4}$$

Proof. Let α_i, $i=1,2,\ldots,2n$ be as in §11.2. Since $\theta\in F^1H^2_{\rm dR}(X)$ we can always write θ in the form $\theta=\alpha_1\wedge\beta_{n+1}+\alpha_2\wedge\beta_{n+2}+\cdots+\alpha_n\wedge\beta_{2n}$ for some $\beta_{n+i}\in H^1_{\rm dR}(X)$. We prove that α_i,β_{n+i}'s form a basis of $H^1_{\rm dR}(X)$. This follows from the equality

$$\theta^n=n!\alpha_1\wedge\beta_{n+1}\wedge\alpha_2\wedge\beta_{n+2}\wedge\cdots\wedge\alpha_n\wedge\beta_{2n}.$$

Therefore, we can make a change of basis and we set $\beta_i=\alpha_{n+i}$. We have

$$\alpha_i\wedge\alpha_j\wedge\theta^{n-1}=\begin{cases}0, & \text{if } |i-j|\neq n\\ \frac{1}{n}\theta^n, & \text{if } j=n+i\end{cases}.$$

We further make a change of of basis replacing α_i, $i=1,2,\ldots,n$ with $\frac{n}{\deg(X)}\alpha_i$ and we get the desired matrix format (11.4). Note that we have used

$$\deg(X)=\mathrm{Tr}(\theta^n). \tag{11.5}$$

□

Proposition 11.2. *An enhanced principally polarized abelian variety X is uniquely determined by*

$$X,[\alpha_1,\alpha_2,\ldots,\alpha_{2n}]$$

such that α_i, $i=1,2,\ldots,2n$ (respectively, $i=1,2,\ldots,n$) form a basis of $H^1_{\rm dR}(X)$ (respectively, $F^1H^1_{\rm dR}(X)$) and $[\langle\alpha_i,\alpha_j\rangle]=\Phi$, where $\langle\cdot,\cdot\rangle$ and θ is defined by (11.2) and (11.3).

Recall Definition 3.7. In the case of abelian varieties we will not need to fix a marked variety, however, if the reader insists he may try to prove the following: A CM abelian variety has a Hodge structure defined over $\bar{\mathfrak{k}}$.

Remark 11.1. An abelian variety is an R-variety in the sense of Definition 2.25 in §2.17. For this we take $m=1$ and $k=0$. Note that the map (2.47) in this case is

$$\delta_0:\Delta\to\mathrm{Hom}\left(H^0(X,\Omega^1_X),\ H^1(X,\Omega^0_X)\right),$$

which is not surjective, and hence it is not an R-variety in the sense of Definition 2.26. Its image is characterized by the equality

$$Q_1(\delta_0(v)(\alpha),\beta)+Q_0(\alpha,\delta_0(v)(\beta))=0,\quad \alpha,\beta\in H^0(X,\Omega^1_X),\quad v\in\Delta,$$

where Q is defined in (2.35).

11.4 Algebraic group

The algebraic group G for polarized abelian varieties is

$$\mathsf{G}=\left\{\begin{bmatrix}k & k'\\ 0 & k^{-\mathsf{tr}}\end{bmatrix}\in \mathrm{GL}(2n,\mathfrak{k})\,\middle|\, kk'^{\mathsf{tr}}=(kk'^{\mathsf{tr}})^{\mathsf{tr}}\right\}\subset \mathrm{Sp}(2n,\mathfrak{k}),$$

where the form of $\mathsf{g}\in\mathsf{G}$ is derived from the facts that g respects the Hodge filtration, and hence $\mathsf{g}^{21}=0$, and $\mathsf{g}^{\mathsf{tr}}\Phi\mathsf{g}=\Phi$. The algebraic group G is of dimension

$$\dim(\mathsf{G})=2n^2-\frac{n(n-1)}{2}. \tag{11.6}$$

It is also called Siegel parabolic subgroup of $\mathrm{Sp}(4,\mathfrak{k})$. The Lie group of G is given by

$$\mathrm{Lie}(\mathsf{G}):=\left\{\begin{bmatrix}k & k'\\ 0 & -k^{\mathsf{tr}}\end{bmatrix}\in \mathrm{Mat}(2n,\mathfrak{k})\,\middle|\, k'^{\mathsf{tr}}=k'\right\}.$$

This is derived from the fact that $\mathfrak{g}\in\mathrm{Lie}(\mathsf{G})$ is upper triangular with respect to the Hodge blocks and $\mathfrak{g}^{\mathsf{tr}}\Phi+\Phi\mathfrak{g}=0$ for $\mathfrak{g}\in\mathrm{Lie}(\mathsf{G})$.

Remark 11.2. The Siegel parabolic group G over the ring $\mathbb{Z}/N\mathbb{Z}$, $N\in\mathbb{N}$ appears in a natural way as quotient of $\mathrm{Sp}(2n,\mathbb{Z})$ by its Siegel congruence subgroup:

$$\Gamma_0^{2n}(N):=\left\{A\in\mathrm{Sp}(2n,\mathbb{Z})\,\middle|\, A\equiv_N\begin{bmatrix}* & *\\ 0 & *\end{bmatrix}\right\}\subset\mathrm{Sp}(2n,\mathbb{Z}).$$

Note that G is formulated in an algebraic context (algebraic de Rham cohomology, etc.), however, $\Gamma_0^{2n}(N)$ is formulated in a topological context (monodromy, homology etc.). For some properties of the Siegel congruence group, see for instance [Shu18].

11.5 Homology group

Let $X_0 \subset \mathbb{P}^n$ be an abelian variety of dimension n over complex numbers. As a complex manifold X_0 is biholomorphic to a complex torus $\mathbb{C}^n/\Lambda$, where $\Lambda = \mathbb{Z}\omega_1 + \mathbb{Z}\omega_2 + \cdots + \mathbb{Z}\omega_{2n}$ is a lattice. We take a basis δ_i, $i = 1, 2, \ldots, 2n$ of $H_1(X_0, \mathbb{Z})$. For instance, δ_i is the loop in X_0 obtained by the identification of the initial and end points of the vector ω_i. A basis of $H_m(X_0, \mathbb{Z})$ is given by

$$\delta_{i_1} \wedge \delta_{i_2} \wedge \cdots \wedge \delta_{i_m}, \quad i_1 < i_2 < \cdots < i_m,$$

which are diffeomorphic to m-dimensional real tori. Note that wedge product in homology is a special feature of tori, and for an arbitrary variety we do not have such a natural structure. We assume that

$$\Delta := \delta_1 \wedge \delta_2 \wedge \cdots \wedge \delta_{2n}$$

is a basis of $H_{2n}(X_0, \mathbb{Z})$ induced by the orientation of X_0 due to its complex structure. This gives us an isomorphism

$$\mathrm{Tr} : H_{2n}(X_0, \mathbb{Z}) \to \mathbb{Z}, \quad \mathrm{Tr}(\alpha) := n, \quad \text{where } \alpha := n\Delta.$$

This data is enough to describe the intersection $\delta_i \cdot \delta_j$ of cycles $\delta_i \in H_{m_1}(X_0, \mathbb{Z})$ and $\delta_j \in H_{m_2}(X_0, \mathbb{Z})$. The intersection $\delta_i \cdot \delta_j$ is zero if the union of ingredient sets of δ_i and δ_j is not $\{\delta_1, \delta_2, \ldots, \delta_{2n}\}$. In particular, this happens for $m_1 + m_2 < 2n$. In other cases it is an element in $H_{m_1+m_2-2n}$ obtained by removing Δ from the ordered set of ingredients of δ_i and δ_j and then taking the wedge product of the rest. We have a natural isomorphism

$$H_m(X_0, \mathbb{Z}) \to H_{2n-m}(X_0, \mathbb{Z}), \quad \delta_i \mapsto \widehat{\delta_i}, \tag{11.7}$$

where $\widehat{\delta_i}$ is obtained by removing δ_i from Δ, without changing the order of its ingredient elements. This resembles a kind of Hard Lefschetz theorem over $\mathbb{Z}$ and without using a polarization element. This map transforms the wedge product into intersection of cycles, that is,

$$\widehat{\delta_i \wedge \delta_j} = \widehat{\delta_i} \cdot \widehat{\delta_j}. \tag{11.8}$$

Let $Y_0 \subset X_0$ be a hyperplane section of X_0 and

$$[Y_0] = \sum_{1 \leq i < j \leq 2n} a_{ij}(-1)^{i+j-1} \widehat{\delta_i \wedge \delta_j} \in H_{2n-2}(X_0, \mathbb{Z}), \quad a_{ij} \in \mathbb{Z}$$

be the induced homology class. In $H_1(X_0,\mathbb{Z})$ we get an anti-symmetric bilinear map

$$H_1(X_0,\mathbb{Z})\times H_1(X_0,\mathbb{Z})\to\mathbb{Z},\quad \langle\delta_i,\delta_j\rangle:=\mathrm{Tr}(\delta_i\wedge\delta_j\wedge[Y_0])=a_{ij}. \tag{11.9}$$

In cohomological terms, we have $u_0:=[Y_0]^{\mathsf{pd}}\in H^2(X_0,\mathbb{Z})$ and (11.9) is obtained from

$$u_0\in H^2(X_0,\mathbb{Z})\cong \mathrm{Hom}(H_1(X,\mathbb{Z})\wedge H_1(X_0,\mathbb{Z}),\mathbb{Z}).$$

The data $(H_1(X_0,\mathbb{Z}),\mathrm{Tr},\langle\cdot,\cdot\rangle)$ is equivalent to the whole homology ring with a polarization element. By abuse of notation we also call (11.9) the polarization. We also define

$$\Gamma_{\mathbb{Z}}:=\mathrm{Aut}\left(H_1(X_0,\mathbb{Z}),\mathrm{Tr},\langle\cdot,\cdot\rangle\right). \tag{11.10}$$

In a basis of $H_1(X_0,\mathbb{Z})$ its matrix is of the format

$$\Psi=\begin{bmatrix}0 & D_n\\ -D_n & 0\end{bmatrix},\quad D_n=\mathrm{diag}(d_1,d_2,\ldots,d_n),\quad d_1\mid d_2\mid\cdots\mid d_n$$

see for instance the references in [LB92, §3.1].

Definition 11.1. We say that X_0 is principally polarized if all the integers d_i's are equal to 1. In other words, the bilinear map $\langle\cdot,\cdot\rangle$ has a symplectic basis.

For principally polarized abelian varieties we have

$$\Gamma_{\mathbb{Z}}=\mathrm{Sp}(2n,\mathbb{Z})=\left\{\begin{bmatrix}a & b\\ c & d\end{bmatrix}\in\mathrm{GL}(2n,\mathbb{Z})\,\middle|\, ab^{\mathsf{tr}}=ba^{\mathsf{tr}},\, cd^{\mathsf{tr}}=dc^{\mathsf{tr}},\, ad^{\mathsf{tr}}-bc^{\mathsf{tr}}=I_n\right\}. \tag{11.11}$$

In the case of abelian varieties, we redefine the notion of Poincaré duals.

Definition 11.2. The Poincaré dual of $\delta\in H_1(X_0,\mathbb{Z})$ is an element $\delta^{\mathsf{pd}}\in H^1(X_0,\mathbb{Q})$ such that

$$\langle\omega,\delta^{\mathsf{pd}}\rangle=\frac{1}{2\pi i}\int_{\delta}\omega,\quad\forall\omega\in H^1_{\mathrm{dR}}(X_0), \tag{11.12}$$

where the bilinear map $\langle\cdot,\cdot\rangle$ is defined in (11.2).

There are two important issues which must be observed. First, since

$$H^1(X_0, \mathbb{Q}) \cong H^{2n-1}(X_0, \mathbb{Q}), \ \delta \mapsto \delta \cup u_0^{n-1}$$

might not be isomorphism using $\mathbb{Z}$ instead of $\mathbb{Q}$, the Poincaré dual may not live in $H^1(X_0, \mathbb{Q})$. Second, the bilinear map (11.9) is not necessarily dual to (11.2), that is, $\langle \delta_1^{\mathsf{pd}}, \delta_2^{\mathsf{pd}} \rangle$ and $\langle \delta_1, \delta_2 \rangle$ might be different. In practice, we will need the first bilinear product and not the second one which is commonly used in the literature. Therefore,

$$\Psi := [\langle \delta_i^{\mathsf{pd}}, \delta_j^{\mathsf{pd}} \rangle] \tag{11.13}$$

has entries in $\mathbb{Q}$. For principally polarized abelian varieties Ψ has entries in $\mathbb{Z}$ and the reason is follows. By the hard Lefschetz theorem, we have an embedding

$$H^1(X_0, \mathbb{Z}) \hookrightarrow H^{2n-1}(X_0, \mathbb{Z}), \quad \delta \mapsto \delta \cup u_0^{n-1}. \tag{11.14}$$

Proposition 11.3. *The abelian variety X_0 is principally polarized if the map (11.14) is surjective. In particular, the Hard Lefschetz theorem for principally polarized abelian varieties is valid over $\mathbb{Z}$, that is, $H^i(X_0, \mathbb{Z}) \to H^{2n-i}(X_0, \mathbb{Z})$, $\delta \mapsto \delta \cup u_0^{n-i}$, $i = 1, 2, \ldots, n$ is an isomorphism of $\mathbb{Z}$-modules.*

Note that the above map for $i = 0$ is not surjective and this is the only case in which the hard Lefschetz theorem fails.

Proof. The second part follows from the first part and the fact that $H^*(X_0, \mathbb{Z})$ is generated by $H^1(X_0, \mathbb{Z})$. □

The conclusion is that for principally polarized varieties we can assume that

$$\Psi := \begin{bmatrix} 0 & -I_n \\ I_n & 0 \end{bmatrix} \tag{11.15}$$

and we may discard the bilinear form (11.9) and replace it with $\langle \delta_i, \delta_j \rangle := \langle \delta_i^{\mathsf{pd}}, \delta_j^{\mathsf{pd}} \rangle$ which is automatically Poincaré dual to (11.2).

11.6 Generalized period domain

Recall the notion of a generalized period domain in Chapter 8. We take a basis α_i, $i = 1, 2, \ldots, 2n$ of $V_0 := H^1_{\mathrm{dR}}(X_0)$ as in §§11.2 and 11.3 and a basis δ_i, $i = 1, 2, \ldots, 2n$ of $V_{0,\mathbb{Z}} := H_1(X_0, \mathbb{Z})$ as in §11.5. Moreover, we only consider principally polarized abelian varieties as in Definition 11.1. For an element of the period domain $\mathsf{\Pi}$ we have

$$\mathsf{P} := \left[\int_{\delta_i} \alpha_j\right] = \begin{bmatrix} x_1 & x_2 \\ x_3 & x_4 \end{bmatrix},$$

where $x_i, i = 1, \ldots, 4$, are $n \times n$ matrices. We use Proposition 4.1 for $m = 1$, in order to find relations among the entries of x_i's. By Proposition 11.3 we have $\Psi = \Psi_{2n-1} = \tilde{\Psi}_{2n-1}$, where $\tilde{\Psi}_m$ is defined in §4.2. We have also $\Psi^{-\mathsf{tr}} = \Psi$, therefore

$$\Psi = \mathsf{P}^{\mathsf{tr}} \Psi \mathsf{P} \tag{11.16}$$

which is the following polynomial relations between periods

$$\begin{bmatrix} 0 & -I_n \\ I_n & 0 \end{bmatrix} = \begin{bmatrix} x_1^{\mathsf{tr}} & x_3^{\mathsf{tr}} \\ x_2^{\mathsf{tr}} & x_4^{\mathsf{tr}} \end{bmatrix} \begin{bmatrix} 0 & -I_n \\ I_n & 0 \end{bmatrix} \begin{bmatrix} x_1 & x_2 \\ x_3 & x_4 \end{bmatrix}$$
$$= \begin{bmatrix} x_3^{\mathsf{tr}} x_1 - x_1^{\mathsf{tr}} x_3 & x_3^{\mathsf{tr}} x_2 - x_1^{\mathsf{tr}} x_4 \\ x_4^{\mathsf{tr}} x_1 - x_2^{\mathsf{tr}} x_3 & x_4^{\mathsf{tr}} x_2 - x_2^{\mathsf{tr}} x_4 \end{bmatrix}.$$

We have also

$$[\langle \alpha_i, \bar{\alpha}_j^x \rangle] = \begin{bmatrix} x_1^{\mathsf{tr}} & x_3^{\mathsf{tr}} \\ x_2^{\mathsf{tr}} & x_4^{\mathsf{tr}} \end{bmatrix} \begin{bmatrix} 0 & -I_n \\ I_n & 0 \end{bmatrix} \begin{bmatrix} \bar{x}_1 & \bar{x}_2 \\ \bar{x}_3 & \bar{x}_4 \end{bmatrix}$$
$$= \begin{bmatrix} x_3^{\mathsf{tr}} \bar{x}_1 - x_1^{\mathsf{tr}} \bar{x}_3 & x_3^{\mathsf{tr}} \bar{x}_2 - x_1^{\mathsf{tr}} \bar{x}_4 \\ x_4^{\mathsf{tr}} \bar{x}_1 - x_2^{\mathsf{tr}} \bar{x}_3 & x_4^{\mathsf{tr}} \bar{x}_2 - x_2^{\mathsf{tr}} \bar{x}_4 \end{bmatrix}.$$

We can redefine $\mathsf{\Pi}$ through x_i's as follows.

Definition 11.3. The generalized period domain $\mathsf{\Pi}$ is the set of all $2n \times 2n$ matrices $\left[\begin{smallmatrix} x_1 & x_2 \\ x_3 & x_4 \end{smallmatrix}\right]$ satisfying the above properties:

$$x_3^{\mathsf{tr}} x_1 = x_1^{\mathsf{tr}} x_3, \; x_3^{\mathsf{tr}} x_2 - x_1^{\mathsf{tr}} x_4 = -I_n,$$

$$x_1, x_2 \in \mathrm{GL}(n, \mathbb{C}),$$

$$\sqrt{-1}(x_3^{\mathsf{tr}} \bar{x}_1 - x_1^{\mathsf{tr}} \bar{x}_3) \text{ is a positive matrix.}$$

We also define $\mathsf{U} := \Gamma_{\mathbb{Z}} \backslash \mathsf{\Pi}$, where $\Gamma_{\mathbb{Z}}$ is defined in (11.11).

The matrix $x := x_1 x_3^{-1}$ is well-defined and invertible and it satisfies the well-known Riemann relations.

Definition 11.4. The set of matrices $x \in \text{Mat}^{n\times n}(\mathbb{C})$ with

$$x^{\mathsf{tr}} = x, \ \text{Im}(x) \text{ is a positive matrix}$$

is called the Siegel upper half-space and is denoted by $\mathbb{H}_n$.

The group $\Gamma_\mathbb{Z}$ acts on $\mathbb{H}$ by

$$\begin{bmatrix} a & b \\ c & d \end{bmatrix} \cdot x = (ax+b)(cx+d)^{-1}, \ \begin{bmatrix} a & b \\ c & d \end{bmatrix} \in \Gamma_\mathbb{Z}, \ x \in \mathbb{H}_n$$

and we have the isomorphism

$$\mathsf{U}/\mathsf{G} \to \Gamma_\mathbb{Z} \backslash \mathbb{H}_n,$$

given by

$$\begin{bmatrix} x_1 & x_2 \\ x_3 & x_4 \end{bmatrix} \to x_1 x_3^{-1}.$$

Remark 11.3. The quotient $\Gamma_\mathbb{Z}\backslash\mathbb{H}_n$ is the moduli of principally polarized abelian varieties of dimension n, and its singular locus is fairly studied in the literature, see for instance [GMZ12] and the references therein. Since the action of $\Gamma_\mathbb{Z}$ in $\sqcap$ is free, the study of the singular locus of $\mathsf{U} := \Gamma_\mathbb{Z}\backslash\sqcap$ is reduced to the study of the singular locus of $\sqcap$ itself. Note that $\sqcap$ is an open set (usual topology) of the affine variety given by (11.16). For small n's the description of singularities of $\sqcap$ can be done by computer. For instance, for $n = 2$ we can use the procedure `slocus` of SINGULAR and verify that $\sqcap$ is smooth. The algebraic moduli T for $n = 2$ is explicitly constructed in [Mov20b].

11.7 Moduli of enhanced polarized tori

For a point x of $\sqcap$ we associate a triple $(A_x, \theta_x, \alpha_x)$ as follows. We have $A_x := \mathbb{C}^n/\Lambda_x$, where Λ_x is the $\mathbb{Z}$-submodule of $\mathbb{C}^n$ generated by the rows of x_1 and x_3. We have cycles $\delta_i \in H_1(A_x, \mathbb{Z})$, $i = 1, 2, \ldots, 2n$, which are defined by the property

$$\begin{bmatrix} \int_{\delta_i} dz_j \end{bmatrix} = \begin{bmatrix} x_1 \\ x_3 \end{bmatrix},$$

where z_j, $j = 1, 2, \ldots, n$, are linear coordinates of $\mathbb{C}^n$. There is a basis $\alpha_x = \{\alpha_1, \alpha_2, \ldots, \alpha_{2n}\}$ of $H^1_{\mathrm{dR}}(A_x)$ such that

$$\left[\int_{\delta_i} \alpha_j\right] = \begin{bmatrix} x_1 & x_2 \\ x_3 & x_4 \end{bmatrix}.$$

The polarization in $H_1(A_x, \mathbb{Z}) \cong \Lambda_x$ (which is defined by $[\langle \delta_i, \delta_j \rangle] = \Psi_0$) is an element $\theta_x \in H^2(A_x, \mathbb{Z}) = \bigwedge_{i=1}^2 \mathrm{Hom}(\Lambda_x, \mathbb{Z})$. It gives the following bilinear map:

$$\langle \cdot, \cdot \rangle : H^1_{\mathrm{dR}}(A_x) \times H^1_{\mathrm{dR}}(A_x) \to \mathbb{C}, \; \langle \alpha, \beta \rangle = \frac{1}{2\pi i} \int_{A_x} \alpha \cup \beta \cup \theta_x^{n-1},$$

which satisfies $[\langle \alpha_i, \alpha_j \rangle] = \Psi_0$.

The triple $(A_x, \theta_x, \alpha_x)$ that we constructed in the previous paragraph does not depend on the action of $\Gamma_{\mathbb{Z}}$ from the left on $\mathsf{\Pi}$. Therefore, for each $x \in \mathsf{U}$ we have constructed such a triple. In fact, U is the moduli space of enhanced principally polarized abelian varieties.

11.8 Moduli of enhanced abelian varieties

The moduli M of principally polarized abelian varieties of dimension n over $\mathfrak{k}$ is a quasi-projective variety over $\mathbb{Q}$, see for instance [MFK94]. It is also denoted by $\mathscr{A}_n$ in the literature. One first construct a family of principally polarized abelian varieties $Y_t \subset \mathbb{P}^N, t \in V$ for a full Hilbert scheme V, and then one proves that under the action of the reductive group $\mathbf{G} = \mathrm{GL}(N+1)$ all principally polarized abelian varieties are stable, and hence, the geometric quotient $\mathbf{G}\backslash V$ exists as a scheme over $\mathbb{Q}$. One can even construct the moduli of abelian schemes over a ring and construct the corresponding moduli stack, which is mainly known as Deligne–Mumford stack, however, due to the lack of motivation we avoid this in the present text and refer the reader to T. J. Fonseca's article [Fon21] and the references therein.

Let $\check{\mathsf{X}} \to \check{\mathsf{T}}$ be the full enhanced family constructed from $Y \to V$ in §§3.6 and 3.7. Theorem 3.5 gives us the coarse moduli space $\mathsf{T} := \mathbf{G}\backslash\check{\mathsf{T}}$ of enhanced principally polarized abelian varieties and we claim the following.

Conjecture 11.1. *There is a universal family $\mathsf{X} \to \mathsf{T}$ of enhanced principally polarized abelian varieties of dimension n.*

Note that the moduli space M is not fine, that is, there is no universal family over M, however, the moduli space of principally polarized abelian varieties with a full level n-structure $n \geq 3$ is fine, see [MFK94, Theorem 7.9; Fon21, §10.3] and so a similar statement as in Conjecture 11.1 in this case is an easy exercise. For the main purposes of the present text we do not need to solve Conjecture 11.1. Proposition 3.10 tells us that the Gauss–Manin connection matrix $\mathsf{A} := \mathsf{A}_m$ for $m = 1$ can be transported to T, and this is all what we need to reproduce the theory of modular vector fields and modular foliation in T. In the rest of this chapter by abuse of terminology, we will talk about a universal family of enhanced principally polarized abelian varieties X/T, knowing that only T and A exist as algebraic objects over $\mathbb{Q}$.

The dimension of the moduli of principally polarized abelian varieties is $\frac{n(n+1)}{2}$ and the dimension of the algebraic group G is given in (11.6). We conclude that the dimension of the moduli T of enhanced principally polarized abelian varieties is

$$\dim(\mathsf{T}) = 2n^2 + n.$$

Let us now consider the moduli space T over complex numbers and identify it with its points. The period map turns out to be

$$\mathsf{P} : \mathsf{T} \to \mathsf{U}, \quad t \mapsto \left[\int_{\delta_j} \alpha_i \right]_{2n \times 2n}, \tag{11.17}$$

where t is identified with (X, α) as in §11.3.

Theorem 11.4. *The period map* (11.17) *is a biholomorphism of complex manifolds.*

Proof. This follows from the fact that the classical period map $\mathsf{M} \to \mathrm{Sp}(2n, \mathbb{Z}) \backslash \mathbb{H}_n$ is a biholomorphism. □

11.9 Modular vector fields

Recall that the moduli space T and the Gauss–Manin connection matrix A are defined over $\mathbb{Q}$.

Theorem 11.5. *There are unique vector fields $\mathsf{v}_{ij}, i, j = 1, 2, \ldots, n, i \leq j$ defined over $\mathbb{Q}$ in the moduli space T of enhanced principally polarized abelian varieties such that*

$$\mathsf{A}_{\mathsf{v}_{ij}} = C_{ij}, \tag{11.18}$$

where $\mathsf{A}_{\mathsf{v}_{ij}}$ is the Gauss–Manin connection matrix composed with the vector field v_{ij} and C_{ij} is the constant matrix define as follows. All the entries of

C_{ij} are zero except $(i, n+j)$ and $(j, n+i)$ entries which are -1. In other words, the Gauss–Manin connection ∇ satisfies

$$\nabla_{\mathsf{v}_{ij}}\alpha_i = -\alpha_{n+j}, \quad \nabla_{\mathsf{v}_{ij}}\alpha_j = -\alpha_{n+i}, \qquad i, j = 1, 2 \ldots, n$$

and $\nabla_{\mathsf{v}_{ij}}\alpha_k = 0$ otherwise. Moreover, the Lie bracket of two such vector field is zero.

The main idea behind the proof of the above theorem is taken from [Mov13]. It also appeared in the author's webpage in 2013. Later, T. J. Fonseca in [Fon21] proved Theorem 11.5 in the framework of moduli stacks.

Proof. Using Theorem 11.4 which says that the period map is a biholomorphism, it is enough to prove the existence and uniqueness of v_{ij} in the period domain U. Note that the statement that v_{ij}'s are defined over $\mathbb{Q}$ follows from this and the fact that A is defined over $\mathbb{Q}$.

Since $\mathsf{U} := \Gamma_{\mathbb{Z}}\backslash\sqcap$, we prove the theorem in $\sqcap$ and moreover we prove that such vector fields are invariant under the action of $\Gamma_{\mathbb{Z}}$. The Gauss–Manin connection matrix in $\sqcap$ is of the form $\mathsf{A} := d\mathsf{P}^{\mathsf{tr}} \cdot \mathsf{P}^{-\mathsf{tr}}$. Therefore, we are looking for vector fields v_{ij} such that

$$d\mathsf{P}(\mathsf{v}_{ij}) = \mathsf{P} \cdot C_{ij}^{\mathsf{tr}}. \tag{11.19}$$

This equality can be used to define the vector field v_{ij} in the space of $2n \times 2n$ matrices. We need to prove that it is tangent to $\sqcap$ and it is also $\Gamma_{\mathbb{Z}}$-invariant. The first statement follows from (11.16) and the equality $C_{ij}^{\mathsf{tr}}\Phi + \Phi C_{ij} = 0$. The second statement follows from Proposition 8.10. The statement on Lie brackets of v_{ij} follows from Proposition 6.17 and the fact that the Lie bracket of matrices C_{ij}'s is zero. □

The vector field v_{ij} is the algebraic incarnation of derivation of Siegel modular forms with respect to τ_{ij}, where $\tilde{\tau} = [\tau_{ij}]$ is an element in the Siegel domain $\mathbb{H}_n$. We define the τ map:

$$\tau : \mathbb{H}_n \to \sqcap, \quad \tau(\tilde{\tau}) := \begin{pmatrix} \tilde{\tau} & -I_n \\ I_n & 0 \end{pmatrix}$$

and call its image in $\sqcap$ the τ locus. The Gauss–Manin connection restricted to the τ locus is given by $d\tau \cdot \tau^{-\mathsf{tr}}$. If we compose it with the vector field $\frac{\partial}{\partial \tau_{ij}}$ then we get the matrix C_{ij}. For simplicity, we write $\tilde{\tau} = \tau$; being clear in the text whether τ refers to a point in $\mathbb{H}_n$ or the τ map.

The Lie algebra of $\mathrm{Sp}(2n, \mathfrak{k})$ consists of $2n \times 2n$ matrices g with coefficients in $\mathfrak{k}$ such that $\mathsf{g}^{\mathsf{tr}}\Phi + \Phi\mathsf{g} = 0$, where Φ is the matrix in (11.4).

It follows that

$$\mathrm{Lie}(\mathrm{Sp}(2n,\mathsf{k})) = \left\{ \begin{bmatrix} a & b \\ c & -a^{\mathsf{tr}} \end{bmatrix} \in \mathrm{Mat}(2n,\mathbb{Q}) \middle| b^{\mathsf{tr}} = b, \quad c^{\mathsf{tr}} = c \right\}.$$

It is a direct sum of two sub-Lie algebra

$$\begin{bmatrix} 0 & 0 \\ c & 0 \end{bmatrix}, \quad c^{\mathsf{tr}} = c,$$

and

$$\begin{bmatrix} a & b \\ 0 & -a^{\mathsf{tr}} \end{bmatrix}, \quad a^{\mathsf{tr}} = a.$$

A basis of the first group is given by the matrices C_{ij} in Theorem 11.5. The second group will produce the constant vector fields v_{g}, $\mathsf{g} \in \mathrm{Lie}(\mathsf{G})$. Theorem 11.5 implies that families of abelian varieties have the constant Gauss–Manin connection.

11.10 Modular foliations

By definition the foliation $\mathscr{F}(2)$ is given by the $\mathbb{Q}$-vector space of vector fields v_{ij}, $1 \leq i \leq j \leq n$. The action of G on $\mathbb{A}^{2n}_{\mathsf{k}} - \{0\}$ has two orbits represented by

$$\mathsf{C}_1 = \begin{pmatrix} 1 \\ 0 \\ \vdots \\ 0 \\ 0 \end{pmatrix}, \quad \mathsf{C}_2 = \begin{pmatrix} 0 \\ 0 \\ \vdots \\ 0 \\ 1 \end{pmatrix}.$$

The modular foliation $\mathscr{F}(\mathsf{C}_1)$ is given by the vector fields

$$\mathsf{v}_{ij}, \quad 1 \leq i \leq j \leq n, \qquad \mathsf{v}_{\mathfrak{g}}, \quad \mathfrak{g}^{\mathsf{tr}}\mathsf{C}_1 = 0, \quad \mathfrak{g} \in \mathrm{Lie}(\mathsf{G}), \tag{11.20}$$

and the modular foliation $\mathscr{F}(\mathsf{C}_2)$ is given by

$$\mathsf{v}_{ij}, \quad 1 \leq i \leq j \leq n-1, \qquad \mathsf{v}_{\mathfrak{g}}, \quad \mathfrak{g}^{\mathsf{tr}}\mathsf{C}_2 = 0, \quad \mathfrak{g} \in \mathrm{Lie}(\mathsf{G}). \tag{11.21}$$

Note that $\mathsf{A}_{\mathsf{v}_{ij}}\mathsf{C}_1 = 0$ is automatically satisfied. Note also that (11.20) consists of $\mathfrak{g} \in \mathrm{Lie}(\mathsf{G})$ whose first column is zero.

In §11.2 we have computed Hodge numbers of abelian varieties. It turns out that both numbers b_n and $\mathsf{h}^{n,0} + \mathsf{h}^{n-1,1} + \cdots, \mathsf{h}^{\frac{n}{2},\frac{n}{2}}$ grow faster than

$\dim(\mathsf{T}) := 2n^2 + n$. This may suggest that for generic $\mathsf{C} \in \mathbb{A}_{\mathfrak{k}}^{\mathsf{b}_n}$ the modular foliation $\mathscr{F}(\mathsf{C})$ is zero-dimensional. However, this is not the case. Recall that for abelian varieties we have the constant Gauss–Manin connection

$$\mathrm{Der}(\mathsf{T}) = \mathscr{C}(\mathsf{X}/\mathsf{T}) \otimes_{\mathfrak{k}} \mathscr{O}_{\mathsf{T}}, \quad \mathscr{C}(\mathsf{X}/\mathsf{T}) \cong \mathrm{Lie}(\mathrm{Sp}(2n, \mathfrak{k})).$$

We consider the canonical left action of G on $\mathbb{A}_{\mathfrak{k}}^{2n}$. By taking the wedge product we gain a new action:

$$\mathsf{G} \times \mathbb{A}_{\mathfrak{k}}^{\mathsf{b}_m} \to \mathbb{A}_{\mathfrak{k}}^{\mathsf{b}_m}, \quad \mathbb{A}_{\mathfrak{k}}^{\mathsf{b}_m} := \bigwedge_{n=1}^{m} \mathbb{A}_{\mathfrak{k}}^{2n}. \tag{11.22}$$

For $m \geq 2$ the corresponding quotient

$$\mathsf{MF}_m := \mathsf{G} \backslash \mathbb{A}_{\mathfrak{k}}^{\mathsf{b}_m}$$

might not be discrete. This is the moduli of modular foliations defined in §6.7. We also need to consider the canonical action of the Lie algebra $\mathrm{Lie}(\mathrm{Sp}(2n, \mathfrak{k}))$ on $\mathbb{A}_{\mathfrak{k}}^{2n}$ and the resulting action on the wedge product space:

$$\mathrm{Lie}(\mathrm{Sp}(2n, \mathfrak{k})) \times \mathbb{A}_{\mathfrak{k}}^{\mathsf{b}_m} \to \mathbb{A}_{\mathfrak{k}}^{\mathsf{b}_m}, \quad (\mathfrak{g}, \mathsf{C}) \mapsto \mathfrak{g} \bullet \mathsf{C}. \tag{11.23}$$

For $\mathsf{C} \in \mathbb{A}_{\mathfrak{k}}^{\mathsf{b}_m}$ we have

$$\mathscr{F}(\mathsf{C}) = \mathscr{F}(\mathsf{v}_{\mathfrak{g}} \in \mathrm{Lie}(\mathrm{Sp}(2n, \mathfrak{k})) | \mathfrak{g} \bullet \mathsf{C} = 0).$$

Therefore, the codimension of the foliation $\mathscr{F}(\mathsf{C})$ is $\dim_{\mathfrak{k}}(\mathrm{Lie}(\mathrm{Sp}(2n, \mathfrak{k}) \bullet \mathsf{C})$ and so it is natural to define the following filtration of $\mathbb{A}_{\mathfrak{k}}^{\mathsf{b}_m}$:

$$\begin{aligned} &\mathsf{MF}_m(a) := \{\mathsf{C} \in \mathbb{A}_{\mathfrak{k}}^{\mathsf{b}_m} | \dim_{\mathfrak{k}}(\mathrm{Lie}(\mathrm{Sp}(2n, \mathfrak{k})) \bullet \mathsf{C}) \leq a\}, \\ &\cdots \subset \mathsf{MF}_m(a) \subset \mathsf{MF}_m(a+1) \subset \cdots \subset \mathsf{MF}_m(2n^2+n) = \mathbb{A}_{\mathfrak{k}}^{\mathsf{b}_m}. \end{aligned} \tag{11.24}$$

The set $\mathsf{MF}_m(a)$ is a closed algebraic subvariety of $\mathbb{A}_{\mathfrak{k}}^{\mathsf{b}_m}$. We take a basis $\mathfrak{g}_i$, $i = 1, 2, \ldots, 2n^2 + n$ of $\mathrm{Lie}(\mathrm{Sp}(2n, \mathfrak{k}))$ and define the $\mathsf{b}_m \times (2n^2 + n)$ matrix whose columns are $\mathfrak{g}_i\mathsf{C}$. The variety $\mathsf{MF}_m(a)$ consists of C's such that the matrix B has rank $\leq a$. Therefore, it is given by the determinants of all $(a+1) \times (a+1)$ submatrices of B. The study of Noether–Lefschetz loci in the case of abelian varieties is partially done in [DL90] and it is of interest to translate such results into the framework of modular foliations.

11.11 Differential Siegel modular forms

In this section, we recover the theory of Siegel modular forms in our framework. The reader is referred to [Kli90, Fre83, Maa71] for more information on Siegel modular forms.

Definition 11.5. The algebra of (algebraic) differential Siegel modular forms is by definition $\mathfrak{k}[\mathsf{T}]$, that is, that algebra of global regular functions on the moduli of enhanced principally polarized abelian varieties of dimension n.

In order to relate this definition to classical Siegel modular forms, we work over the field of complex numbers and consider the holomorphic map $\mathbb{H}_n \to \mathsf{T}$ which is the composition of the maps:

$$\mathbb{H}_n \to \Pi \to \mathsf{U} \stackrel{\mathsf{P}^{-1}}{\to} \mathsf{T}.$$

The first map is the τ maps given by

$$\tau \to \begin{bmatrix} \tau & -I_n \\ I_n & 0 \end{bmatrix}$$

and the second is the canonical map. Note that the period map P is a biholomorphism.

Definition 11.6. The algebra of (holomorphic) differential Siegel modular forms is by definition the pull-back of the algebra of algebraic differential Siegel modular forms $\mathfrak{k}[\mathsf{T}]$ by the map $\mathbb{H}_n \to \mathsf{T}$.

In order to recover the algebra of Siegel modular forms in the framework of enhanced principally polarized abelian varieties we need to introduce the following moduli.

Definition 11.7. We define S the moduli of (A, ω, θ), where A is a principally polarized abelian variety of dimension n and with the polarization θ, and ω is a holomorphic differential n-form on A.

In a similar way as in §11.8 we know that S is a moduli scheme over $\mathfrak{k}$. We have a canonical morphism of $\mathfrak{k}$ schemes

$$\mathsf{T} \to \mathsf{S}, \tag{11.25}$$

$$(X, [\alpha_1, \alpha_2, \ldots, \alpha_{2n}],\ \theta) \mapsto (X, \alpha_1 \wedge \alpha_2 \wedge \cdots \wedge \alpha_{2n}, \theta). \tag{11.26}$$

This map is surjective and hence we have an inclusion $\mathfrak{k}[\mathsf{S}] \subset \mathfrak{k}[\mathsf{T}]$. Recall that the algebraic group G acts on the space T and hence on $\mathfrak{k}[\mathsf{T}]$. If we impose a functional equation for $f \in \mathfrak{k}[\mathsf{T}]$ with respect to the action of G then we may decompose $\mathfrak{k}[\mathsf{T}]$ into graded pieces. We will do this just in the case of S. It inherits from T an action of the multiplicative group $\mathbb{G}_m$.

Definition 11.8. The algebra of (holomorphic) Siegel modular forms is by definition the pull-back of $\mathfrak{k}[\mathsf{S}]$ by the composition $\mathbb{H}_n \to \mathsf{T} \to \mathsf{S}$. A function $f \in \mathfrak{k}[\mathsf{S}]$ is called an (algebraic) Siegel modular form of weight $k \in \mathbb{N}$ if

$$f(t \bullet \mathsf{g}) = f(t) \cdot \mathsf{g}^{-k}, \quad \forall \mathsf{g} \in \mathbb{G}_m. \tag{11.27}$$

Its pull-back to $\mathbb{H}_n$ is called a (holomorphic) Siegel modular form.

The following proposition justifies the name Siegel modular form.

Proposition 11.6. *A holomorphic function $f : \mathbb{H}_n \to \mathbb{C}$ is a Siegel modular form of weight $k \in \mathbb{N}$ if and only if*

$$\det(c\tau + d)^{-k} f\left((a\tau + b) \cdot (c\tau + d)^{-1}\right) = f(\tau), \quad \forall \begin{bmatrix} a & b \\ c & d \end{bmatrix} \in \mathrm{Sp}(2n, \mathbb{Z}). \tag{11.28}$$

The algebra of Siegel modular forms can be defined over $\mathbb{Q}$ and in this way we have the following theorem.

Theorem 11.7. *The moduli space S is a quasi-affine open subset of the affine scheme*

$$\mathrm{Spec}\,(\textit{the algebra of Siegel module forms in } \mathbb{H}_n \textit{ for } \mathrm{Sp}(2n, \mathbb{Z}))\,. \tag{11.29}$$

Proof. For a large $k \in \mathbb{N}$, a basis of Siegel modular forms $f_0, f_1, \ldots, f_N$ of weight k gives us an embedding $\mathrm{Sp}(2n, \mathbb{Z})\backslash\mathbb{H}_n \hookrightarrow \mathbb{C}^{N+1}$. □

The case $n = 1$ corresponds to classical modular forms and it is discussed in Chapter 9. For $n = 2$, the algebra of genus 2 Siegel modular forms is

$$\mathbb{C}\left[\mathscr{E}_4, \mathscr{E}_6, \mathscr{C}_{10}, \mathscr{C}_{12}, \mathscr{C}_{35}\right] / \left(\mathscr{C}_{35}^2 = \mathrm{P}(\mathscr{E}_4, \mathscr{E}_6, \mathscr{C}_{10}, \mathscr{C}_{12})\right),$$

where $\mathscr{E}_4, \mathscr{E}_6, \mathscr{C}_{10}, \mathscr{C}_{12}, \mathscr{C}_{35}$ are generalization of Eisenstein series and P is an explicit polynomial, see [Igu62].

For $n = 1$ the algebra of classical modular forms (without growth condition) is $\mathbb{Q}[E_4, E_6, \frac{1}{\Delta}]$, where $\Delta := \frac{1}{1728}(E_4^3 - E_6^2)$. In this case, we have the invertible element Δ in this algebra and its logarithmic derivative is E_2:

$$\frac{\Delta'}{\Delta} = E_2, \quad ' = q\frac{d}{dq}. \tag{11.30}$$

For $n \geq 2$ all the non-constant Siegel modular forms have zeros in $\mathbb{H}_n$ and hence the only invertible Siegel modular forms are constants, see [Wei87]. Moreover, by a result of Bertrand and Zudilin in the articles [BZ01, BZ03] we know that field generated by Siegel modular forms and their derivations has transcendental degree $2n^2+n$ over $\mathbb{C}$. Since this is exactly the dimension of the moduli space T, it turns out that any meromorphic function in T is algebraic over the field generated by $\mathfrak{k}[\mathsf{S}]$ and the derivation of its elements along the vector fields v_{ij}. It might be still too early to expect the following conjecture. In a private communication, T. J. Fonseca has also shown doubts about its trueness.

Conjecture 11.2. *For $n \geq 2$ the algebra of differential Siegel modular forms is generated by the algebra of Siegel modular forms and their derivations under $\frac{\partial}{\partial \tau_{ij}}$.*

One may look for a weaker version of the above conjecture which is easier to state it in the algebraic framework.

Conjecture 11.3. *The $\mathfrak{k}$-algebra $\mathfrak{k}[\mathsf{T}]$ is generated by $\mathfrak{k}[\mathsf{S}]$ and the derivation of its elements along the vector fields v_{ij} and vector fields $\mathsf{v}_{\mathfrak{g}}$, $\mathfrak{g} \in \mathrm{Lie}(\mathsf{G})$.*

Our geometric point of view gives a natural context for the study of differential equations of Siegel modular forms, for some explicit differential equations Resnikof's articles [Res70a, Res70b]. If Conjecture 11.2 is false, one may try study the degree of the field of differential Siegel modular forms over the field generated by Siegel modular forms and their derivations. Our construction of modular vector fields in §6.14 might be useful for this study.

The Shimura varieties parameterize certain Hodge structures, and a smaller class of them called Shimura varieties of PEL-type, are the moduli of abelian varieties with polarization, endomorphism, and level structure, see [Del71b, Milb, Ker14] and the references therein. Constructing T as a moduli stack in the case of a Shimura variety seems to be doable as the literature in this case is abundant. One special case is the moduli of abelian varieties with real multiplication. In this case one has the theory of Hilbert–Blumenthal modular forms, and the differential equations of such modular forms have been studied in [Pel05]. The construction of functions on T and the corresponding modular vector fields will give us the algebra of differential Hilbert–Blumenthal modular forms. For the existence and uniqueness of such vector fields in this case see [Fon21].

Chapter 12

Hypersurfaces

I see the process of mathematical creation as a kind of recognizing a preexisting pattern. When you study something-topology, probability, number theory, whatever-first you acquire a general vision of the vast territory, then you focus on a part of it. Later you try to recognize "what is there?" and "what has already been seen by other people?". So you can read other papers and finally start discerning something nobody has seen before you (Y. Manin in The Berlin Intelligencer, 1998, pp. 16–19).

12.1 Introduction

In this chapter, we focus on enhanced smooth hypersurfaces of degree d and dimension n, and their moduli space T. The algebra $\mathfrak{k}[\mathsf{T}]$ of global regular functions in T only in the case of elliptic curves $(n, d) = (1, 3)$, quartic surfaces $(n, d) = (2, 4)$, and cubic fourfolds $(n, d) = (4, 3)$, will be related to automorphic forms and their derivations. In these cases, we will describe the differential equations of such automorphic forms in terms of modular vector fields. Beyond these cases it is not clear whether $\mathfrak{k}[\mathsf{T}]$ can be as useful as any theory of automorphic forms. Instead, we will focus on the study of Noether–Lefschetz and Hodge loci in the moduli space T. Using Griffiths theorem on de Rham cohomology of hypersurfaces, we construct explicit coordinates on T and we show that most of the hypersurfaces have isolated Hodge cycles and hence the corresponding modular foliations are trivial. A simplified version of T is introduced in [Mov17c] and the content of this chapter can be considered as a better and more complete presentation of this article.

12.2 Fermat hypersurface as marked variety

Recall that for some special varieties, the Hodge decomposition might be defined over the base field, see Definition 3.7. This is going to be the case of Fermat varieties:

$$X = X_n^d : \quad x_0^d + x_1^d + \cdots + x_{n+1}^d = 0. \tag{12.1}$$

We remind the description of the de Rham cohomology and Hodge filtration of hypersurfaces in [Gri69] and in particular the Fermat variety in [Mov19, Chapter 15]. Let

$$I := \{0, 1, \dots, d-2\}^{n+1},$$
$$A_\beta := \sum_{i=1}^{n+1} \frac{\beta_i + 1}{d}, \quad \beta \in I.$$

In the affine chart $x_0 = 1$

$$L : \quad x_1^d + x_2^d + \cdots + x_{n+1}^d = -1,$$

a basis of the de Rham cohomology $H_d^n(X)_0 \subset H_{\mathrm{dR}}^n(L)$ is given by

$$\omega_\beta := x^\beta \left(\sum_{i=1}^{n+1} \frac{(-1)^{i-1}}{d} x_i dx_1 \wedge dx_2 \wedge \cdots \wedge \widehat{dx_i} \wedge \cdots \wedge dx_{n+1} \right), \quad \beta \in I, \ A_\beta \notin \mathbb{N}, \tag{12.2}$$

where $x^\beta := x_1^{\beta_1} x_2^{\beta_2} \cdots x_{n+1}^{\beta_{n+1}}$ is a monomial. A basis of $F^{n+1-k} H_{\mathrm{dR}}^n(X)_0$ is given by

$$\omega_\beta, \qquad A_\beta < k, \qquad A_\beta \notin \mathbb{N}.$$

Proposition 12.1. *The Hodge decomposition of the Fermat variety is defined over rational numbers.*

Proof. For the Fermat variety $X = X_n^d$, it is enough to prove that the primitive de Rham cohomology $H_d^n(X)_0$ has the Hodge decomposition defined over $\mathbb{Q}$. The integral formula in [Mov19, Proposition 15.1], implies that

$$\overline{\omega_\beta} = \omega_{d-2-\beta} \quad \text{in } H_{\mathrm{dR}}^n(L), \tag{12.3}$$

where

$$d-2-\beta := (d-2-\beta_1, d-2-\beta_2, \ldots).$$

Since $A_{d-2-\beta} = n+1-A_\beta$, a basis of $\overline{F^{n+1-k}H^n_d(X)_0}$ is given by

$$\omega_\beta, \quad A_\beta > n+1-k, \quad A_\beta \notin \mathbb{N}.$$

We conclude that a basis of $H^{n+1-k,k-1} := F^{n+1-k} \cap \overline{F^{n-k}}$ is given by ω_β, $k-1 < A_\beta < k$. □

Proposition 12.2. *The cup product in the basis ω_β, $\beta \in I$, $A_\beta \notin \mathbb{N}$ of $H^n_{\mathrm{dR}}(X^d_n)_0$ is given by*

$$\omega_\beta \cup \omega_{\beta'} = \begin{cases} \Phi_{\beta,\beta'} \cdot \theta^n, & \text{for } A_\beta = n+1-A_{\beta'}, \text{ with } \Phi_{\beta,\beta'} \in \mathbb{Q}, \\ 0 & \text{otherwise,} \end{cases} \tag{12.4}$$

$$\omega_\beta \cup \theta^{\frac{n}{2}} = 0. \tag{12.5}$$

Proof. The equality (12.5) follows from

$$\int_X \omega_\beta \cup \theta^{\frac{n}{2}} = \int_{Z_\infty} \omega_\beta = 0,$$

where Z_∞ is an intersection of a linear $\mathbb{P}^{\frac{n}{2}+1} \subset \mathbb{P}^{n+1}$ with X. Since ω_β, $k-1 < A_\beta < k$ forms a basis of the piece $H^{n+1-k,k-1}$ of the Hodge decomposition of $H^n_{\mathrm{dR}}(X)_0$, if $A_\beta + A_{\beta'} \neq n+1$ then $\omega_\beta \cup \omega_{\beta'} = 0$. The rationality of $\Phi_{\beta,\beta'}$ follows from the fact that ω_β, θ and the cup product are defined over $\mathbb{Q}$. □

The cup product matrix in the basis $\{\omega_\beta,\ \beta \in I,\ A_\beta \notin \mathbb{N},\}$ is of the Hodge block form

$$\Phi := [\Phi^{i,j}] = \begin{pmatrix} 0 & 0 \; 0 \cdots & 0 & \Phi^{0,n} \\ 0 & 0 \; 0 \cdots & \Phi^{1,m-1} & 0 \\ \vdots & \vdots \; \vdots \; \ddots & \vdots & \vdots \\ \Phi^{n,0} & 0 \; 0 \cdots & 0 & 0 \end{pmatrix}, \tag{12.6}$$

where $\Phi^{n-i,i}$ is an $\mathsf{h}^{i,n-i} \times \mathsf{h}^{n-i,i}$ matrix with rational entries. One can use the explicit construction of the cup product in algebraic de Rham cohomology, see for instance [Mov20a], and Carlson–Griffiths description of the algebraic de Rham cohomology of hypersurfaces in [CG80, p. 7], in order to compute the entries Φ explicitly. For now, we make a change of basis in ω_β's such that for n even, Φ is written as the following type of anti-diagonal matrix

$$\Phi = \begin{pmatrix} 0 & 0 & 0 & 0 & J \\ 0 & 0 & 0 & J & 0 \\ 0 & 0 & I & 0 & 0 \\ 0 & J & 0 & 0 & 0 \\ J & 0 & 0 & 0 & 0 \end{pmatrix} \tag{12.7}$$

(a sample for $n = 4$). Here, J is the anti-diagonal matrix with 1 in its anti-diagonal entries and zero elsewhere, and I is the identity matrix. For this we have to work over $\mathbb{Q}(\sqrt{a})$ for some rational number $a \in \mathbb{Q}$, see Proposition 12.3. For n odd after a change of basis we have

$$\Phi = \begin{pmatrix} 0 & 0 & 0 & J \\ 0 & 0 & J & 0 \\ 0 & -J & 0 & 0 \\ -J & 0 & 0 & 0 \end{pmatrix} \tag{12.8}$$

(a sample for $n = 3$).

For a smooth hypersurface defined over $\mathbb{Q}$ and of even dimension n, we have a bilinear map

$$\begin{aligned} H^{\frac{n}{2},\frac{n}{2}}(X)_0 \times H^{\frac{n}{2},\frac{n}{2}}(X)_0 &\to \mathbb{Q}, \\ \langle \omega_1, \omega_2 \rangle &\mapsto \frac{\omega_1 \cup \omega_2}{\theta^n}. \end{aligned} \tag{12.9}$$

Proposition 12.3. *There is a positive non-zero rational number a which depends on d and n, such that for X over $\mathfrak{k} := \mathbb{Q}(\sqrt{a})$, $H^{\frac{n}{2},\frac{n}{2}}(X)_0$ has a basis α_i with $[\langle \alpha_i, \alpha_j \rangle]$ being the identity matrix.*

Proof. The bilinear map $\langle \cdot, \cdot \rangle$ is non-degenerate, and by second Hodge–Riemann bilinear relations, we have $\langle \omega, \omega \rangle > 0$ for all non-zero ω. In order to find an orthogonal basis we start with a non-zero $\omega_1 \in H^{\frac{n}{2},\frac{n}{2}}(X)_0$ and replace it with $a_1\omega_1$, where $a_1 = (\langle \omega_1, \omega_1 \rangle)^{-\frac{1}{2}}$. Then we repeat this with the space orthogonal to ω_1. □

12.3 The algebraic group

Recall the general definition of the algebraic group G in §3.3. Since for a hypersurface, only the middle cohomology is non-trivial, we set $\mathsf{b} = \mathsf{b}_n$, and so $\mathsf{b}_0 := \mathsf{b} - 1$ is the dimension of the primitive cohomology of X_0 (and not the zeroth Betti number of X_0). We identify $\mathsf{g} \in \mathsf{G}$ with its representation in the nth primitive cohomology of $X_0 := X_n^d$ and hence with $\mathsf{b}_0 \times \mathsf{b}_0$ matrices:

$$\mathsf{G} := \{\mathsf{g} \in \mathrm{GL}(\mathsf{b}_0, \mathfrak{k}) | \mathsf{g} \text{ block upper triangular and } \mathsf{g}^{\mathsf{tr}} \Phi \mathsf{g} = \Phi\}. \quad (12.10)$$

This implies that we have a surjective morphism of groups

$$\mathsf{G} \to \mathrm{O}(\mathsf{h}_0^{\frac{n}{2},\frac{n}{2}}),$$
$$\mathsf{g} \mapsto \mathsf{g}^{\frac{n}{2},\frac{n}{2}},$$

where $\mathsf{h}_0^{\frac{n}{2},\frac{n}{2}} := \mathsf{h}^{\frac{n}{2},\frac{n}{2}} - 1$ and

$$\mathrm{O}(\mathsf{h}_0^{\frac{n}{2},\frac{n}{2}}) = \mathrm{O}(\mathsf{h}_0^{\frac{n}{2},\frac{n}{2}}, \mathfrak{k}) := \{\mathsf{g} \in \mathrm{GL}(\mathsf{h}^{\frac{n}{2},\frac{n}{2}}0, \mathfrak{k}) | \mathsf{g}^{\mathsf{tr}} \cdot \mathsf{g} = \text{identity matrix}\} \quad (12.11)$$

is the orthogonal group. We have also a canonical immersion $\mathrm{O}(\mathsf{h}_0^{\frac{n}{2},\frac{n}{2}}) \hookrightarrow \mathsf{G}$ which follows from the fact that the Hodge decomposition of the Fermat variety is defined over $\bar{\mathfrak{k}}$ (see Proposition 12.1), and hence we have the inclusion (3.25).

12.4 Enhanced hypersurfaces

For a smooth hypersurface $X/\mathfrak{k}$ in $\mathbb{P}^{n+1}$, we use Lefschetz theorems and we know that only the middle cohomology is non-trivial in the sense that

$$H_{\mathrm{dR}}^m(X) \cong \begin{cases} \mathfrak{k}, & m \neq n \text{ is even,} \\ 0, & m \neq n \text{ is odd.} \end{cases}$$

For the middle cohomology we use Lefschetz decomposition and we conclude that for n an odd number $H_{\mathrm{dR}}^n(X)_0 = H_{\mathrm{dR}}^n(X)$ and for n an even number

$$H_{\mathrm{dR}}^n(X)_0 := \frac{H_{\mathrm{dR}}^n(X)}{\mathfrak{k} \cdot \theta^{\frac{n}{2}}} \hookrightarrow H_{\mathrm{dR}}^n(L).$$

The primitive cohomology can be also embedded in $H_{\mathrm{dR}}^n(X)$. For this we redefine

$$H_{\mathrm{dR}}^n(X)_0 := \{\omega \in H_{\mathrm{dR}}^n(X) | \langle \omega, \theta^{\frac{n}{2}} \rangle = 0\}.$$

Definition 12.1. An enhanced hypersurface is a pair (X, α), where X is a smooth hypersurface and $\alpha = [\alpha_1, \alpha_2, \ldots, \alpha_{\mathsf{b}_0}]$ is a basis of $H_{\mathrm{dR}}^n(X)_0$ with the following conditions:

(1) The basis α is compatible with the Hodge filtration of $H_{\mathrm{dR}}^n(X)_0$.
(2) The intersection matrix $[\langle \alpha_i, \alpha_j \rangle]$ is the constant matrix (12.6).

Let $V_n^d \subset \mathfrak{k}[x]_d$ be the affine variety parameterizing non-singular hypersurfaces. We take $\pi : Y \to V_n^d$ the full family of hypersurfaces over V_n^d. Recall the construction of the full enhanced family X/T in Theorems 3.3 and 3.4, and in particular, in the proof of these theorems. We use the Griffiths theorem on the de Rham cohomology of hypersurfaces, see [Mov19], for construction of such an enhanced family. From now on we are going to work with a full enhanced family $\mathsf{X} \to \mathsf{T}$ of hypersurfaces.

Remark 12.1. Hypersurfaces with $k := \frac{n+2}{d} \in \mathbb{N}$ have Hodge numbers $0, \ldots, 0, 1, \mathsf{h}^{k+1, m-k-1}, \ldots$ and they are R-varieties in the sense of Definition 2.26. For instance, a smooth cubic tenfold in $\mathbb{P}^{11}$ has Hodge numbers $0, 0, 0, 1, 220, 925, 220, 1, 0, 0, 0$ and its Moduli is of dimension $\dim(H^1(X, \Theta_X)) = 220$, see [Mov19, Chapter 15] and Chapter 12. It might be interesting to classify all odd-dimensional hypersurfaces such that

$$H_1(X, \Theta_X) \to \{A \in \mathrm{Hom}(H^{\frac{n-1}{2}}(X, \Omega_X^{\frac{n+1}{2}}), H^{\frac{n+1}{2}}(X, \Omega_X^{\frac{n-1}{2}})) \\ |Q(\delta(v)(\alpha), \beta) + Q(\alpha, \delta(v)(\beta)) = 0, \quad \alpha, \beta \in H^{\frac{n-1}{2}}(X, \Omega^{\frac{n+1}{2}})\}$$

is an isomorphism, where Q is defined in (2.35). This would give us other type of R-varieties in the context of hypersurfaces.

12.5 Modular foliations of Hodge type

Recall the definition of constant periods of Hodge type and the corresponding modular foliation $\mathscr{F}(\mathsf{C})$ in §6.7.

Proposition 12.4. *All the modular foliations $\mathscr{F}(\mathsf{C})$ of Hodge type with C as in (6.15) are isomorphic.*

Proof. The subgroup of the algebraic group G sending constant periods of Hodge type (6.15) to themselves contains the orthogonal group $\mathsf{O}(\mathsf{h}_0^{\frac{n}{2}, \frac{n}{2}})$. Since the orthogonal group act transitively on the unitary ball, the action of G on constant periods of Hodge type is transitive. Now, the affirmation follows from Proposition 6.3. □

The conclusion is that we have just one modular foliation of Hodge type $\mathscr{F}(\mathsf{C})$.

Definition 12.2. We define $\mathscr{F}_d^n = \mathscr{F}(\mathsf{C})$, where C has zeros every where except in the first entry of the middle Hodge block, where it is 1 in this entry, and call it the modular foliation for hypersurfaces.

Conjecture 12.1. *For $n \geq 4$ an even number and $d > \frac{2(n+1)}{(n-2)}$ the modular foliation $\mathscr{F}_n^d$ is trivial in the sense of Definition 6.4.*

This conjecture is motivated by the study of Hodge cycles of the Fermat variety. For n, d as in Conjecture 12.1, the generic Hodge cycle of the Fermat variety is conjectured to be isolated, see [Mov19, §16.8], and so it does produce a leaf of $\mathscr{F}_n^d$ which is an orbit of $\mathsf{Stab}(\mathsf{G}, \mathsf{C})$. Despite this fact, note that inside $\mathrm{Sing}(\mathscr{F}_n^d)$ we might have a non-trivial foliation.

12.6 Moduli of enhanced hypersurfaces

In this section, we want to construct the moduli of enhanced hypersurfaces. Before this we recall some useful information regarding the classical moduli of hypersurfaces. We closely follow Mukai's book [Muk03, Chapter 5]. For many missing definitions and proofs see this and [MFK94].

Let $\mathsf{k}[x]_d = \mathsf{k}[x_0, x_1, \ldots, x_{n+1}]_d$ be the space of homogeneous degree d polynomials in $n + 2$ variables $x_0, x_1, \ldots, x_{n+1}$ and with coefficients in k. Note that the projectivization of this space is the Hilbert scheme $\mathrm{Hilb}_P(\mathbb{P}_{\mathsf{k}}^{n+1})$, where P is the Hilbert polynomial of hypersurfaces in (2.19). For $f \in \mathsf{k}[x]_d$ we denote by X the hypersurface in $\mathbb{P}^{n+1}$ given by $f = 0$. We consider the action of the reductive group $\mathbf{G} := \mathrm{GL}(n + 2)$ on V_n^d:

$$\mathbf{G} \times \mathsf{k}[x]_d \to \mathsf{k}[x]_d, \quad (\mathbf{g}, f) \mapsto \mathbf{g} \cdot f := \text{ the polynomial } f \text{ evaluated at } x^{\mathrm{tr}}\mathbf{g}.$$

The following theorem has been proved in Mukai [Muk03, Theorem 5.23, p. 170] and it is attributed to Jordan in [Jor80] and Matsumura and Monsky in [MM64]. In Mumford, Fogarty and Kirwan's book [MFK94, §2, p. 80], it is attributed to Kodaira and Spencer [KS58, Lemma 14.2].

Theorem 12.5. *Any non-singular hypersurface X of degree ≥ 3 and dimension $n \geq 2$ given by a homogeneous polynomial $f \in \mathsf{k}[x_0, x_1, \ldots, x_{n+1}]_d$ is invariant under at most finitely many $\mathbf{g} \in \mathrm{GL}(n + 2)$.*

Let $V_n^d \subset \mathsf{k}[x]_d$ be the affine variety parameterizing non-singular hypersurfaces:

$$V_n^d := \mathrm{Spec}\left(\mathsf{k}\left[t, \frac{1}{\Delta(t)}\right]\right), \tag{12.12}$$

where $t = (t_\alpha)$ is a collection of all parameters with t_α for each monomial x^α of degree d and $\Delta = \Delta(t)$ is the discriminant function on t. By the above theorem all the points in V_n^d have a finite stabilizer. The variety V_n^d

is affine and so all its points are stable for the action of $\mathrm{GL}(n+2)$ on V_n^d, see [Muk03, Corollary 5.14, p. 166 and Corollary 5.24, p. 171].

Definition 12.3. The quotient

$$\mathsf{M}_n^d := \mathrm{GL}(n+2)\backslash V_n^d \tag{12.13}$$

has a canonical structure of an algebraic variety over $\mathfrak{k}$ and it is the moduli of smooth hypersurfaces of degree d and dimension n.

In the literature we also find the action of $\mathrm{SL}(n+1)$ on $\mathfrak{k}[x]_d$ without mentioning the geometric meaning of this. We next explain this, and in particular its geometric interpretation in the case:

$$k := \frac{n+2}{d} \in \mathbb{N}. \tag{12.14}$$

For $f \in \mathfrak{k}[x]$, $f \neq 0$ consider the following meromorphic differential $(n+1)$-form in $\mathbb{A}_{\mathfrak{k}}^{n+2}$:

$$\Omega_f := \frac{\sum_{i=0}^{n+1} (-1)^i x_i \widehat{dx_i}}{f^k}. \tag{12.15}$$

For the action of $\mathbf{g} \in \mathrm{GL}(n+1)$ in $\mathfrak{k}[x]_d$ we have

$$\mathbf{g}^* \Omega_f := \det(\mathbf{g}) \cdot \Omega_{\mathbf{g}\cdot f}. \tag{12.16}$$

The differential form Ω_f induces an $(n+1)$-form in $\mathbb{P}_{\mathfrak{k}}^{n+1}$ if and only if (12.14) occurs. The residue of Ω_f in $X \subset \mathbb{P}_{\mathfrak{k}}^{n+1}$ is a basis of the one-dimensional space $H^{n-k+1,k-1}$ of the Hodge decomposition of X. For $k=1$, X is a Calabi–Yau variety and such a residue is a holomorphic differential n-form in X. It follows from (12.14) and the Griffiths theorem on the cohomology of hypersurfaces that $H^{p,q}=0$ for $q < k-1$. We are interested in the moduli S of pairs (X, α), where X is a hypersurface of degree d and dimension n and $0 \neq \alpha \in H^{n-k+1,k-1}(X)$. If we write (X,α) in a coordinate system this is as follows. The group $\mathrm{GL}(n+2)$ acts on the set of pairs $(f, a\Omega_f)$, $f \in \mathfrak{k}[x]_d$, $a \in \mathbb{G}_m$ in a natural way:

$$\mathbf{g} \cdot (f, a\Omega_f) := (\mathbf{g}\cdot f, a\mathbf{g}^*\Omega_f),$$

and we are interested in the corresponding quotient. It turns out that it is the same as the the quotient of the action of the group $\mathrm{SL}(n+2)$ on

$\mathfrak{k}[x]_d$. Moreover, the trivial $\mathbb{G}_m$-action on pairs (X, α) (multiplying α with a constant) is translated to the right action of $\mathbb{G}_m$:

$$\mathfrak{k}[x]_d \times \mathbb{G}_m \to \mathfrak{k}[x]_d,$$
$$(f, \mathsf{g}) \mapsto f \bullet \mathsf{g} := \text{ the polynomial } f \text{ evaluated at } (\mathsf{g}x_0, \mathsf{g}x_1, \ldots, \mathsf{g}x_{n+1}).$$

It commutes with the action of $\mathrm{SL}(n+2)$ on $\mathfrak{k}[x]_d$, and so we have the action of $\mathbb{G}_m$ on the quotient space $\mathrm{SL}(n+2)\backslash\mathfrak{k}[x]_d$.

Theorem 12.6. *The quotient*

$$\mathsf{S} := \mathrm{SL}(n+2)\backslash V_n^d \tag{12.17}$$

is an affine variety over $\mathfrak{k}$ *and the* $\mathbb{G}_m$*-action on* S *is also defined over* $\mathfrak{k}$.

The affine variety S is called the moduli of holomorphic differential forms. This naming makes sense for the case (12.14) with $k = 1$, however, we will use it in general. Note that we can recover the classical moduli of hypersurfaces by taking the quotient:

$$\mathsf{M}_n^d = \mathsf{S}/\mathbb{G}_m.$$

For a proof of Theorem 12.6 see [Muk03, §5.2]. By Hilbert's theorem the ring $\mathfrak{k}[t]^{\mathrm{SL}(n+2)}$ of polynomials in t and invariant under the action of $\mathrm{SL}(n+2)$ is finitely generated. We take homogeneous polynomials $F_0, F_1, \ldots, F_s \in \mathfrak{k}[t]^{\mathrm{SL}(n+2)}$ and consider the map

$$\mathsf{S} \to \mathbb{A}_{\mathfrak{k}}^s, \quad t \mapsto (F_0(t), F_1(t), \ldots, F_s(t)),$$

which realizes S as a quasi-affine subvariety of $\mathbb{A}_{\mathfrak{k}}^{s+1}$. Moreover, the action of $\mathbb{G}_m$ in S is compatible with the action of $\mathbb{G}_m$ in $\mathbb{A}_{\mathfrak{k}}^{s+1}$:

$$(f_0, f_1, \ldots, f_s) \bullet \mathsf{g} := (\mathsf{g}^{d_0} f_0, \mathsf{g}^{d_1} f_1, \ldots, \mathsf{g}^{d_s} f_s),$$

where $d_i := \deg(F_i)$. It turns out that the classical moduli space M_n^d is a subvariety of the weighted projective variety $\mathbb{P}^{(d_0, d_1, \ldots, d_s)}$. In [Muk03] and many other texts, it seems that S has been used as an auxiliary object for the study of M_n^d, however, in the present text and for cases (12.14) it is the little brother of our main moduli space T. The case of elliptic curves in Theorem 12.6, that is, $n = 1$, $d = 3$, $k = 1$, is a classical statement.

The geometric quotient $\mathrm{SL}(3)\backslash V_1^3$ exists. It is

$$\mathsf{S} = \mathrm{Spec}\left(\mathfrak{k}\left[t_2, t_3, \frac{1}{27t_3^2 - t_2^3}\right]\right),$$

and the action of $\mathbb{G}_m$ on S is given by $(t_2, t_3) \bullet \mathsf{g} := (\mathsf{g}^{-4}t_2, \mathsf{g}^{-6}t_3)$. For this, we take a tangent line at an inflection point of the elliptic curve X and by an action of $\mathrm{SL}(3)$, put the inflection point at $[0 : 1 : 0]$ and the line at $x_2 = 0$. In Mukai [Muk03, Example 5.26, p. 172], this simple observation is attributed to Aronhold in [Aro50].

Let $H_n^d \to V_n^d$ be the nth algebraic de Rham cohomology bundle, that is, its fiber over the point f is $H_{\mathrm{dR}}^n(X)$, where X is the hypersurface in $\mathbb{P}^{n+1}$ given by $f = 0$. Let us define $\tilde{\mathsf{T}}$ to be the set of all enhanced hypersurfaces (X, α) with X as before, see §12.4. This is the total space of all possible choices of α in the fibers of $H_n^d \to V_n^d$. The group $\mathrm{SL}(n+1)$ acts in $\tilde{\mathsf{T}}$ in a natural way and we have

Theorem 12.7. *The moduli of enhanced hypersurfaces*

$$\mathsf{T} := \tilde{\mathsf{T}}/\mathrm{SL}(n+1) \tag{12.18}$$

is an algebraic variety over $\mathfrak{k}$*, with the action of the algebraic group* G *on* T.

Proof. In our way of constructing enhanced families in §3.6, the group $\mathrm{SL}(n+2)$ acts in each chart U_0, and all the points in U_0 have finite stabilizer. Therefore, $\mathrm{SL}(n+2)\backslash U_0$ is an affine variety with the action of the algebraic group G, all defined over $\mathfrak{k}$. The quotient T is obtained by gluing these affine charts. □

Note that T is not necessarily affine. When we choose sections of the cohomology bundle, we can at most say that they form a basis in fibers for a Zariski open neighborhood in the base space V_n^d, that is, they might not form a basis in all fibers. This forces us to study T in charts.

We note that the parameter space of hypersurfaces is full in the sense of Definition 2.27.

Proposition 12.8. *Let* V_n^d *be the parameter space of smooth hypersurfaces of dimension* n *and degree* d *in* $\mathbb{P}_{\mathfrak{k}}^{n+1}$*. The Kodaira–Spencer map*

$$\mathbf{T}_0 V_n^d \to H^1(X, \Theta_X)_0 \tag{12.19}$$

is surjective and its kernel is given by vector fields $\mathsf{v}_{\mathfrak{g},t}$, $\mathfrak{g} \in \mathrm{Lie}(\mathbf{G})$.

Proof. For hypersurfaces we have $H^1(X, \Theta_X)_0 = H^1(X, \Theta_X)$, except for $(n, d) = (2, 4)$. In this exceptional case $H^1(X, \Theta_X)_0$ is of codimension one in $H^1(X, \Theta_X)$. For the second statement, see [Voi03, Lemma 6.15]. □

12.7 Hypersurfaces with a finite group action

In this section, we take a finite group G acting on $\mathbb{P}^{n+1}_{\mathfrak{k}}$. In practice, this will be a subgroup of the automorphism group of the classical Fermat variety $X = X^d_n$ of dimension n and degree d given in (12.1). The group S_{n+2} of all permutations in $n+2$ elements $\{0, 1, \ldots, n+1\}$ acts on X^d_n in a natural way. An element in $b \in \mathsf{S}_{n+2}$ acts on X^d_n by permuting the coordinates:

$$(x_0, x_1, \ldots, x_{n+1}) \mapsto (x_{b_0}, x_{b_1}, \ldots, x_{b_{n+1}}).$$

Multiplication of the coordinates by dth roots of unity provides other automorphisms of the Fermat variety. Let

$$\mu_d^{n+2}/\mu_d := \underbrace{\mu_d \times \mu_d \times \cdots \times \mu_d}_{(n+2)\text{-times}}/\mathrm{diag}(\mu_d), \tag{12.20}$$

where

$$\mu_d := \{1, \zeta_d, \ldots, \zeta_d^{d-1}\} \tag{12.21}$$

is the group of dth roots of unity and $\mathrm{diag}(\mu_d)$ is the image of the diagonal map

$$\mu_d \to \mu_d^{n+2}, \quad \zeta \mapsto (\zeta, \zeta, \ldots, \zeta).$$

The group μ_d^{n+2}/μ_d acts on X^d_n by multiplication of coordinates:

$$(\zeta_0, \zeta_1, \ldots, \zeta_{n+1}), (x_0, x_1, \ldots, x_{n+1}) \mapsto (\zeta_0 x_0, \zeta_1 x_1, \ldots, \zeta_{n+1} x_{n+1}). \tag{12.22}$$

Let us define the free product group

$$G^d_n := \left(\mu_d^{n+2}/\mu_d\right) * \mathsf{S}_{n+2}, \tag{12.23}$$

which is a subgroup of the automorphism group of the Fermat variety X^d_n. Let $G \subset G^d_n$ be any finite subgroup. Our main examples for G are

$$G := \{\zeta \in \mu_d^{n+2}/\mu_d | \zeta_0 \zeta_1 \cdots \zeta_{n+1} = 1\}, \tag{12.24}$$

for the case $d = n+2$ and the permutation group $G := \mathsf{S}_{n+1}$. The group G acts on the space V^d_n of smooth hypersurfaces V^d_n in a canonical way and we define

$$V_G := \{t \in V_n^d | g \cdot t = t\}, \tag{12.25}$$

that is, V_G parameterizes hypersurfaces X with $G \subset \mathrm{Aut}(X)$. By definition the Fermat point $0 \in V_n^d$ is in V_G. For the group (12.24) the corresponding family of hypersurfaces is given by

$$X_t: \ t_0 x_0^{n+2} + t_1 x_1^{n+2} + \ldots + t_{n+1} x_{n+1}^{n+2} - t_{n+3} x_0 x_1 \ldots x_n = 0, \tag{12.26}$$

which is called the Dwork family. For the permutation group $G = \mathsf{S}_{n+2}$ we will consider the case $d = 3$. The corresponding family of hypersurfaces is given by

$$X_t: \quad t_0(x_0^3 + \cdots) + t_1(x_0^2 x_1 + \cdots) + t_3(x_0 x_1 x_2 + \cdots) = 0, \tag{12.27}$$

where $\cdots$ means the sum of all possible monomials obtained from the leading monomial by permuting the variables. We call X_t the Deligne's family, as working with hypersurfaces with large automorphism group is proposed by P. Deligne (personal communication, February 20, 2016).

An automorphism of a smooth projective variety leaves the Hodge filtration invariant and hence it is natural to consider the invariant part of $H^n_{\mathrm{dR}}(X)$

$$H^n_{\mathrm{dR}}(X)^G := \{\omega \in H^n_{\mathrm{dR}}(X) | \sigma^* \omega = \omega \quad \forall \sigma \in G\} \tag{12.28}$$

and the induced Hodge filtration. This is also called the invariant cohomology of X.

Proposition 12.9. *A basis of $H^n_{\mathrm{dR}}(X)^G$ for a member $X = X_t$ of the Dwork and Deligne families and for t in a neighborhood of the Fermat point are given by*

$$\frac{(x_0 x_1 \cdots x_{n+1})^{k-1} \Omega}{f^k}, \qquad k = 1, 2, \ldots, n+1,$$
$$\frac{(x_0 x_1 \cdots x_{3k-n-2} + \cdots) \Omega}{f^k}, \qquad k = 1, 2, \ldots, n+1,$$

respectively, where $\cdots$ means the sum of all possible monomials obtained from the leading monomial by permuting the variables. It is compatible with the Hodge filtration. For the Dwork family $\dim(H^n_{\mathrm{dR}}(X)^G) = n+1$ *and the*

Hodge numbers are

$$\underbrace{1, 1, 1, \ldots, 1}_{(n+1)\text{-times}},$$

and for Deligne family $\dim(H^n_{\mathrm{dR}}(X)^G) = n + 1 - 2\left[\frac{n+1}{3}\right]$ *and the Hodge numbers are*

$$\underbrace{0, 0, \ldots, 0}_{\left[\frac{n+1}{3}\right]\text{-times}}, \quad \underbrace{1, 1, \ldots, 1}_{n+1-2\left[\frac{n+1}{3}\right]\text{-times}}, \underbrace{0, 0, \ldots, 0}_{\left[\frac{n+1}{3}\right]\text{-times}}.$$

Proof. This follows from Griffiths theorem on the cohomology of hypersurfaces in [Gri69], see also [Mov19]. □

12.8 Automorphic forms for hypersurfaces

Let S be the moduli space of holomorphic differential forms constructed in Theorem 12.6. We denote by $\mathfrak{k}[\mathsf{S}]$ the ring of regular functions on S.

Definition 12.4. A regular function $f \in \mathfrak{k}[\mathsf{S}]$ is an automorphic form of weight k if

$$f(t \bullet \mathsf{g}) = \mathsf{g}^{-k} f(t), \quad \forall t \in \mathsf{S}, \quad \mathsf{g} \in \mathbb{G}_m.$$

The following theorem says that in which cases Definition 12.4 has to do with automorphic forms on some Hermitian symmetric domain. Recall the definition of Griffiths period domain in Chapter 8.

Theorem 12.10. *The Griffiths period domain D arising from the middle primitive cohomology of a smooth hypersurface is a Hermitian symmetric domain if and only if (n, d) is in the Table* 12.1.

Table 12.1. Hypersurfaces with a Hermitian symmetric domain D.

(n,d)	Hodge numbers of the primitive cohomology	Dimension of the moduli
$(1, d \geq 3)$	$\frac{(d-1)(d-2)}{2}, \frac{(d-1)(d-2)}{2}$	$\frac{(d+1)(d+2)}{2} - 9$
$(2,4)$	$1, 19, 1$	19
$(3,3)$	$5, 5$	10
$(3,4)$	$30, 30$	45
$(4,3)$	$1, 20, 1$	20
$(5,3)$	$21, 21$	35

Proof. The Hodge numbers of a hypersurface satisfy $\mathsf{h}^{n,0} \leq \mathsf{h}^{n-1,1} \leq \cdots \leq \mathsf{h}^{n-[\frac{n}{2}],[\frac{n}{2}]}$ and the Griffiths transversality implies that the corresponding period domain is Hermitian symmetric if and only if the Hodge numbers are of the format $0,\ldots,0,1,a,1,0\ldots,0$ for n even, and of the format $0,\ldots,0,a,a,0\ldots,0$ for n odd. Classical formulas for the Hodge number of hypersurfaces imply the desired statement. □

The dimension of the moduli space of hypersurfaces is

$$\dim(\mathsf{M}_n^d) = \binom{d+n+1}{n+1} - (n+2)^2.$$

The dimension of the Hermitian symmetric domain for n odd in Table 12.1 is $\frac{a(a+1)}{2}$, where $a = \mathsf{h}^{\frac{n+1}{2},\frac{n-1}{2}}$ is the non-zero Hodge number. Therefore, in these cases and except for $(n,d) = (1,3),(1,4)$, the period map is not surjective. Similar to the one dimensional case in which a curve is replaced with its Jacobian and the Jacobian is replaced with a polarized abelian variety so that we get a biholomorphic period map, we can also repeat this process in other cases using Griffiths' intermediate Jacobian. In the present text, we are only interested in the even-dimensional cases which are $(n,d) = (2,4),(4,3)$. In the following sections, we are going discuss these cases in more details.

12.9 Period domain

The cohomologies of a smooth hypersurface in $\mathbb{P}^{n+1}$ in dimensions below and over n is generated by the polarization $\theta \in H^2_{\rm dR}(X)$, and hence, our version of the Griffiths period domain in Definition 8.4 is the same as Griffiths' original definition. The cases with Hodge numbers in Theorem 12.10 are of special interest. The generalized period domain for Hodge numbers $0,\ldots,0,m,m,0\ldots,0$ is discussed in Chapter 11. In this section, we discuss the case of Hodge numbers $1,m,1$. We make a slight modification of the material in Chapter 8 as follows.

Let X_0 be a projective smooth variety of even dimension n over $\mathbb{C}$ such that its middle cohomology has Hodge numbers $0,\ldots,1,a,1,0\ldots,0$. By the Poincaré duality and Hodge index theorem, the homology $H_n(X_0,\mathbb{Z})$ together with the intersection of cycles is unimodular and it is of signature $(3,a-1)$ (respectively, $(a,2)$) for $\frac{n}{2}$ an odd number (respectively, even number). The lattice of Hodge cycles $\mathrm{Hodge}_n(X_0,\mathbb{Z})$ is also non-degenerate and of signature $(1, b)$ for some $b \leq a - 1$ (respectively, $(b, 0)$ for some $b \leq a$) for $\frac{n}{2}$ an odd number (respectively, even number). We take a non-degenerate sublattice $L \subset \mathrm{Hodge}_n(X_0, \mathbb{Z})$ of signature $(1, a-1-m)$ for $\frac{n}{2}$ an odd number (respectively, $(a - m, 0)$ for $\frac{n}{2}$ an even number) and define

$$V_{0,\mathbb{Z}} := \{\delta \in H_2(X_0, \mathbb{Z}) | \langle \delta, L \rangle = 0\}.$$

Since L is non-degenerate, $L \oplus L^{\perp}$ has finite index in $H_n(X_0, \mathbb{Z})$, and hence, the $\mathbb{Z}$-module $V_{0,\mathbb{Z}}$ is equipped with a non-degenerate, not necessarily unimodular, symmetric bilinear form inherited from $H_n(X_0, \mathbb{Z})$, and of signature $(2, m)$ for $\frac{n}{2}$ odd and signature $(m, 2)$ for $\frac{n}{2}$ an even number. Let $L^{\mathsf{pd}} \subset H^n_{\mathrm{dR}}(X_0)$ be the set of Poincaré duals of elements of L. In the cohomology side, we define:

$$V_0 := (L^{\mathsf{pd}})^{\perp} = \{\delta \in H^2_{\mathrm{dR}}(X_0) | \langle \delta, L^{\mathsf{pd}} \rangle = 0\}.$$

The integration map

$$V_{0,\mathbb{Z}} \hookrightarrow V_0^{\vee}, \quad \delta \mapsto \int_{\delta}$$

is well-defined and we proceed the same discussion as in Chapter 8 with this $V_{0,\mathbb{Z}}$ and V_0. We take basis $\delta_1, \ldots, \delta_{m+2}$ of $V_{0,\mathbb{Z}}$ and $\alpha_1, \alpha_2, \ldots, \alpha_{m+2}$ of V_0 with

$$\Psi = [\langle \delta_i, \delta_j \rangle], \quad \Phi = [\langle \alpha_i, \alpha_j \rangle].$$

The generalized period domain Π consists of $(m+2) \times (m+2)$ matrices:

$$\mathsf{P} := \begin{bmatrix} \mathsf{P}^{00} & \mathsf{P}^{01} & \mathsf{P}^{02} \\ \mathsf{P}^{10} & \mathsf{P}^{11} & \mathsf{P}^{12} \\ \mathsf{P}^{20} & \mathsf{P}^{21} & \mathsf{P}^{22} \end{bmatrix}$$

(written in the Hodge block notation corresponding to $1 + m + 1$) such that

$$\Phi = \mathsf{P}^{\mathsf{tr}} \Psi^{-\mathsf{tr}} \mathsf{P} \tag{12.29}$$

and

$$\begin{bmatrix} \mathsf{P}^{00} \\ \mathsf{P}^{10} \\ \mathsf{P}^{20} \end{bmatrix}^{\mathsf{tr}} \Psi^{-\mathsf{tr}} \begin{bmatrix} \overline{\mathsf{P}}^{00} \\ \overline{\mathsf{P}}^{10} \\ \overline{\mathsf{P}}^{20} \end{bmatrix} > 0. \tag{12.30}$$

The group

$$\Gamma_{\mathbb{Z}} := \mathrm{Aut}(V_{\mathbb{Z}}, \langle \cdot, \cdot \rangle) \cong \{A \in \mathrm{GL}(m+2, \mathbb{Z}) \mid A^{\mathsf{tr}} \Psi A = \Psi\}$$

written in the basis δ_i, $i = 1, 2, \ldots, m+2$ acts from the left on Π with the usual multiplication of matrices and $\mathsf{U} := \Gamma \backslash \Pi$.

We use $x = [x_1, x_2, \ldots, x_{m+2}]$ to denote the first column of P. The Griffiths period domain is just the projectivization of the space of first columns of P:

$$D = \{[x] \in \mathbb{P}^{m+1} \mid x\Psi^{-\mathsf{tr}} x^{\mathsf{tr}} = 0,\ x\Psi^{-\mathsf{tr}} \bar{x}^{\mathsf{tr}} > 0\ \}. \tag{12.31}$$

In the literature, one mainly find the following format of the period domain D. Let $W_{\mathbb{Z}} = V_{0,\mathbb{Z}}^{\vee}$ be a non-degenerate lattice of rank $m+2$. We have

$$D := \mathbb{P}(\{\omega \in W_{\mathbb{C}} | \langle \omega, \omega \rangle = 0,\ \langle \omega, \bar{\omega} \rangle > 0\}).$$

Both definitions are related by writing

$$\omega = \sum_{i=1}^{m+2} x_i \delta_i^{\mathsf{pd}},\ x_i \in \mathbb{C}.$$

Note that $[\langle \delta_i^{\mathsf{pd}}, \delta_j^{\mathsf{pd}} \rangle] = \Psi^{-\mathsf{tr}}$. The group $\Gamma_{\mathbb{Z}}$ acts from the left on D. The quotient $\Gamma_{\mathbb{Z}} \backslash D$ is the moduli of polarized Hodge structures of type $1, m, 1$.

12.10 K3 surfaces

For a detailed exposition of $K3$ surfaces the reader is referred to [Dol96]. A complex compact smooth surface X with the trivial canonical bundle Ω_X^2 and with $H^1(X, \mathbb{Q}) = 0$ is called a $K3$ surface. Using a result of Siu in [Siu83] we know that a $K3$ surface is Kähler and hence its only non-trivial cohomology carries a Hodge decomposition. Using Serre duality we know the Hodge numbers are $(\mathsf{h}^{20}, \mathsf{h}^{11}, \mathsf{h}^{02}) = (1, 20, 1)$. The homology $H_2(X, \mathbb{Z})$ is torsion free and together with the intersection of cycles is isomorphic to the $K3$ lattice

$$H \oplus H \oplus H \oplus (-E_8) \oplus (-E_8),$$

where H is the two-dimensional lattice with the gram matrix $\begin{bmatrix} 0 & -1 \\ -1 & 0 \end{bmatrix}$ and E_8 is the eight-dimensional lattice with the gram matrix

$$\begin{bmatrix} 2 & 0 & -1 & 0 & 0 & 0 & 0 & 0 \\ 0 & 2 & 0 & -1 & 0 & 0 & 0 & 0 \\ -1 & 0 & 2 & -1 & 0 & 0 & 0 & 0 \\ 0 & -1 & -1 & 2 & -1 & 0 & 0 & 0 \\ 0 & 0 & 0 & -1 & 2 & -1 & 0 & 0 \\ 0 & 0 & 0 & 0 & -1 & 2 & -1 & 0 \\ 0 & 0 & 0 & 0 & 0 & -1 & 2 & -1 \\ 0 & 0 & 0 & 0 & 0 & 0 & -1 & 2 \end{bmatrix} \tag{12.32}$$

see [Ser78]. By the Lefschetz $(1,1)$-theorem, the elements of the lattice $\mathrm{NS}(X) := H^2(X,\mathbb{Z}) \cap H^{1,1}$ are Poincaré dual to the Néron–Severi group of divisors in X. Moreover, since $H^1(X,\mathbb{Q}) = 0$ this is also the Picard group which is the group of line bundles in X. For an algebraic $K3$ surface, the signature of $\mathrm{NS}(X)$ is $(1,a)$ for some $a \leq 19$, and hence, the Picard rank of the surface is $1+a$. For instance, for a generic quartic in $\mathbb{P}^3$ $\mathrm{NS}(X) = \mathbb{Z} \cdot u$ with $u \cdot u = 4$.

Let L be an even non-degenerate lattice of signature $(1, 19-m)$, $m \geq 0$. Our main example for this is

$$L = \langle 2n \rangle, \quad n \geq 2 \tag{12.33}$$

which means that $L = \mathbb{Z} \cdot u$ with $\langle u, u \rangle = 2n$. A lattice polarization on the $K3$ surface X is given by a primitive embedding

$$i : L \hookrightarrow \mathrm{NS(X)}, \tag{12.34}$$

whose image contains a pseudo-ample class, that is, a numerically effective class with positive self-intersection. This gives us a line bundle l in X such that four linearly independent sections s_0, s_1, s_2, s_3 of l give us a generically one-to-one map $X \to \mathbb{P}^3$ whose image is given by a polynomial of degree 4. After the works of many authors we have the following theorem.

Theorem 12.11. *The coarse moduli space $\mathscr{M}$ of L-polarized $K3$ surfaces is a quasi-projective variety over $\mathbb{C}$ and the period map*

$$\mathscr{M} \to \Gamma_{\mathbb{Z}} \backslash D$$

is a biholomorphism of analytic spaces.

See [Dol96, p. 11]. It is natural to expect that $\mathscr{M}$ is a quasi-projective variety defined over $\mathbb{Q}$ or $\mathbb{Z}[\frac{1}{N}]$. This is only done in the case (12.33) for which one uses a Theorem of Viehweg, see §13.4. Let S be the moduli space of pairs (X, α_1), where X is an L-polarized $K3$ surfaces and α_1 is a holomorphic 2-form on X. In many examples the following conjecture is true.

Conjecture 12.2. *There is a universal family $\check{\mathsf{X}} \to \mathsf{S}$ of L-polarized $K3$ surfaces enhanced with a holomorphic differential 2-form.*

An immediate translation of Theorem 12.11 into the context of enhanced lattice polarized $K3$ surfaces is that the generalized period map

$$\mathsf{T} \to \mathsf{U} \tag{12.35}$$

is a biholomorphism. Moreover, if the above conjecture is true then we have a universal family $\mathsf{X} \to \mathsf{T}$ of enhanced L-polarized $K3$ surfaces.

Theorem 12.12. *Let X/T be a universal family of of enhanced L-polarized K3 surfaces. There are unique vector fields v_k, $k = 1, 2, \ldots, \mathsf{h}_0^{1,1}$ in T such that*

$$\mathsf{A}_{\mathsf{v}_k} = \begin{pmatrix} 0 & \delta_k^j & 0 \\ 0 & 0 & -\delta_k^i \\ 0 & 0 & 0 \end{pmatrix}. \tag{12.36}$$

Proof. The proof in the period domain U is similar to the case of Abelian varieties, see Theorem 11.5 and it is left to the reader. For the algebraic description of modular vector fields in T for L in (12.33) see §13.6. □

12.11 Cubic fourfolds

For missing details in this section see Hassett's article [Has00]. The generalized period map $\mathsf{P} : \mathsf{T} \to \mathsf{U}$ in the case of cubic fourfolds becomes an open immersion, and so, the modular vector fields constructed in Proposition 8.10 can be transported to T. We get a similar statement as in Theorem 12.12 in this case.

Theorem 12.13. *Let X/T be a universal family of enhanced cubic fourfolds. There are unique vector fields v_k, $k = 1, 2, \ldots, \mathsf{h}_0^{2,2}$ in T such that*

$$\mathsf{A}_{\mathsf{v}_k} = \begin{pmatrix} 0 & \delta_k^j & 0 \\ 0 & 0 & -\delta_k^i \\ 0 & 0 & 0 \end{pmatrix}. \tag{12.37}$$

Chapter 13

Calabi–Yau Varieties

Every time I gave a counterexample [to Calabi Conjecture], it failed in a very delicate manner, so I felt it cannot be that delicate unless God had fooled me; so it had to be right now. I changed my mind completely, and then I prepared everything to try to solve it (S.-T. Yau in Kavli IPMU News No. 33 March 2016).

13.1 Introduction

In this chapter, we will prove the existence and uniqueness of modular vector fields for Calabi–Yau varieties of arbitrary dimension. This has been formulated in Property 6.4 in Chapter 6. For Calabi–Yau threefolds this has been proved in [AMSY16] using certain manipulations of periods in mirror symmetry, and so, the proof might not be accessible for a general mathematics reader or it might not be considered a polished mathematical proof. Our proof in this chapter simplifies many arguments used in [AMSY16], and it works for Calabi–Yau varieties of arbitrary dimension and defined over any field of characteristic zero. In even dimension, we aim to use modular vector fields in order to study Hodge cycles.

Modular vector fields are algebraic incarnation of derivations with respect to the so-called flat coordinate system on the moduli of Calabi–Yau varieties. Such a coordinate system was first introduced in the physics literature for Calabi–Yau threefolds, see [CdlOGP91b, BCOV93, BCOV94], and in the mathematics literature it was used in the proof of the Bogomolov–Tian–Todorov theorem, see for instance Todorov's expository article [Tod03]. Recall from Chapter 9 that in the case of elliptic curves such a coordinate is the variable τ in the upper half-plane and $\frac{\partial}{\partial\tau}$ in the moduli of enhanced elliptic curves is incarnated as the Ramanujan vector field.

13.2 Preliminaries

For the main purposes of the present text we will only need the following definition of a Calabi–Yau variety.

Definition 13.1. A smooth projective variety $X \subset \mathbb{P}^N_{\mathfrak{k}}$ of dimension n over $\mathfrak{k}$ is called a Calabi–Yau variety if its canonical line bundle Ω^n_X is trivial.

Throughout the text, we will not need the extra condition

$$H^i(X, \mathcal{O}_X) = 0, \quad 0 < i < n.$$

Only the condition $H^2(X, \mathcal{O}_X) = 0$ for $n \geq 3$ might be useful for some discussions. For instance, from this we deduce that the primitive deformation space $H^1(X, \Theta_X)_0$ of X (see Definition 2.42) is equal to the usual one $H^1(X, \Theta_X)$. For Calabi–Yau twofolds ($K3$ surfaces) note that the dimension of $H^2(X, \mathcal{O}_X)$ is one, and so, we cannot drop 0 from our cohomology notations. There are two important aspects of a Calabi–Yau variety.

(1) The vector space $H^0(X, \Omega^n)$ is one dimensional and it is generated by a holomorphic nowhere vanishing differential n-form ω. In particular, the Hodge numbers satisfy

$$\mathsf{h}^{n,0} = 1 = \mathsf{h}^{0,n}.$$

(2) Using Serre duality we have

$$H^1(X, \Theta_X) \cong H^{n-1}(X, \Omega^n_X \otimes \Omega^1_X) = H^{n-1}(X, \Omega^1_X). \tag{13.1}$$

In particular, the dimension of the deformation space of X is equal to the Hodge number $\mathsf{h}^{n-1,1}$.

The Serre duality also manifests itself in the IVHS for smooth Calabi–Yau varieties. Recall the map in (2.40).

Proposition 13.1. *For a smooth projective Calabi–Yau variety over* $\mathfrak{k}$, *the following part of IVHS*

$$\delta_{n,0} : H^1(X, \Theta_X)_0 \to \operatorname{Hom}\left(H^0(X, \Omega_X^n),\ H^1(X, \Omega_X^{n-1})_0\right) \quad (13.2)$$

is an isomorphism of $\mathfrak{k}$*-vector spaces.*

Note that for $n \geq 3$ we have $H^1(X, \Omega_X^{n-1})_0 = H^1(X, \Omega_X^{n-1})$.

Proof. We know that $H^0(X, \Omega_X^n)$ is one-dimensional, and so, the right-hand side of (13.2) is canonically isomorphic to $H^1(X, \Omega_X^{n-1})_0$. We give a proof in the complex context $\mathfrak{k} = \mathbb{C}$. Since the canonical bundle of X is trivial, we can take a global holomorphic nowhere vanishing n-form ω in X. In a local holomorphic coordinate system $(z_1, z_2, \ldots, z_n)$, we can write $\omega = dz_1 \wedge dz_2 \wedge \cdots \wedge dz_n$ and the map

$$i.\omega : \Theta_X \to \Omega_X^{n-1},\ \mathsf{v} \mapsto i_{\mathsf{v}}\omega$$

is an isomorphism of holomorphic coherent sheaves, because for

$$\mathsf{v} = \sum_{i=1}^{n} \mathsf{v}_i(z)\frac{\partial}{\partial z_i}, \quad \text{we have} \quad i_{\mathsf{v}}\omega = \sum_{i=1}^{n} \mathsf{v}_i(z)(-1)^{i-1}\widehat{dz_i}.$$

This induces an isomorphism in H^1. By Serre's GAGA the same isomorphism is valid for Cech cohomologies with Zariski topology of X. The affirmation for the primitive part of H^1 follows from the fact that the following diagram commutes:

$$\begin{array}{ccc} H^1(X, \Theta_X) & \overset{i.\omega}{\to} & H^1(X, \Omega_X^{n-1}) \\ {\scriptstyle i.\theta}\downarrow & & \downarrow{\scriptstyle \theta\wedge\cdot} \\ H^2(X, \mathcal{O}_X) & \overset{\cdot\omega}{\to} & H^2(X, \Omega_X^n). \end{array}$$

This in turn follows from

$$0 = i_{\mathsf{v}}(\theta \wedge \omega) = i_{\mathsf{v}}\theta \wedge \omega - \theta \wedge i_{\mathsf{v}}\omega,$$

for $\mathsf{v} \in H^1(X, \Theta_X)$. Note that $\theta \in H^1(X, \Omega_X^1)$ and so it contributes (-1) in the above formula. It must not be confused with its C^∞ counterpart which is a 2-form. □

By Proposition 13.1, a Calabi–Yau variety is an R-variety with $m = n$ and $k = 0$ in the sense of §2.17.

13.3 R-varieties and modular vector fields

Recall the notion of an R-variety in §2.17. We are interested in R-varieties because they give us some important information about the corresponding modular vector fields.

Proposition 13.2. *Let Y/V be an* R*-family of smooth projective varieties in the sense of Definition 2.26 and let* X/T *be the corresponding enhanced family constructed in §§3.6 and 3.7. Assume that the Kodaira–Spencer map* $\mathbf{T}_tV \to H^1(Y_t, \Theta_{Y_t})_0$ *is surjective for all* $t \in V$*. Then the map* $\mathsf{Y}^{k-1,k}_{m,\cdot} : \mathfrak{M}_m(\mathsf{X}/\mathsf{T}) \to \mathrm{Mat}(\mathsf{h}_0^{m-k+1,k-1} \times \mathsf{h}_0^{m-k,k}, \mathcal{O}_\mathsf{T})$ *defined in* (6.26) *is a surjective morphism of* $\mathcal{O}_\mathsf{T}$*-modules.*

A slight modification of the proof of Proposition 13.2 might give us a similar statement for R-varieties in the sense of Definition 2.25.

Proof. By Theorem 6.23, we know that the Kodaira–Spencer map at a point $t \in V$ is the composition

$$\mathbf{T}_tV \xrightarrow{f} \mathbf{T}_t\mathsf{T} \overset{\delta_{m,k}}{\to} \mathrm{Hom}(H^k(X, \Omega_X^{m-k})_0,\ H^{k+1}(X, \Omega_X^{m-k-1})_0),$$

where $X = X_t$. Since $\mathsf{Y}^{k-1,k}_{m,\cdot}$ is just $\delta_{m,k}$ with its image written in the standard basis of the enhanced family X/T (Proposition 6.26), the result follows from the surjectivity of the Kodaira–Spence map. □

Let i and j be integers with $1 \leq i \leq \mathsf{h}_0^{m-k+1,k-1}$ and $1 \leq j \leq \mathsf{h}_0^{m-k,k}$ and let us take the $\mathsf{h}^{m-k+1,k-1} \times \mathsf{h}^{m-k,k}$ matrix M_{ij} such that its entries are all zero except its (i,j)-entry which is 1. Proposition 13.2 gives us a vector field v_{ij} in T with $\mathsf{Y}^{k-1,k}_{m,\mathsf{v}_{ij}} = M_{ij}$. This vector field might not be unique. We are going to discuss the uniqueness issue in the case of Calabi–Yau varieties, see Theorem 13.4. We will discuss the moduli of enhanced Calabi–Yau varieties using the available results on the classical moduli, however, the construction of vector fields v_{ij} as above suggests that we might be able to construct the moduli of enhanced R-varieties in the following way which seems to be completely new. Let us consider the action of a reductive group $\mathbf{G}$ for the enhanced R-family X/T as in §3.7. Starting from v_{ij} constructed above, we have to construct modular vector fields $\mathsf{v} \in \mathfrak{M}(\mathsf{X}/\mathsf{T})$ in T which are $\mathbf{G}$-invariant. By Proposition 6.20 all the entries of $\mathsf{Y}^{i-1,i}_{m,\mathsf{v}}$ are constant along the orbits of $\mathbf{G}$. Further derivations of these functions along $\mathbf{G}$-invariant vector fields produce more functions constant along the orbits of $\mathbf{G}$. We might hope that all these functions would be enough to construct the quotient $\mathbf{G}\backslash\mathsf{T}$ as an affine scheme.

13.4 Universal family of enhanced Calabi–Yau varieties

The classical moduli of Calabi–Yau varieties X is fairly well-understood both in the complex and algebraic context. In the complex side we have the following theorem.

Theorem 13.3 (Bogomolov–Tian–Todorov). *The moduli space* M *of complex polarized Calabi–Yau manifolds* X *is smooth and of dimension* $N := \dim H^1(X, \Omega^{n-1})_0$. *Moreover, a holomorphic coordinate system for* M *around a point* $t_0 \in M$, $X_{t_0} = X$ *is given by* $\tau := (\tau_1, \tau_2, \ldots, \tau_N)$, *where*

$$\tau_i := \frac{\int_{\delta_{i,t}} \omega}{\int_{\delta_{1,t}} \omega} \tag{13.3}$$

for some cycles $\delta_{i,t} \in H_n(X_t, \mathbb{Z})$ *with* $t \in (\mathsf{M}, t_0)$ *and* ω *is the holomorphic* n*-form on* X_t.

For a summary of results in this direction see Todorov's expository article [Tod03], see also [IM10] for an algebraic proof of this theorem. For the affirmation concerning construction of τ_i's, see Tian's article [Tia87, Corollary 1, p. 664]. Note that Tian uses a period domain which is different from the Griffiths period domain and it essentially encodes the periods of ω, see [Tia87, p. 637]. In the Physics literature, see [CdlOGP91b, BCOV93, BCOV94], and for many examples of Calabi–Yau varieties, we find a precise description of how to choose the topological cycles $\delta_{i,t}$. This description is necessary if we want to find relations between Theorem 13.4 and periods of Calabi–Yau varieties. This is only done in the case of Calabi–Yau varieties of dimension ≤ 3, see [AMSY16].

In the algebraic side, E. Viehweg has constructed the coarse moduli space of polarized smooth Calabi–Yau varieties as a quasi-projective variety, see [Vie95]. His method is based on Mumford–Hilbert geometric invariant theory (GIT). He has shown that smooth Calabi–Yau varieties $X_t \subset \mathbb{P}^N, t \in V$ for an irreducible component V of a Hilbert scheme, are stable under the action of the reductive group $\mathbf{G} = \mathrm{GL}(N+1)$, and hence, the geometric quotient $\mathbf{G}\backslash V$ exists as a scheme over $\mathfrak{k}$. According to [Tod03, pp. 700 and 707], the first moduli is just a local holomorphic chart of the second moduli for $\mathfrak{k} = \mathbb{C}$.

In order to construct universal families of enhanced Calabi–Yau varieties we are going to take an irreducible component V of a full Hilbert scheme of Calabi–Yau varieties $X \subset \mathbb{P}^N_{\mathfrak{k}}$ defined in §2.18. Our main example for this is the parameter space V of smooth hypersurfaces of dimension n and degree $d = n+2$ in $\mathbb{P}^{n+1}_{\mathfrak{k}}$ or the Dwork family discussed in §12.7. In Proposition 12.8, we have proved that the parameter space of hypersurfaces is full. Let $\mathsf{X} \to \mathsf{T}$ be the full enhanced family constructed from $Y \to V$ in §§3.6 and 3.7. Theorem 3.5 gives us the coarse moduli space $\check{\mathsf{T}} := \mathbf{G}\backslash\mathsf{T}$ of enhanced Calabi–Yau varieties and we claim the following conjecture.

Conjecture 13.1. *There is a universal family $\check{\mathsf{X}} \to \check{\mathsf{T}}$ of enhanced Calabi–Yau varieties.*

Despite the fact that we have not solved the above conjecture, Proposition 3.10 tells us that the Gauss–Manin connection matrices A_m can be transported to $\check{\mathsf{T}}$, and this is all what we need to reproduce the theory of modular vector fields and modular foliations in $\check{\mathsf{T}}$. In the rest of this chapter by abuse of terminology, we will talk about a universal family of enhanced Calabi–Yau varieties X/T, knowing that only T and A_m's exists as algebraic objects over $\mathfrak{k}$ (we will use $\check{\mathsf{T}}$ instead of T, etc.). It must be remarked that according to Todorov [Tod03, Theorem 33], there exists a universal family over a finite covering of the classical moduli space of polarized smooth Calabi–Yau varieties.

13.5 Modular vector fields for Calabi–Yau varieties

We are now ready to state our main theorem on modular vector fields for Calabi–Yau varieties.

Theorem 13.4. *Let X/T be a universal family of enhanced smooth projective Calabi–Yau varieties of dimension n. There exist unique global vector fields v_j, $j = 1, 2, \ldots, \mathsf{h}_0^{n-1,1}$ in T and unique $\mathsf{h}_0^{m-i+1,i-1} \times \mathsf{h}_0^{m-i,i}$ matrices $\mathsf{Y}^{i-1,i}_{m,\mathsf{v}}$, $i = 1, 2, \ldots, m$, $m = 1, 2, \ldots, 2n-1$ with entries as global regular functions in T such that*

$$\mathsf{A}_{m,\mathsf{v}_j} = \mathsf{Y}_{m,j} := \begin{pmatrix} 0 & \mathsf{Y}^{01}_{m,j} & 0 & \cdots & 0 \\ 0 & 0 & \mathsf{Y}^{12}_{m,j} & \cdots & 0 \\ \vdots & \vdots & \vdots & \ddots & \vdots \\ 0 & 0 & 0 & \cdots & \mathsf{Y}^{m-1,m}_{m,j} \\ 0 & 0 & 0 & \cdots & 0 \end{pmatrix} \tag{13.4}$$

with

$$\mathsf{Y}^{01}_{n,j} = [0,0,\ldots,0,1,0,\ldots,0], \quad 1 \textit{ is in the } j\textit{th place}, \tag{13.5}$$

$$\mathsf{Y}^{i-1,i}_{m,j} = (-1)^{m-1}(\mathsf{Y}^{m-i,m-i+1}_{m,j})^{\mathsf{tr}}, \tag{13.6}$$

$$\mathsf{v}_j(\mathsf{Y}^{i-1,i}_{m,k}) = \mathsf{v}_k(\mathsf{Y}^{i-1,i}_{m,j}), \tag{13.7}$$

$$\mathsf{Y}^{i-1,i}_{m,j}\mathsf{Y}^{i,i+1}_{m,k} = \mathsf{Y}^{i-1,i}_{m,k}\mathsf{Y}^{i,i+1}_{m,j}. \tag{13.8}$$

Proof. The existence is already proved in Proposition 13.2 in which we have used Theorem 6.23. This also gives the matrix format (13.5). The intersection matrix Φ_n is taken in the standard format (3.24). Therefore, (13.6), (13.7) and (13.8) follows respectively from (6.28), (6.31) and (6.32). For (6.31) we need to prove that $[\mathsf{v}_j, \mathsf{v}_k] = 0$. In summary, we only need to prove that v_i's are unique and commute with each other. If we have given two such vector fields v_j and $\check{\mathsf{v}}_j$ then for $\mathsf{v} := \mathsf{v}_j - \check{\mathsf{v}}_j$ we have $\mathsf{Y}^{01}_{n,\mathsf{v}} = 0$. Moreover, if we have two such vector fields v_j and v_k with different j,k, it follows from (6.31) applied for $i = 1$ that $\mathsf{Y}^{0,1}_{n,[\mathsf{v}_j,\mathsf{v}_k]} = 0$ because $\mathsf{Y}^{01}_{n,\mathsf{v}_j}$ and $\mathsf{Y}^{01}_{n,\mathsf{v}_k}$ are constant matrices. Therefore, for $\mathsf{v} := [\mathsf{v}_j, \mathsf{v}_k]$ we have $\mathsf{Y}^{0,1}_{n,\mathsf{v}} = 0$.

We prove that if v is a vector field in T with $\mathsf{Y}^{0,1}_{n,\mathsf{v}} = 0$ then v must be necessarily the zero vector field. We have $\nabla_{\mathsf{v}}\alpha_1 = 0$, where α_1 is the first element (a basis of H^{n0}) of the standard basis of $H^n_{\mathrm{dR}}(X_{t_0})$. We proceed the proof over complex numbers, that is, $\mathfrak{k} = \mathbb{C}$. Let $\gamma : (\mathbb{C},0) \to \mathsf{T}$ be an arbitrary integral curve of the vector field v with $\gamma(0) = t_0$. In order to prove that $\mathsf{v} = 0$ it is enough to prove that γ is a constant map. It follows that all the integrals

$$\int_{\delta_x} \alpha_1, \ \delta_x \in H_n(X_{\gamma(x)}, \mathbb{Z})$$

do not depend on x and hence are constant numbers. Let $\pi : \mathsf{T} \to \mathsf{M} := \mathsf{T}/\mathsf{G}$ be the canonical map to the classical moduli space of Calabi–Yau varieties. By the Bogomolov–Tian–Todorov theorem, we know that M is smooth of dimension $N := \mathsf{h}_0^{n-1,1}$, moreover, a coordinate system for M around the point $\pi(t_0)$ is given by $\tau := (\tau_1, \tau_2, \ldots, \tau_N)$, where $\tau_i := \frac{\int_{\delta_{i,t}} \alpha_1}{\int_{\delta_{1,t}} \alpha_1}$ for certain cycles $\delta_{i,t} \in H_n(X_t, \mathbb{Z})$ with $t \in (\mathsf{T}, t_0)$. We conclude that $\pi \circ \gamma$ is a constant map, and hence if we identify $t \in \mathsf{T}$ with $(X_t, \alpha_{1,t}, \ldots)$ then $X_{\gamma(x)} = X_{t_0}$ does not depend on x. Now, we use $\nabla_{\mathsf{v}}\alpha^i_m = \mathsf{Y}^{i,i+1}_{m,\mathsf{v}}\alpha^{i+1}_m$. We take the pull-back of this equation by γ and regard all the involved quantities depending on x. The operation ∇_{v} turns out to be the usual derivation $\frac{\partial}{\partial x}$. We write the

Taylor series $\alpha^i_m = \sum_{j=0}^{\infty} x^j \alpha^i_{m,j}$, where the entries of $\alpha^i_{m,j}$'s are elements in $H^m_{\rm dR}(X_{t_0})$ independent of x. The conclusion is that the Gauss–Manin connection ∇_{v} leaves the Hodge filtration invariant and hence $\mathsf{Y}^{i,i+1}_{m,\mathsf{v}}$ are all zero, and hence, the Gauss–Manin connection matrix A_m of X/T composed with v is identically zero and the elements of the standard basis α of $H^*_{\rm dR}(X_t)$ do not depend on x. This means that γ is a constant map. □

As a corollary of Theorem 13.4, we have the following theorem.

Theorem 13.5. *For a universal family of enhanced smooth projective Calabi–Yau varieties* X/T, *the vector fields* v_i, $i = 1, 2, \ldots, \mathsf{h}_0^{n-1,1}$ *are linearly independent at each point* $t \in \mathsf{T}$ *and the* $\mathscr{O}_{\mathsf{T}}$*-module* $\mathfrak{M}(\mathsf{X}/\mathsf{T})$ *of modular vector fields is freely generated by* v_i*'s. In particular, the modular foliation* $\mathscr{F}(2)$ *has no singularities and it is of dimension* $\mathsf{h}_0^{n-1,1}$.

Proof. The first statement follows from (13.5). The argument for the second statement is as follows. Let $\mathsf{v} \in \mathfrak{M}(\mathsf{X}/\mathsf{T})$ be a modular vector field and let $\mathsf{Y}^{01}_{n,\mathsf{v}} = [a_1, a_2, \ldots, a_N]$. For $\check{\mathsf{v}} = \mathsf{v} - \sum_{i=1}^{N} a_i \mathsf{v}_i$ we have $\mathsf{Y}^{01}_{n,\check{\mathsf{v}}} = 0$ and the same argument as in the last step of the proof of Theorem 13.4 shows that $\check{\mathsf{v}} = 0$. The last statement follows from the fact that any vector field tangent to the foliation $\mathscr{F}(2)$ is modular. □

After a partial compactification of T one might get singularities for the foliation $\mathscr{F}(2)$, see for instance the case of elliptic curves in Chapter 9.

The cases of Calabi–Yau 1, 2, 3 and 4-folds are of special interest. The Calabi–Yau 1-folds are elliptic curves and Theorem 13.4 is reduced to the existence of the Ramanujan vector field discussed in §9.3. In the next sections for our matrices (13.4) we are going to use the notation of [AMSY16]. We define δ^j_k (respectively, δ^i_k) to be the $1 \times \mathsf{h}_0^{n-1,1}$ (respectively, $\mathsf{h}_0^{n-1,1} \times 1$) matrix with zeros everywhere except at its $(1, k)$th (respectively, $(k, 1)$th) entry which is one.

13.6 Modular vector fields for K3 surfaces

The Calabi–Yau two folds are K3 surfaces. In this case, it is also fruitful to enhance X with some fixed curves in its Neron–Severi group. This is well-known under the name "lattice polarization" and it is discussed in §12.10. A version of the following theorem has been proved in [Ali17] using ideas from period manipulations of mirror symmetry. For Dwork family of K3 surfaces it is also proved in [MN18].

Theorem 13.6. *Let* X/T *be a universal family of K3 surfaces. There are unique vector fields* v_k, $k = 1, 2, \ldots, \mathsf{h}_0^{1,1}$ *in* T *such that*

$$\mathsf{A}_{\mathsf{v}_k} = \begin{pmatrix} 0 & \delta_k^j & 0 \\ 0 & 0 & -\delta_k^i \\ 0 & 0 & 0 \end{pmatrix}. \tag{13.9}$$

Note that for $K3$ surfaces there is no non-constant Y and they belong to the class of varieties with the constant Gauss–Manin connection, see §6.12. The intersection matrix in this case is taken to be

$$\Phi := \begin{pmatrix} 0 & 0 & 1 \\ 0 & I & 0 \\ 1 & 0 & 0 \end{pmatrix},$$

where $I = I_{\mathsf{h}_0^{1,1}\times\mathsf{h}_0^{1,1}}$ is the identity matrix. We get the following generalization of $\mathfrak{sl}_2$ Lie algebra. This is namely the $\mathfrak{k}$-vector space generated by vector fields

$$\mathsf{v}_{\mathsf{g}},\ \mathsf{g} \in \mathrm{Lie}(\mathsf{G}), \quad \mathsf{v}_k,\ k = 1, 2, \ldots, \mathsf{h}_0^{1,1}.$$

After composing with the transpose of the Gauss–Manin connection matrix we get the following representation of this Lie algebra:

$$\left\{ \mathfrak{g} \in \mathrm{Mat}(N \times N, \mathfrak{k}) \middle| \mathfrak{g} = \begin{pmatrix} * & * & * \\ * & * & * \\ 0 & * & * \end{pmatrix}, \quad \mathfrak{g}^{\mathrm{tr}}\Phi + \Phi\mathfrak{g} = 0 \right\}, \tag{13.10}$$

where $N := 1 + \mathsf{h}_0^{1,1} + 1$ and it contains $\mathrm{Lie}(\mathsf{G})$. Note that $\mathfrak{g}^{20} = \mathfrak{g}_{N,1} = 0$ follows from the $(N, 1)$-entry of the equality $\mathfrak{g}^{\mathrm{tr}}\Phi + \Phi\mathfrak{g} = 0$.

13.7 Modular vector fields for Calabi–Yau threefolds

Theorem 13.4 in the case of Calabi–Yau threefolds is inspired from many period manipulations in mirror symmetry.

Theorem 13.7 (Alim–Movasati–Scheidegger–Yau [AMSY16]). *Let* X/T *be a universal family of enhanced smooth Calabi–Yau threefolds. There are unique vector fields* v_k, $k = 1, 2, \ldots, \mathsf{h}_0^{2,1}$ *in* T *and unique regular* *functions in* T, Y_{ijk}, $i,j,k = 1,2,\ldots,\mathsf{h}_0^{2,1}$ *symmetric in* i,j,k *such that*

$$\mathsf{A}_{\mathsf{v}_k} = \begin{pmatrix} 0 & \delta_k^j & 0 & 0 \\ 0 & 0 & \mathsf{Y}_{kij} & 0 \\ 0 & 0 & 0 & \delta_k^i \\ 0 & 0 & 0 & 0 \end{pmatrix} \tag{13.11}$$

with

$$\mathsf{v}_{i_1}\mathsf{Y}_{i_2i_3i_4} = \mathsf{v}_{i_2}\mathsf{Y}_{i_1i_3i_4}. \tag{13.12}$$

Note that in this case we have taken

$$\Phi := \begin{pmatrix} 0 & 0 & 0 & 1 \\ 0 & 0 & I & 0 \\ 0 & -I & 0 & 0 \\ -1 & 0 & 0 & 0 \end{pmatrix}, \tag{13.13}$$

and if $\mathsf{h}^{1,0} = 0$ then we do not need to use the primitive cohomology notation and so $\mathsf{h}_0^{2,1} = \mathsf{h}^{2,1}$.

13.8 Modular vector fields for Calabi–Yau fourfolds

For applications in Hodge theory and in particular the study of Hodge loci, we highlight Theorem 13.4 in the case of Calabi–Yau fourfolds. There has been some interest in Calabi–Yau fourfolds in the physics literature for which we refer ro [HMY17] and the references therein. It might be possible to generalize the period manipulations of periods in the case of Calabi–Yau threefolds to higher dimensions, and in this we reprove Theorem 13.4 and its particular case of Calabi–Yau fourfolds.

Theorem 13.8. *Let* X/T *be a universal family of enhanced smooth projective Calabi–Yau fourfolds. There are unique vector fields* v_k, $k = 1,2,\ldots,\mathsf{h}_0^{3,1}$ *in* T *and unique regular functions in* T

$$\mathsf{Y}_{kij}, \quad i,k = 1,2,\ldots,\mathsf{h}_0^{3,1}, \ j = 1,2,\ldots,\mathsf{h}_0^{2,2}$$

symmetric in i,k *such that*

$$\mathsf{A}_{\mathsf{v}_k} = \begin{pmatrix} 0 & \delta_k^j & 0 & 0 & 0 \\ 0 & 0 & \mathsf{Y}_{kij} & 0 & 0 \\ 0 & 0 & 0 & -\mathsf{Y}_{kji} & 0 \\ 0 & 0 & 0 & 0 & -\delta_k^i \\ 0 & 0 & 0 & 0 & 0 \end{pmatrix} \tag{13.14}$$

with

$$\mathsf{v}_r \mathsf{Y}_{kij} = \mathsf{v}_k \mathsf{Y}_{rij}, \tag{13.15}$$

for $r, i, k = 1, 2, \ldots, \mathsf{h}_0^{3,1}$, $j = 1, 2, \ldots, \mathsf{h}_0^{2,2}$.

Let C be a period vector of Hodge type as in (6.15):

$$\mathsf{C} = \begin{pmatrix} 0 \\ 0 \\ \mathsf{C}^{\frac{n}{2}} \\ 0 \\ 0 \end{pmatrix}, \quad \mathsf{C}^{\frac{n}{2}} = \begin{pmatrix} 1 \\ 0 \\ \vdots \\ 0 \end{pmatrix}. \tag{13.16}$$

For simplicity we have assumed that all the entries of the middle Hodge block $\mathsf{C}^{\frac{n}{2}}$ are also zero except for the first entry which is 1.

Proposition 13.9. *The foliation* $\mathscr{F}(2) \cap \mathscr{F}(\mathsf{C})$ *is of dimension zero and its singular set is given by*

$$\mathrm{Sing}(\mathscr{F}(2) \cap \mathscr{F}(\mathsf{C})) = \{\det[\mathsf{Y}_{ij1}]_{\mathsf{h}_0^{3,1} \times \mathsf{h}_0^{3,1}} = 0\}.$$

Proof. A vector field $\sum_{k=1}^{\mathsf{h}_0^{3,1}} f_k \mathsf{v}_k$ is tangent to the mother foliation $\mathscr{F}(2)$ if

$$\sum_{k=1}^{\mathsf{h}_0^{3,1}} f_k \mathsf{A}_{\mathsf{v}_k} \mathsf{C} = 0.$$

The special format of C implies that all the Hodge blocks of $\mathsf{A}_{\mathsf{v}_k}\mathsf{C}$ are zero except for the middle one which is $[\mathsf{Y}_{ki1}]_{\mathsf{h}_0^{3,1} \times 1}$. □

13.9 Periods of Calabi–Yau threefolds

Apart from the polynomial relations between the entries of P_m described in Proposition 4.1 we do not know any other relation which might be used in the definition of the period domains Π, U in order to make it smaller dimension. For Calabi–Yau threefolds with constant Yukawa couplings the following discussion might be developed into a precise formulation of extra polynomial relations. The main sources for this section are [AMSY16, Ali17]. For the special case of mirror quintic and in general Calabi–Yau threefolds with Hodge numbers of the third cohomology all equal to one, similar discussions are developed in [Mov17b] with different notations.

Let M be a moduli space of Calabi–Yau threefolds X with $h = \mathsf{h}^{2,1}(X)$. Let also $\mathbb{H}$ be a connected component of the moduli of (X, δ), where $X \in \mathsf{M}$ and δ stands for a symplectic basis $\delta_1, \delta_2, \ldots, \delta_{h+1}, \delta_{h+2} \ldots, \delta_{2h+2}$ of $H_3(X, \mathbb{Z})$, that is

$$[\delta_i \cdot \delta_j] = \begin{bmatrix} 0 & I_{(h+1)\times(h+1)} \\ -I_{(h+1)\times(h+1)} & 0 \end{bmatrix}.$$

Note that this moduli space might have infinitely many components, for instance, this is the case of mirror quintic, see the discussion in [Mov17b, §4.6].

Proposition 13.10 ([AMSY16, Proposition 3]). *For (X, δ) in an open dense subset of $\mathbb{H}$ there is a unique enhanced variety (X, α) such that the period matrix is of the form*

$$\left[\int_{\delta_j} \alpha_i\right] = \tau^{\mathsf{tr}} := \begin{bmatrix} 1 & \tau_j & F_j & 2F_0 - \sum_{l=1}^{h} \tau_l F_l \\ 0 & \delta_j^i & F_{ij} & F_i - \sum_{l=1}^{h} \tau_l F_{li} \\ 0 & 0 & \delta_j^i & -\tau_i \\ 0 & 0 & 0 & -1 \end{bmatrix} \tag{13.17}$$

for some holomorphic function F_0 in τ which is called the prepotential and $F_i := \frac{\partial F_0}{\partial \tau_i}$.

The functions τ_i are those which are used in Bogomolov–Tian–Todorov theorem (Theorem 13.3). In the first line of the equality we claim that

$$\left(\frac{\int_{\delta_1}\omega}{\int_{\delta_1}\omega}, \frac{\int_{\delta_2}\omega}{\int_{\delta_1}\omega}, \frac{\int_{\delta_3}\omega}{\int_{\delta_1}\omega}, \ldots, \frac{\int_{\delta_{h+1}}\omega}{\int_{\delta_1}\omega}, \frac{\int_{\delta_{h+2}}\omega}{\int_{\delta_1}\omega}, \ldots, \frac{\int_{\delta_{2h+1}}\omega}{\int_{\delta_1}\omega}, \frac{\int_{\delta_{2h+2}}\omega}{\int_{\delta_1}\omega}\right)$$
$$= \left(1, \tau_1, \tau_2, \ldots, \tau_h, \frac{\partial F_0}{\partial \tau_1}, \ldots, \frac{\partial F_0}{\partial \tau_h}, 2F_0 - \sum_{i=1}^{h} \tau_i \frac{\partial F_0}{\partial \tau_i}\right)$$

for a holomorphic $(3,0)$-form ω on X. Evidently, due to the quotient, this does not depend on the choice of ω. Proposition 13.10 gives us a meromorphic map

$$\mathsf{t} : \mathbb{H} \dashrightarrow \mathsf{T}, \; (X, \delta) \mapsto (X, \alpha). \tag{13.18}$$

For (X, δ) we first choose an arbitrary enhancement $(X, \tilde{\alpha})$. For the period matrix $\tilde{\mathsf{P}} := [\int_{\delta_i} \tilde{\alpha}_j]$ there is a unique $\mathsf{g} \in \mathsf{G}$ such that $\tilde{\mathsf{P}}\mathsf{g}$ is of the form (13.17). For this we have divided over many periods $\int_{\delta_i} \alpha_j$ which might be zero, and the meromorphic points of the map (13.18) is due to these zeros.

For further details in the case of mirror quintic see [Mov17b, §§4.6 and 4.7]. We call (13.18) the special map. Here, the adjective "special" is taken from the special geometry of mirror symmetry.

Proposition 13.11 ([AMSY16, Proposition 5]). *For the period matrix in* (13.17) *we have*

$$d\tau^{\mathsf{tr}} \cdot \tau^{-tr} = \begin{bmatrix} 0 & d\tau_j & 0 & 0 \\ 0 & 0 & \sum_{k=1}^{h} F_{ijk} d\tau_k & 0 \\ 0 & 0 & 0 & d\tau_i \\ 0 & 0 & 0 & 0 \end{bmatrix}. \tag{13.19}$$

Proof. This follows from (13.17) after performing explicit matrix multiplication. □

Recall from §13.7 the modular vector fields v_k, $k = 1, 2, \ldots, h$ in the moduli T of enhanced Calabi–Yau threefolds.

Theorem 13.12. *The modular vector fields* v_k*'s are tangent to the image of the map* t *in* (13.18) *and* $\frac{\partial}{\partial \tau_k}$ *is mapped to* v_k *under the derivation of* t.

Proof. This follows from Proposition 13.11, Theorem 13.7 and the uniqueness of the vector field v_k. □

13.10 Calabi–Yau threefolds with constant Yukawa couplings

Recall the algebraic Yukawa couplings in §13.7. It might be useful to describe the functional equation of Y_{ijk} with respect to the action of G. By Proposition 3.6 we know that the pull-back of the Gauss–Manin connection matrix A under the isomorphism $\mathsf{g} : \mathsf{T} \to \mathsf{T}$, $t \mapsto t \bullet \mathsf{g}$ is $\mathsf{g}^{\mathsf{tr}} \cdot \mathsf{A} \cdot \mathsf{g}^{-\mathsf{tr}}$. If v_k is a modular vector field in T and $\mathsf{g}_* \mathsf{v}_k$ is its push-forward then we have $\mathsf{g}^{\mathsf{tr}} \cdot \mathsf{Y} \cdot \mathsf{g}^{-\mathsf{tr}} = \mathsf{A}(\mathsf{g}_* \mathsf{v}_k)$ which is not in the desired format of the Yukawa coupling matrix, and hence, $\mathsf{g}_* \mathsf{v}_k$ is not modular. We might work out further in order to get the functional equation of Y_{ijk}. It seems that such a functional equation does only depends on the Hodge numbers and not the underlying Calabi–Yau varieties, see for instance the case of Calabi–Yau threefolds with $(h^{3,0}, h^{2,1}) = (1, 1)$ in [Mov17b, §§2.4 and 7.8]. This means that the algebraic Yukawa couplings Y_{ijk} might not be constant, even though the Yukawa coupling which are mainly used in physics and are written in terms of period might be constant.

Definition 13.2. We say that a Calabi–Yau threefold (or in fact its moduli) has constant Yukawa couplings if there are constants c_{ijk} symmetric in $i, j, k = 1, 2, \ldots, \mathsf{h}^{2,1}$ such that the variety

$$\check{\mathsf{T}}: \quad \mathsf{Y}_{ijk} - c_{ijk} = 0, \quad i, j, k = 1, 2, \ldots, \mathsf{h}^{21} \tag{13.20}$$

is non-empty and the modular vector fields v_l, $l = 1, 2, \ldots, \mathsf{h}^{21}$ are tangent to $\check{\mathsf{T}}$. In other words, the quantities $\mathsf{v}_l(\mathsf{Y}_{ijk})$ restricted to $\check{\mathsf{T}}$ are zero.

Let us consider Calabi–Yau threefolds such that prepotential F_0 is a polynomial of degree at most 3 in $\tau_1, \tau_2, \ldots, \tau_h$. By Proposition 13.11, the pull-back of the algebraic Yukawa couplings Y_{ijk} by t is F_{ijk} which is a constant c_{ijk}. This implies that the image of t is inside $\check{\mathsf{T}}$ given by (13.20). This might suggest that the algebraic Yukawa couplings are constant in the sense of Definition 13.2. For Calabi–Yau threefolds with constant Yukawa couplings, there are many polynomial relations among the entries of the period matrix $\mathsf{P}(t)$, $t \in \mathsf{T}$, and this indicates that one might be able to define smaller dimensional period domains.

Conjecture 13.2. *A moduli of Calabi–Yau threefolds is a quotient of a Hermitian symmetric domain (constructed from periods) by an arithmetic group if and only if the corresponding algebraic Yukawa couplings are constants in the sense of Definition 13.2.*

The conjecture must be formulated precisely, as the term "constructed from periods" is ambiguous. Note that Conjecture 13.2 is analogous to Conjecture 6.2 which is formulated for arbitrary projective varieties. Due to the enumerative meaning of Yukawa couplings Y_{ijk} in the case of Calabi–Yau threefolds we have reformulated it again. The main idea is that Calabi–Yau threefolds with constant Yukawa couplings are like K3 surfaces and abelian varieties. For them there is a new period domain which is much smaller than the Griffiths period domain and such that the period map is surjective. There is a list of examples for which Conjecture 13.2 is true or expected. This includes C. Schoen's fiber product of two rational elliptic surfaces in [Sch88] and Borcea–Voisin's examples [Bor97, Voi93], see also [EK93, Roh09, GvG10]. In a personal communication (February 5, 2018) E. Scheidegger informed the author most of the references cited in this paragraph. In many of these examples a common feature is that they do not admit a maximal unipotent monodromy. This can be rigorously proved for a class of Calabi–Yau threefolds which we discuss in §13.11. Another class of Calabi–Yau varieties which might fit into our discussion is those which are cyclic covers branched over a $(2n+2)$ hyperplane arrangements in a general position in $\mathbb{P}^n$. A particular family of such Calabi–Yau varieties can be also constructed as a quotient of a product of higher genus curves, see [SXZ13, §3; SZ10]. This is informed to the author by Tsung-Ju Lee (personal communication, January 15, 2019). For these classes of Calabi–Yau varieties it seems that some of the Yukawa couplings are constant and some are not. In the case of K3 surfaces, where we do not have Yukawa couplings, the description of the moduli space as a quotient of a Hermitian symmetric domain by an arithmetic group, and the corresponding automorphic forms needed for inverting the mirror map, has been discussed in [HLTY18]. It is also interesting to investigate the relation of Conjecture 13.2 with the description of Hermitian symmetric domains in [Yau93].

The following statement which is similar to Proposition 13.10 might be helpful for understanding Conjecture 13.2. For lattice polarized $K3$ surfaces we define $\mathbb{H}$ in a similar way as for Calabi–Yau threefolds.

Proposition 13.13 ([Ali17, Proposition 7.1]). *For (X,δ) in an open dense subset of $\mathbb{H}$ there is a unique enhanced lattice polarized $K3$ surface (X,α) such that the period matrix is of the form*

$$\left[\int_{\delta_j}\alpha_i\right] = \begin{bmatrix} 1 & \tau_j & F_0 \\ 0 & \delta^i_j & \tau_i \\ 0 & 0 & 0 \end{bmatrix} \tag{13.21}$$

for some degree two homogeneous polynomial F_0 in τ_i's.

Proposition 13.10 for Calabi–Yau threefolds with constant Yukawa couplings and Proposition 13.13 suggest that there might be some relation between these two contexts. For instance, there might be a birational map between $X_1\times X_2\times X_3/G_1$ and $Y_1\times Y_2/G_2$, where X_i's (respectively, Y_i's) are three lattice polarized $K3$ surfaces (respectively, two Calabi–Yau threefolds) and G_1 (respectively, G_2) is a finite group acting on their product.

13.11 Calabi–Yau equations

Conjecture 13.2 seems to be non-trivial even in the case of Hodge numbers $\mathsf{h}^{i,3-i}$ all equal to one. In this section, we discuss the classification of such Calabi–Yau threefolds with constant Yukawa couplings in the sense that the prepotential F_0 becomes a polynomial of degree 3, see the comments after Definition 13.2. We follow the notations used in [Mov17b] which is different (in particular those related to periods) from the notation of an arbitrary Calabi–Yau threefolds used in this chapter.

Let us consider a family $X_z, z \in \mathbb{P}^1$ of Calabi–Yau threefolds with $\mathsf{h}^{1,2} = \mathsf{h}^{2,1} = 1$ and with possibly finitely many singular fibers.

Definition 13.3. The periods $\int_{\delta_z} \omega^{3,0}$, $\delta_z \in H_3(X_z, \mathbb{Z})$ generate a $\mathbb{C}$-vector space of dimension ≤ 4, and we assume that this dimension is exactly four, (for many examples of Calabi–Yau threefolds such that this vector space is of dimension two see [Cv17]). This implies that they generate the solution space of a Picard–Fuchs equation of order four:

$$L := \theta^4 - \sum_{i=0}^{3} a_i(z)\theta^i = 0, \tag{13.22}$$

where $\theta = z\frac{\partial}{\partial z}$ and a_i's are rational functions in z. The Picard–Fuchs equation (13.22) is called a Calabi–Yau equation (of order 4).

We define:

$$a_4 := \frac{1}{4}a_3^2 + a_2 - \frac{1}{2}\theta a_3, \quad a_5(z) := e^{\frac{1}{2}\int a_3(z)\frac{dz}{z}}. \tag{13.23}$$

Griffiths transversality implies the following equality:

$$a_1 = -\frac{1}{2}a_2a_3 - \frac{1}{8}a_3^3 + \theta a_2 + \frac{3}{4}a_3\theta a_3 - \frac{1}{2}\theta^2 a_3. \tag{13.24}$$

See [Mov17b, Proposition 15 and Remark 9]. The algebraic condition (13.24) can be written as

$$a_1 = -\frac{1}{2}a_3a_4 + \theta a_4 \tag{13.25}$$

where a_4 is given in (13.23). The condition (13.24) is not changed when a solution of L (equivalently $\omega^{3,0}$) is multiplied with a holomorphic function f. In particular, for $f = a_5$ in (13.23) we get a new differential equation with $a_3 = 0$. In this case we have $a_4 = a_2$ and (13.24) becomes

$$a_1 = \theta a_2. \tag{13.26}$$

Proposition 13.14. *A Calabi–Yau equation L with a constant Yukawa coupling is the third symmetric power of a second-order Fuchsian linear* *differential equation:*

$$\check{L}: \theta^2 - b_1(z)\theta - b_0(z) = 0, \tag{13.27}$$

that is, $L = \mathrm{sym}^3 \check{L}$.

Proof. Let us assume that the Yukawa coupling of X_z, $z \in \mathbb{P}^1$ is constant. It turns out that the prepotential F_0 a polynomial of degree ≤ 3 in τ_0 and

$$\tau_1 := \frac{\partial F_0}{\partial \tau_0}, \quad \tau_3 = \frac{\partial^2 F_0}{\partial \tau_0^2}, \ \mathsf{Y} = \frac{\partial^3 F_0}{\partial \tau_0^3}, \tag{13.28}$$

$$\tau_2 = \int \left(\frac{\partial F_0}{\partial \tau_0} - \tau_0 \frac{\partial^2 F_0}{\partial \tau_0^2} \right) = 2F_0 - \tau_0 \frac{\partial F_0}{\partial \tau_0}, \tag{13.29}$$

where we are using the following τ-locus format

$$\tau = \begin{pmatrix} \tau_0 & 1 & 0 & 0 \\ 1 & 0 & 0 & 0 \\ \tau_1 & \tau_3 & 1 & 0 \\ \tau_2 & -\tau_0\tau_3 + \tau_1 & -\tau_0 & 1 \end{pmatrix} \tag{13.30}$$

used in [Mov17b, Chapter 4]. By our assumption in Definition 13.3 the degree of F_0 in τ_0 is exactly 3. We have $\tau_0 := \frac{x_{11}}{x_{21}}$, $\tau_1 = \frac{x_{31}}{x_{21}}$ and $\tau_2 = \frac{x_{41}}{x_{21}}$ for four linearly independent solutions x_{i1}, $i = 1, 2, 3, 4$ of L. If we write $x_{21} = y_0^3$ and $x_{11} = y_0^2 y_1$ we conclude that the solution space of L is generated by y_0^3, $y_0^2 y_1$, $y_0 y_1^2, y_1^3$, where y_0, y_1 form a basis of solutions of a second-order differential equation (13.27). This implies that $L = \mathrm{sym}^3 \check{L}$. In a personal communication (June 14, 2018), S. Reiter kindly reminded the author that the third symmetric product $\mathrm{sym}^3 \check{L}$ of an arbitrary second-order linear differential equation $\check{L}$ is given by:

$$\begin{aligned} \mathrm{sym}^3 \check{L} : \theta^4 &- 6b_1\theta^3 + \left(11b_1^2 - 4\theta b_1 - 10b_0\right)\theta^2 \\ &+ \left(-6b_1^3 + 7b_1\theta b_1 + 30b_0 b_1 - 10\theta b_0 - \theta^2 b_1\right)\theta \\ &- 18b_1^2 b_0 + 6b_0\theta b_1 + 15(\theta b_0)b_1 + 9b_0^2 - 3\theta^2 b_0. \end{aligned} \tag{13.31}$$

This implies that b_0, b_1 are rational functions in z and $\check{L}$ is Fuchsian. □

Proposition 13.15. *The third symmetric product of an arbitrary second-order linear differential equation satisfies* (13.24).

Proof. This can be verified directly from the formula (13.31). Another proof is as follows. The second-order Fuchsian differential equation (13.27) can be transformed into SL-form by substituting y by fy, where $0 \neq f$ satisfies $\theta f = \frac{1}{2}b_1(z)f$. Then by [IKSY91, p. 166]

$$\theta^2 y = p(z)y, \, p(z) = b_0 + \tfrac{1}{4}b_1^2 - \tfrac{1}{2}\theta b_1. \tag{13.32}$$

For $b_1 = 0$, equation (13.31) is just

$$\mathrm{sym}^3 \check{L} : \theta^4 - 10b_0\theta^2 - 10\theta b_0\theta + 9b_0^2 - 3\theta^2 b_0.$$

This satisfies automatically (13.26). □

Example 13.1. This example is take from [Bv95, p. 525]. The product of three elliptic cubic curves in $\mathbb{P}^2$ is an abelian threefold. The periods of its mirror satisfy the Picard–Fuchs equation

$$\begin{aligned} L := \theta^4 &- 3z(6 + 29\theta + 56\theta^2 + 54\theta^3 + 27\theta^4) + 81z^2(27\theta^2 + 54\theta + 40)(\theta + 1)^2 \\ &- 2187z^3(3\theta + 5)(3\theta + 4)(\theta + 2)(\theta + 1). \end{aligned}$$

In a personal communication (February 17, 2018), Duco van Straten informed the author about the following Picard–Fuchs equation with a constant Yukawa coupling.

$$\begin{aligned} L := \theta^4 &- 12z(4\theta + 1)(4\theta + 3)(18\theta^2 + 18\theta + 7) \\ &+ 2^4 3^6 z^2 (4\theta + 1)(4\theta + 3)(4\theta + 5)(4\theta + 7). \end{aligned}$$

We would like to classify all Calabi–Yau equations, and once this is done, we may try to classify the underlying Calabi–Yau threefolds with Hodge numbers $\mathsf{h}^{i,3-i} = 1$. This might be as hard as the Hodge conjecture and construction of algebraic cycles. We have to collect properties for the Picard–Fuchs equation L which has been derived from the underlying geometry. This job with non-constant Yukawa coupling and conjecturally integral instanton numbers is undertaken by Almkvist, Enckevort, van Straten and Zudilin in [AvEvSZ10], and up to the writing of the present text they were able to find approximately 400 such equations. Calabi–Yau equations L with constant Yukawa coupling are not included in their list. In Proposition 13.15, we have already seen that $L = \mathrm{sym}^3\check{L}$, where $\check{L}$ is a second-order differential equation. One needs more data from the underlying geometry in order to characterize Calabi–Yau equations with constant Yukawa coupling. Here are some suggestions.

Let $\check{L}$ be the a second-order differential equation as in (13.27). The monodromy $\check{\Gamma} \subset \mathrm{GL}(2, \mathbb{C})$ of $\check{L}$ satisfies

$$\Gamma = A\left(\mathrm{sym}^3 \check{\Gamma}\right) A^{-1} \subset \mathrm{Sp}(4, \mathbb{Z})$$

for some 4×4 matrix A with $\det(A) \neq 0$. Even if $\check{\Gamma} \subset \mathrm{SL}(2, \mathbb{Z})$ we may have A with this property which has not rational entries. We have to classify all degree three F_0 such that

$$\mathrm{Im}(\tau_0 \bar{\tau}_1 + \bar{\tau}_2) < 0, \tag{13.33}$$

$$\mathrm{Re}(\tau_1(-\tau_0\tau_3 + \tau_1) - \tau_2\tau_3 - (\tau_0(-\tau_0\tau_3 + \tau_1) - \tau_2)\bar{\tau}_3) - |-\tau_0\tau_3 + \tau_1|^2 < 0. \tag{13.34}$$

These are the positivity conditions of the Hodge structure $H^3(X_z) = H^{30} \oplus H^{21} \oplus H^{12} \oplus H^{03}$, see [Mov17b, §4.3]. Further, we may assume that the family of Calabi–Yau threefolds X_z (or the holomorphic function $\int_{\delta_z} \omega^{3,0}$) is not a pull-back under a rational map $p : \mathbb{P}^1 \to \mathbb{P}^1$.

Since the Yukawa coupling is constant, one may try to find some constrains for Calabi–Yau equations from genus 1 instanton numbers. Such numbers are encoded in the following expression:

$$\mathsf{F}_1^{\mathrm{hol}} := -\frac{1}{2}\log\left(\left(\theta\frac{x_{11}}{x_{21}}\right)^{-1} x_{21}^{-3-h^{11}+\frac{\chi}{12}} \prod_i (b_i - z)^{*_i}\right). \tag{13.35}$$

Here, b_i's are singularities of L and $*_i$'s are some constant ambiguities. The number $h^{1,1}$ and χ are $(1,1)$-Hodge number and Euler number of X_z, respectively. In the absence of the geometric object X_z they might be also considered as ambiguities. When X_z has a singularity, let us say $z = 0$, with maximal unipotent monodromy, we have a recipe to define q-coordinate around $z = 0$ and inside the log one has a q-expansion of the form (up to multiplication with a constant):

$$\begin{aligned} &q^{a_0}\left(\prod_{n=1}^{\infty}(1-q^n)^{\sum_{r|n} d_r}\right)^{a_1}\left(\prod_{s=1}^{\infty}(1-q^s)^{n_s}\right)^{a_2} \\ &= q^{a_0}\left(\prod_{n=1}^{\infty}(1-q^n)^{\sum_{r|n} d_r}\right)^{a_1}. \end{aligned} \tag{13.36}$$

Since the Yukawa coupling is constant, we have $n_s = 0$ for all $s \geq 1$ which is used in the above equality. However, for Calabi–Yau equations with constant Yukawa coupling we do not have maximal unipotent monodromy.

This follows from Proposition 13.14 which implies that the differential Galois group of L is a proper subgroup of $\mathrm{Sp}(4,\mathbb{C})$, for further details see [Mov17b, §7.6]. Even though, we have $L = \mathrm{Sym}^3\check{L}$ and one might use the q-coordinate constructed from the solution of $\check{L}$ around a point $z = 0$ with maximal unipotent monodromy for $\check{L}$. In this case one can develop modular form theories attached to $\check{L}$ in a similar style as for Gauss hypergeometric equation, see [DGMS13]. It turns out that for a class of Gauss hypergeometric equations the corresponding modular form theory has a basis with integral Fourier coefficients if and only if the corresponding monodromy group is arithmetic, and hence, we have a finite list of them, see [Mov17b, Appendix C; MS14]. This might be the case for second-order differential equations $\check{L}$ computed from Calabi–Yau equations.

Appendix

A Geometric Introduction to Transcendence Questions on Values of Modular Forms

TIAGO J. FONSECA

A.1. Introduction

One of the most striking arithmetical applications of Ramanujan's relations between the normalized Eisenstein series E_2, E_4, E_6 (see [Ram16; NP01, Chapter 1]), namely

$$\frac{1}{2\pi i}\frac{dE_2}{d\tau} = \frac{E_2^2 - E_4}{12}, \quad \frac{1}{2\pi i}\frac{dE_4}{d\tau} = \frac{E_2E_4 - E_6}{3}, \quad \frac{1}{2\pi i}\frac{dE_6}{d\tau} = \frac{E_2E_6 - E_4^2}{2} \tag{R}$$

is the following algebraic independence theorem proved by Nesterenko in 1996.

Theorem A.1 ([Nes96]). *For every* $\tau \in \mathbb{H} = \{z \in \mathbb{C} \mid \operatorname{Im} z > 0\}$, *we have*

$$\operatorname{trdeg}_{\mathbb{Q}}\mathbb{Q}(e^{2\pi i\tau}, E_2(\tau), E_4(\tau), E_6(\tau)) \geq 3.$$

This means that among the four complex numbers $e^{2\pi i\tau}$, $E_2(\tau)$, $E_4(\tau)$, and $E_6(\tau)$ there are always three of them which are algebraically independent over $\mathbb{Q}$.

Nesterenko's result is remarkable both in its short and powerful statement as in its proof method. In the next sections, we explain some applications of Nesterenko's theorem and the main ideas of its proof. For other accounts of Nesterenko's proof, the reader may consult, besides the original paper [Nes96], the collective volumes [NP01, FGK05]. Here, we shall emphasize the special role played by the dynamics of the algebraic differential equations (R) in the guise of Nesterenko's "D-property".

To help the reader with no background in Transcendental Number Theory, we shall start with a brief overview of some of its main concepts and results. Let us point out, however, that some major results such as Baker's theorems or Wüstholz analytic subgroup theorem are not discussed. For more complete and better written introductions to this same subject, we refer to the classic [Bak75], or to the more recent [MR14].

In the last section, we give a very short introduction to "periods" and discuss some general transcendence questions related to Nesterenko's theorem. Periods are complex numbers given by integrals in algebraic geometry which have recently gained much attention of the Number Theory community due to their deep connections with the theory of motives. This places Nesterenko's theorem in a larger context and allows us not only to better appreciate its content, but also to dream and speculate on future generalizations.

A.2. A biased overview of transcendence theory

Transcendence theory is one of the oldest, and reputedly one of the most difficult, domains of Mathematics. Here, we can only scratch the surface.

A.2.1. *First notions*

In mathematics, "transcendental" is the antonym of "algebraic". Accordingly, a complex number α is said to be *transcendental* if it is not algebraic — that is, $P(\alpha) \neq 0$ for every $P \in \mathbb{Q}[X] \setminus \{0\}$. Similarly, we say that a function f (seen either as a formal Laurent series in $\mathbb{C}((t))$ or as a meromorphic function on some open domain of $\mathbb{C}$) is transcendental if it is not an algebraic function: $P(t, f(t)) \not\equiv 0$ for every $P \in \mathbb{C}[X, Y] \setminus \{0\}$.

The following definition generalizes these notions.

Definition A.1. Let $k \subset K$ be a field extension. We say that elements $\alpha_1, \ldots, \alpha_n$ of K are *algebraically independent* over k, or that the set $\{\alpha_1, \ldots, \alpha_n\} \subset K$ is algebraically independent over k, if

$$P(\alpha_1, \ldots, \alpha_n) \neq 0$$

for every polynomial $P \in k[X_1, \ldots, X_n] \setminus \{0\}$. When $n = 1$, we rather say that α_1 is *transcendental* over k.

Some of the most traditional choices of fields k and K are as follows:

	k	K
arithmetic case	$\mathbb{Q}$	$\mathbb{C}$
functional case	$\mathbb{C}$	$\mathbb{C}((t_1, \ldots, t_m))$

These give origin to the two main branches of transcendence theory: arithmetic transcendence (or transcendental number theory), and functional transcendence. Albeit essentially distinct, these two branches are subtly, and often mysteriously, intertwined.

Remark A.1. One could also replace the field of formal Laurent series $\mathbb{C}((t_1, \ldots, t_m))$ above by the field of meromorphic functions on some open domain of $\mathbb{C}^m$. In the one variable case, it is common to consider $\mathbb{C}(t)$ as the base field k. Our framework includes this one, since $f_1, \ldots, f_n \in \mathbb{C}((t))$ are algebraically independent over $\mathbb{C}(t)$ if and only if $f_0 = t, f_1, \ldots, f_n$ are algebraically independent over $\mathbb{C}$.

Closely related to the notion of algebraic independence is the quantitative notion of transcendence degree of a field extension.

Definition A.2. Let $k \subset K$ be a field extension. A subset S of K is algebraically independent over k if every finite subset of S is algebraically independent over k. The *transcendence degree* of K over k, denoted by $\mathrm{trdeg}_k\ K$, is the maximal cardinality of a subset of K algebraically independent over k.

For instance, let us assume that $K = k(\alpha_1, \ldots, \alpha_n)$. Then $\mathrm{trdeg}_k\ K \leq n$. To assert that $\mathrm{trdeg}_k\ K \leq n-1$ is equivalent to assert that there exists a non-trivial algebraic relation, with coefficients in k, between $\alpha_1, \ldots, \alpha_n$. If $1 \leq r \leq n$, to assert that $\mathrm{trdeg}_k\ K \geq r$ is equivalent to assert that some subset of r elements of $\{\alpha_1, \ldots, \alpha_n\}$ is algebraically independent over k.

Example A.1 (Exponential function). Consider t as a formal variable and let

$$e^t = \sum_{m=0}^{\infty} \frac{t^m}{m!} \in \mathbb{C}((t))$$

be the exponential power series. Then t and e^t are algebraically independent over $\mathbb{C}$. In other words, e^t is transcendental over $\mathbb{C}(t)$. We prove this by contradiction. If e^t were algebraic over $\mathbb{C}(t)$, then there would exist a minimal integer $d \geq 1$ for which there are polynomials $P_0, \ldots, P_d \in \mathbb{C}[X]$ satisfying

$$\sum_{j=0}^{d} P_j(t)e^{jt} = 0.$$

By differentiating with respect to t and by subtracting the resulting equation from d times the original equation, we obtain

$$\sum_{j=0}^{d} (P_j'(t) + (j-d)P_j(t))e^{jt} = 0,$$

so that the leading coefficient is now $P_d'(t)$. By induction, we would obtain polynomials $Q_0, \ldots, Q_{d-1} \in \mathbb{C}[X]$ (not all zero; check!) such that $\sum_{j=0}^{d-1} Q_j(z)e^{jt} \equiv 0$, thereby contradicting the minimality of d.

In general, the notion of algebraic independence admits the following scheme-theoretic interpretation. Let $k \subset K$ be a field extension, and consider a K-point $\alpha = (\alpha_1, \ldots, \alpha_n) \in \mathbb{A}_k^n(K)$. Then, $\alpha_1, \ldots, \alpha_n$ are algebraically independent over k if and only if the image of

$$\alpha : \operatorname{Spec} K \longrightarrow \mathbb{A}_k^n$$

is dense in $\mathbb{A}_k^n$ for the Zariski topology — that is, if $Y \subset \mathbb{A}_k^n$ is a closed k-subvariety such that $\alpha \in Y(K)$, then $Y = \mathbb{A}_k^n$.

This point of view allows us to give concrete geometric significance to functional transcendence. For instance, algebraic independence of one variable functions corresponds to Zariski-density of parameterized curves. To fix ideas, let U be a neighborhood of $0 \in \mathbb{C}$, and let $\varphi_1, \ldots, \varphi_n$ be holomorphic functions on U. We thus obtain a holomorphic curve

$$\varphi = (\varphi_1, \ldots, \varphi_n) : U \longrightarrow \mathbb{C}^n = \mathbb{A}_{\mathbb{C}}^n(\mathbb{C}).$$

To say that $\varphi_1, \ldots, \varphi_n$ (seen as formal power series) are algebraically independent over $\mathbb{C}$ is equivalent to say that the image of the curve φ is Zariski-dense in $\mathbb{A}_{\mathbb{C}}^n$.

Example A.2 (Exponential function, revisited). It follows from the above discussion that the transcendence of e^t over $\mathbb{C}(t)$ is equivalent to the

Zariski-density of the image of the holomorphic curve

$$\varphi : \mathbb{C} \longrightarrow \mathbb{C}^2, \qquad z \longmapsto (z, e^z).$$

Geometrically, this can be proved as follows. Assume that there exists an irreducible algebraic curve $C \subset \mathbb{A}^2_{\mathbb{C}} = \operatorname{Spec} \mathbb{C}[X, Y]$ containing the image of φ. Since $e^{2\pi i n} = 1$ for every $n \in \mathbb{Z}$, C intersects the line $V(Y-1)$ at an *infinite* number of points: $(2\pi i n, 1) \in \mathbb{C}^2$ for $n \in \mathbb{Z}$. As C and $V(Y-1)$ are both irreducible, this is only possible if $C = V(Y-1)$ (by a weaker form of Bézout's theorem). This would imply that $e^z \equiv 1$, which is clearly absurd.

A.2.2. *Arithmetic transcendence and Diophantine approximation*

It is widely acknowledged that functional transcendence is easier than arithmetic transcendence. Indeed, while results concerning functional transcendence date back to the founding fathers of calculus (see [And17, Footnote 1, p. 2]), the first transcendence proof of explicitly defined numbers appears in Liouville's landmark paper [Lio51].

The fundamental insight of Liouville was to establish a link between arithmetic transcendence and Diophantine approximation.

Theorem A.2 (Liouville). *If a real number α is algebraic of degree $d > 1$ over $\mathbb{Q}$, then there exists $c = c(\alpha) > 0$ such that*

$$\left|\alpha - \frac{p}{q}\right| > \frac{c}{q^d}$$

for every rational number of the form p/q, with $p, q \in \mathbb{Z}$ coprime and $q > 0$.

Proof. Let $P \in \mathbb{Z}[X]$ be an irreducible polynomial of degree d such that $P(\alpha) = 0$, and take p/q as above with $|\alpha - p/q| < 1$. By considering the Taylor expansion of P at α, we obtain

$$|P(p/q)| = \left| \sum_{i=1}^{d} \frac{P^{(i)}(\alpha)}{i!} (p/q - \alpha)^i \right| \le M |p/q - \alpha|,$$

where $M = \sum_{i=1}^{n} |P^{(i)}(\alpha)/i!|$. Since P is irreducible, we have $P(p/q) \neq 0$; since it is of degree d and has integral coefficients, we have $q^d P(p/q) \in \mathbb{Z} \setminus \{0\}$, so that $|P(p/q)| \geq 1/q^d$. We can thus take $c = \min\{1, (2M)^{-1}\}$. □

For instance, the above result gives the transcendence of $\alpha = \sum_{n=0}^{\infty} 10^{-n!}$.

Remark A.2 (The fundamental theorem of transcendence). The seemingly innocuous observation, used in the above proof, that the absolute value of a non-zero integer is at least 1 is at the heart of virtually *every* proof in arithmetic transcendence (cf. [Mas16]).

Liouville's theorem can be regarded as a general transcendence criterion. Some other general algebraic independence criteria in terms of Diophantine approximation exist. Nesterenko's proof of his theorem on values of Eisenstein series, to be discussed below, relies on the following particular case of Philippon's sophisticated criteria [Phi86].

For a polynomial $P \in \mathbb{C}[X_1, \ldots, X_n]$, we denote by $\|P\|_\infty$ the maximum of the absolute values of its coefficients.

Theorem A.3 (Philippon). *Let $n \geq 2$ be an integer and $\alpha_1, \ldots, \alpha_n$ be complex numbers. Suppose that there exists an integer $r \geq 2$ and real constants $a > b > 0$ such that, for every sufficiently large positive integer d, there exists a polynomial $Q_d \in \mathbb{Z}[X_1, \ldots, X_n] \setminus \{0\}$ of degree $\deg Q_d = O(d \log d)$ satisfying*

$$\log \|Q_d\|_\infty = O(d(\log d)^2)$$

and

$$-ad^r \leq \log |Q_d(\alpha_1, \ldots, \alpha_n)| \leq -bd^r .$$

Then

$$\operatorname{trdeg}_{\mathbb{Q}} \mathbb{Q}(\alpha_1, \ldots, \alpha_n) \geq r - 1.$$

Geometrically, we may interpret the hypotheses of the above result as an approximation condition of α by hypersurfaces in $\mathbb{A}^n_{\mathbb{Q}}$ in terms of their degree and their "arithmetic complexity".

A.2.3. *Schneider–Lang and Siegel–Shidlovsky*

The connection between transcendence and Diophantine approximation suggests a closer inspection on values of analytic functions. Indeed, properties of such functions such as growth conditions, and differential or functional equations, can provide additional tools to the study of the approximation properties of their values.

Historically, this vague idea culminated in two precise and general theorems, those of Schneider–Lang and Siegel–Shidlovsky, which deal with entire or meromorphic functions over $\mathbb{C}$ satisfying some algebraic differential equation, a growth condition, a functional transcendence statement, and some hypotheses of arithmetic nature.

We next state both theorems and derive some consequences, without saying anything about their proofs. The idea here is simply to help the reader to put Nesterenko's theorem in perspective (see Appendix A.4.3. below).

A.2.3.1. *Schneider–Lang*

Given a real number $\rho > 0$, we say that the *order* of an entire function f on $\mathbb{C}$ is $\leq \rho$ if there exist real numbers $a, b > 0$ such that

$$|f(z)| \leq a e^{b|z|^\rho}$$

for every $z \in \mathbb{C}$. A meromorphic function on $\mathbb{C}$ is of order $\leq \rho$ if it can be written as a quotient of two entire functions of order $\leq \rho$.

Theorem A.4 (Schneider–Lang, cf. [Wal74, Theorem 3.3.1]). *Let $\rho_1, \rho_2 > 0$ be real numbers, $K \subset \mathbb{C}$ be a number field, $n \geq 2$ be an integer, and $f_1, \ldots, f_n$ be meromorphic functions on $\mathbb{C}$ such that the ring $K[f_1, \ldots, f_n]$ is stable under the derivation $\frac{d}{dz}$. Let us further assume that:*

(1) *f_1 and f_2 are algebraically independent over K;*
(2) *f_i is of order $\leq \rho_i$, for $i = 1, 2$.*

Then, if S denotes the set of $\alpha \in \mathbb{C}$ such that, for every $1 \leq j \leq n$, α is not a pole of f_j and $f_j(\alpha) \in K$, we have:

$$\operatorname{card}(S) \leq (\rho_1 + \rho_2)[K : \mathbb{Q}].$$

In essence, this theorem simply asserts that $f_1, \ldots, f_n$ can only take too many simultaneous algebraic values if there is an algebraic relation between them.

Remark A.3. The Schneider–Lang criterion also admits geometric generalizations replacing differential equations by algebraic foliations. See [Her, Gas13].

As a first corollary, we can recover the following classical result which brings together the pioneering works of Hermite and Lindemann on the transcendence of e and π.

Corollary A.1 (Hermite–Lindemann). *For every $z \in \mathbb{C} \setminus \{0\}$, we have*

$$\operatorname{trdeg}_{\mathbb{Q}} \mathbb{Q}(z, e^z) \geq 1.$$

Proof. Let $f_1(z) = z$ and $f_2(z) = e^z$. Clearly, both f_1 and f_2 are of order ≤ 1 and the ring $\mathbb{Q}[f_1, f_2]$ is stable under $\frac{d}{dz}$. We have already seen in Example A.1 that f_1 and f_2 are algebraically independent over $\mathbb{C}$. By absurd, suppose that there exists $\alpha \in \mathbb{C} \setminus \{0\}$ such that both α and e^α are algebraic, and set $K = \mathbb{Q}(\alpha, e^\alpha)$. Then S would contain the infinite set $\{n\alpha \mid n \in \mathbb{Z}\}$, thereby contradicting Schneider–Lang's theorem. □

Taking $z = 2\pi i$, we obtain the transcendence of π; taking $z = 1$, we obtain the transcendence of e.

In the same spirit, we can prove Schneider's theorem characterizing "bi-algebraic" points for the j-invariant, seen as a holomorphic function on the Poincaré upper half-plane $\mathbb{H} = \{\tau \in \mathbb{C} \mid \operatorname{Im} \tau > 0\}$.

Corollary A.2 (Schneider). *If $\tau \in \mathbb{H}$ is not quadratic imaginary, then*

$$\operatorname{trdeg}_{\mathbb{Q}} \mathbb{Q}(\tau, j(\tau)) \geq 1.$$

Note that, if $\tau \in \mathbb{H}$ is quadratic imaginary, i.e., $\mathbb{Q}(\tau)$ is an imaginary algebraic extension of $\mathbb{Q}$ of degree 2, then it follows from the classical theory of complex multiplication of elliptic curves that $j(\tau)$ is algebraic (see, for instance, [Sil94, Chapter II]).

Proof. Suppose that τ and $j(\tau)$ are both algebraic. Since $j(\tau)$ is algebraic, which means that the elliptic curve corresponding to τ can be defined over the field of algebraic numbers $\overline{\mathbb{Q}} \subset \mathbb{C}$, there exists a lattice $\Lambda = \mathbb{Z}\omega_1 + \mathbb{Z}\omega_2 \subset \mathbb{C}$ such that $\tau = \frac{\omega_2}{\omega_1}$ and $g_2(\Lambda), g_3(\Lambda) \in \overline{\mathbb{Q}}$. Here, $g_2(\Lambda)$ and $g_3(\Lambda)$ are the "invariants" of the Weierstrass elliptic function $\wp_\Lambda$, so that we have the differential equation

$$\wp'_\Lambda(z)^2 = 4\wp_\Lambda(z)^3 - g_2(\Lambda)\wp_\Lambda(z) - g_3(\Lambda).$$

Set

$$(f_1(z), f_2(z), f_3(z), f_4(z)) = (\wp_\Lambda(z), \wp_\Lambda(\tau z), \wp'_\Lambda(z), \wp'_\Lambda(\tau z))$$

and let K be a number field containing τ, $j(\tau)$, and the field of definition of all 2-torsion points on the elliptic curve E over $\mathbb{Q}(j(\tau))$ such that $E(\mathbb{C}) = \mathbb{C}/\Lambda$. Then $K[f_1, f_2, f_3, f_4]$ is stable under derivation and

$$f_j\left(n\omega_1 + \frac{1}{2}\omega_1\right) \in K$$

for every $j = 1, \ldots, 4$ and $n \in \mathbb{Z}$. Since each f_j is of finite order (cf. [MR14, Chapter 10]), it follows from Schneider–Lang's theorem that f_1 and f_2 cannot be algebraically independent over K. This implies that there exists $m \in \mathbb{Z}$ such that $m\omega_2$ is a period of $f_2(z) = \wp_\Lambda(\tau z)$, so that there exists $a, b \in \mathbb{Z}$ satisfying

$$m\omega_2 = a\frac{\omega_1}{\tau} + b\frac{\omega_2}{\tau}.$$

Since $\tau = \omega_2/\omega_1$, we get $m\tau^2 - b\tau - a = 0$. □

Exercise A.4. *The Gelfond–Schneider theorem asserts that for algebraic numbers $\alpha \notin \{0, 1\}$ and $\beta \notin \mathbb{Q}$, α^β is transcendental. Derive this result from Schneider–Lang's theorem. Hint: consider the functions e^z and $e^{\beta z}$.*

A.2.3.2. *Siegel–Shidlovsky*

We say that a power series

$$f(z) = \sum_{n=0}^{\infty} \frac{a_n}{n!} z^n$$

defines a *Siegel E-function* if:

(1) there exists a number field $K \subset \mathbb{C}$ such that $a_n \in K$ for every $n \geq 0$;
(2) for every $\epsilon > 0$, we have $\max_\sigma |\sigma(a_n)| = O(n^{\epsilon n})$, where σ runs through the set of all field embeddings of K into $\mathbb{C}$;
(3) for every $\epsilon > 0$, there exists a sequence of strictly positive numbers $(q_n)_{n\geq 0}$ such that $q_n = O(n^{\epsilon n})$ and $q_n a_j$ is an algebraic integer of K for every $0 \leq j \leq n$.

The second condition above implies that f defines an entire function on $\mathbb{C}$. The prototype of an E-function is the exponential function e^z, but other remarkable examples include some special cases of hypergeometric functions, such as Bessel's function

$$J_0(z) = \sum_{n=0}^{\infty} \frac{(-1)^n}{n!^2} \left(\frac{z}{2}\right)^{2n}.$$

Theorem A.5 (Siegel–Shidlovsky; cf. [Bak75, Chapter 11]). *Let $n \geq 1$ be an integer and $f_1, \ldots, f_n$ be entire functions on $\mathbb{C}$ whose Taylor coefficients at the origin all lie in a same number field $K \subset \mathbb{C}$ and for which there exist rational functions $g_{ij} \in K(z)$, $1 \leq i, j \leq n$, such that*

$$\frac{df_i}{dz} = \sum_{j=1}^{n} g_{ij} f_j$$

for every $1 \leq i \leq n$. If, moreover:

(1) *$f_1, \ldots, f_n$ are algebraically independent over $K(z)$, and*
(2) *each f_i is a Siegel E-function,*

then, for every non-zero algebraic number $\alpha \in \mathbb{C}$ which is not contained in the set of poles of g_{ij}, we have

$$\operatorname{trdeg}_K K(f_1(\alpha), \ldots, f_n(\alpha)) = n.$$

Besides the arithmetic and the growth conditions, Siegel–Shidlovsky's and Schneider–Lang's theorems differ in an essential aspect: in Siegel–Shidlovsky the differential equation must be linear over $K(z)$. This more restrictive hypothesis yields, on the other hand, a stronger result of algebraic independence, while in Schneider–Lang we can only obtain transcendence. Observe however that the hypotheses of both results are structurally very similar.

Example A.3. The Bessel function J_0 introduced above satisfies the linear equation

$$z^2 \frac{d^2 J_0}{dz^2} + z \frac{dJ_0}{dz} + z^2 J_0 = 0.$$

Applying Siegel–Shidlovsky's theorem, we obtain that for every algebraic $\alpha \in \mathbb{C}$, $J_0(\alpha)$ and $J_0'(\alpha)$ are algebraically independent over $\mathbb{Q}$.

Remark A.5. As Schneider–Lang's theorem (see Remark A.3 above), there are also geometric generalizations of Siegel–Shidlovsky's theorem; see [Ber12, Gas10].

A.3. The theorem of Nesterenko

Both in Schneider–Lang and in Siegel–Shidlovsky results, functions (holomorphic or meromorphic) are assumed to be defined on the whole complex plane $\mathbb{C}$. This rules out any immediate application of these methods to functions defined on proper domains of $\mathbb{C}$, such as modular functions.

Note that Schneider's theorem (Corollary A.2 above) is indeed a theorem on the values of some modular function, but its proof relies fundamentally on elliptic functions. Schneider himself, in his famous memoir [Sch57, p. 138], asks if it is possible to recover his result through a *direct* study of the j-function.

Schneider's question remains unanswered, but we dispose nowadays of other transcendence results on modular functions with truly modular proofs. The following statement was conjectured by Mahler [Mah69b] and proved by Barré-Sirieix, Diaz, Gramain and Philibert [BDGP96].

Theorem A.6. *For every $\tau \in \mathbb{H}$,*

$$\operatorname{trdeg}_{\mathbb{Q}} \mathbb{Q}(e^{2\pi i\tau}, j(\tau)) \geq 1.$$

This implies that $j(\tau)$ is transcendental whenever $e^{2\pi i\tau}$ is algebraic. Observe the appearance of the "modular parameter" $q = e^{2\pi i\tau}$ instead of τ. Shortly after, Nesterenko generalized the above result in his famous theorem on values of Eisenstein series [Nes96].

Theorem A.7. *For every $\tau \in \mathbb{H}$, we have*

$$\operatorname{trdeg}_{\mathbb{Q}} \mathbb{Q}(e^{2\pi i\tau}, E_2(\tau), E_4(\tau), E_6(\tau)) \geq 3.$$

Taking, for instance, $\tau = i$, we obtain the algebraic independence of $e^{\pi}, \pi, \Gamma(1/4)$ over $\mathbb{Q}$. We refer to Nesterenko's original paper [Nes96] for more applications. In Appendix A.4.2. below, we shall interpret Nesterenko's result in terms of periods of elliptic curves.

A.3.1. *Nesterenko's D-property and a zero lemma*

Let X be a smooth affine variety over $\mathbb{C}$ equipped with a vector field $v \in \Gamma(X, T_{X/\mathbb{C}})$. We say that a closed subvariety Y of X is *v-invariant* if v restricts to a vector field on Y. In other words, if v is seen as a derivation on the ring of regular functions $\mathcal{O}_X(X)$, and if I_Y denotes the ideal of $\mathcal{O}_X(X)$ corresponding to Y, then Y is v-invariant if $v(I_Y) \subset I_Y$.

Example A.4. Consider the vector field

$$v = \frac{\partial}{\partial x} + y\frac{\partial}{\partial y}$$

on $\mathbb{A}^2_{\mathbb{C}} = \operatorname{Spec} \mathbb{C}[x, y]$. It is clear that $V(y)$ is a v-invariant subvariety of $\mathbb{A}^2_{\mathbb{C}}$. Let us prove that this is the only one. For any v-invariant subvariety Y, we can write $Y = V(P)$ where

$$P(x, y) = P_0(x) + P_1(x)y + \cdots + P_n(x)y^n, \qquad P_n(x) \neq 0$$

is an irreducible polynomial dividing

$$v(P) = \frac{\partial P}{\partial x} + y\frac{\partial P}{\partial y}.$$

Since $\deg P = \deg_x P_n + n = \deg v(P)$, we must have $v(P) = \lambda P$ for some constant $\lambda \in \mathbb{C} \setminus \{0\}$, that is

$$P_j'(x) + jP_j(x) = \lambda P_j(x) \qquad (j = 0, \ldots, n).$$

It is not hard to conclude from our hypotheses that $\lambda = 1$, $P_1(x) = P_1(0)$ is constant $\neq 0$, and $P = P_1(0)y$. In other words, $Y = V(y)$.

Definition A.3. Let X be a smooth affine variety over $\mathbb{C}$, $v \in \Gamma(X, T_{X/\mathbb{C}})$ be a vector field on X, and

$$\varphi : U \subset \mathbb{C} \longrightarrow X(\mathbb{C})$$

be a non-constant holomorphic integral curve[1] of v defined on some connected open neighborhood U of $0 \in \mathbb{C}$. We say that φ satisfies *Nesterenko's D-property* if there exists a constant $c > 0$ such that, for every v-invariant closed subvariety Y of X, there exists a regular function f on X vanishing on Y and satisfying

$$\operatorname{ord}_0(f \circ \varphi) \leq c.$$

Lemma A.1. *For X, v and φ as in the above definition (with or without D-property), let Y be the Zariski closure of $\varphi(U) \subset X(\mathbb{C})$ (i.e., Y is the smallest subvariety of X such that $\varphi(U) \subset Y(\mathbb{C})$). Then Y is v-invariant. In particular, any φ satisfying Nesterenko's D-property must have Zariski-dense image in X.*

[1] By this, we simply mean that the derivative of φ at every $z \in U$ is a multiple of $v_{\varphi(z)}$.

Proof. Let f be a regular function on X vanishing on Y; we must prove that $g := v(f)$ also vanishes on Y. Let λ be the holomorphic function on the open subset of U where $v_{\varphi(z)} \neq 0$ such that $\varphi'(z) = \lambda(z) v_{\varphi(z)}$, so that

$$\lambda(z)(g \circ \varphi)(z) = (f \circ \varphi)'(z) = 0.$$

Since U is connected and φ is not constant, we obtain $g \circ \varphi \equiv 0$. Since the image of φ is dense in Y, we conclude that g vanishes on Y. □

Conversely, if the image of φ is Zariski-dense in X and if there exists a finite number of subvarieties $Y_1, \ldots, Y_m \subset X$ for which every v-invariant subvariety Y of X containing $\varphi(0)$ is contained in some Y_i, then φ satisfies the D-property. This is how Nesterenko's D-property is verified in practice.

Example A.5. We have already seen in Example A.2 that the image of the integral curve

$$\varphi : \mathbb{C} \longrightarrow \mathbb{A}^2_{\mathbb{C}}(\mathbb{C}), \qquad z \longmapsto (z, e^z)$$

of the vector field v of Example A.4 is Zariski-dense in $\mathbb{A}^2_{\mathbb{C}}$. Since v admits at most a finite number of invariant subvarieties (actually, there is only one!), we conclude that φ satisfies the D-property.

Remark A.6. Alternatively, one could remark that the Zariski-density of the image of $z \longmapsto (z, e^z)$ implies the Zariski-density of the image of

$$\varphi_a : z \longmapsto (z + a, e^z)$$

for any $a \in \mathbb{C}$. This "stronger" statement immediately implies Nesterenko's D-property for φ since *any* leaf of the holomorphic foliation on $D(y) = \mathbb{A}^2_{\mathbb{C}} \setminus V(y)$ induced by v can be parameterized by some φ_a (cf. Lemma A.1).

Remark A.7. A famous theorem of Darboux and Jouanolou (see, for instance, [Dar78a, Ghy00]) implies that any vector field v on a smooth algebraic surface X admitting a holomorphic integral curve with Zariski-dense image has at most a finite number of v-invariant subvarieties. Thus, when $\dim X = 2$, Nesterenko's D-property is actually equivalent to the image of φ being Zariski-dense in X.

The *raison d'être* of the D-property is that it gives a sufficient condition to an integral curve to satisfy certain "zero estimates" which are useful in Diophantine approximation. Here is a precise statement.

Theorem A.8 (Zero Lemma). *Let X be an open affine subscheme of $\mathbb{A}^n_{\mathbb{C}}$, $v \in \Gamma(X, T_{X/\mathbb{C}}) \setminus \{0\}$ be a vector field on X, $U \subset \mathbb{C}$ be a neighborhood* *of* 0, *and* $\varphi : U \longrightarrow X(\mathbb{C})$ *be a holomorphic map satisfying the differential equation*

$$z\frac{d\varphi}{dz} = v \circ \varphi.$$

If φ *satisfies the D-property, then there exists a constant* $C > 0$ *such that, for every polynomial* $P \in \mathbb{C}[x_1, \ldots, x_n] \setminus \{0\}$, *we have*

$$\mathrm{ord}_0(P \circ \varphi) \leq C(\deg P)^n.$$

The above result was proved in this geometric form by Binyamini [Bin14] and is based on Nesterenko's original result in [Nes96, Paragraph 5]. It also admits more general versions (see [Fon19, Appendix B]).

Binyamini's approach is based on intersection theory of analytic cycles. We isolate the main technical details in the form of the following lemma.

Lemma A.2 (cf. [Bin14]). *With notation as in Theorem A.8, there exists an additive function* mult_φ, *the* intersection multiplicity with φ at $p = \varphi(0)$, *which takes an effective algebraic cycle* Z *of* X *and associates a natural number (or* $+\infty$*)*

$$\mathrm{mult}_\varphi(Z) \in \mathbb{N} \cup \{+\infty\}$$

satisfying the following properties:

(1) $\mathrm{mult}_\varphi(Z)$ *only depends on the analytic germ of* Z *at* p.
(2) *If* $Z = V(P) \cap X$, *for some* $P \in \mathbb{C}[x_1, \ldots, x_n] \setminus \{0\}$, *then* $\mathrm{mult}_\varphi(Z) = \mathrm{ord}_0(P \circ \varphi)$.
(3) *If* $Z = p$, *then* $\mathrm{mult}_\varphi(Z) = 1$.
(4) *For any closed subvariety* Y *of* X *for which* $p \in Y$, *and any polynomial* $P \in \mathbb{C}[x_1, \ldots, x_n] \setminus \{0\}$ *vanishing identically on* Y, *we have* $\mathrm{mult}_\varphi(Y) \leq \mathrm{ord}_0(P \circ \varphi) \cdot \mathrm{mult}_p(Y)$.[2]
(5) *For any closed subvariety* Y *of* X, *and any polynomial* $P \in \mathbb{C}[x_1, \ldots, x_n] \setminus \{0\}$ *vanishing identically on* Y *for which* $v(P)$ *does not vanish identically on* Y, *we have* $\mathrm{mult}_\varphi(Y) \leq \mathrm{mult}_\varphi(Y \cdot V(v(P)))$, *where* $Z_1 \cdot Z_2$ *denotes the intersection product of algebraic cycles.*
(6) *There is an integer* $n_0 \geq 0$ *such that, for every closed subvariety* Y *of* X *not contained in a* v*-invariant subvariety of* X, *if* $d \geq 1$ *is*

[2]Here, $\mathrm{mult}_p(Y)$ denotes the Samuel multiplicity of the variety Y at the closed point $p \in Y$. It is given by $S(T) = \frac{\mathrm{mult}_p(Y)}{d!}T^d + O(T^{d-1})$, where $d = \dim Y$, and $S \in \mathbb{Q}[T]$ is the unique polynomial such that $S(n) = \mathrm{length}(\mathcal{O}_{Y,p}/\mathfrak{m}_p^{n+1})$ for every $n \in \mathbb{N}$.

the smallest integer for which there exists $P \in \mathbb{C}[x_1, \ldots, x_n] \setminus \{0\}$ *of degree* d *vanishing identically on* Y*, then* $\min\{n \mid v^n(P) = v(v(\cdots(v(P))\cdots))$ *does not vanish identically on* $Y\} \leq n_0$.

Let us now sketch Binyamini's argument assuming the above lemma.

Proof of Theorem A.8. Let $P \in \mathbb{C}[x_1, \ldots, x_n] \setminus \{0\}$ be a polynomial of degree $d \geq 1$. We want to show that $\mathrm{ord}_0(P \circ \varphi) \leq Cd^n$ for some constant $C > 0$ not depending on d or P.

Set $Z^1 = V(P) \cap X$. The idea is to construct, by induction, cycles Z^k of codimension k, for $2 \leq k \leq n$, satisfying

(1) $\mathrm{mult}_\varphi(Z^k) \leq \mathrm{mult}_\varphi(Z^{k+1}) + c \deg Z^k$, where c is the constant of the D-property, and
(2) For every $1 \leq k \leq n-1$, $\deg Z^{k+1} \leq (d+c_0) \deg Z^k$, for some constants $c_0, c_1 > 0$ (not depending on d or P).

Once this is done, we have

$$\begin{aligned}
\mathrm{ord}_0(P \circ \varphi) &= \mathrm{mult}_\varphi(Z^1)\\
&\leq \mathrm{mult}_\varphi(Z^2) + c \deg\ Z^1\\
&\leq \mathrm{mult}_\varphi(Z^3) + c(\deg\ Z^2 + \deg\ Z^1)\\
&\ \ \vdots\\
&\leq \mathrm{mult}_\varphi(\underbrace{Z^n}_{\text{0-cycle}}) + c(\deg\ Z^{n-1} + \cdots + \deg\ Z^1)\\
&\leq c(\deg\ Z^n + \deg\ Z^{n-1} + \cdots + \deg\ Z^1)\\
&\leq c((d+c_0)^n + (d+c_0)^{n-1} + \cdots + (d+c_0))\\
&\leq cn(1+c_0)^n d^n,
\end{aligned}$$

so that we may take $C = cn(1+c_0)^n$.

Let us now see how the sequence of cycles Z^k is constructed, and where the D-property comes in. By induction, suppose that Z^k has been constructed; let us construct Z^{k+1}. We write

$$Z^k = Z^k_n + Z^k_c,$$

where, by definition, the irreducible components of Z^k_c are those of Z^k which are contained in some v-invariant subvariety of X. Now, write $Z^k_n = \sum_i m_i Y_i$, and, for every i, let $d_i \geq 1$ be the smallest integer for which there exists a polynomial $P_i \in \mathbb{C}[x_1, \ldots, x_n] \setminus \{0\}$ of degree d_i vanishing identically on Y_i (note that $d_i \leq d$). Let $n_i = \min\{n \mid v^n(P_i)|_{Y_i} \not\equiv 0\}$. Since, by definition, Y_i is not contained in a v-invariant subvariety, n_i is finite. Then we define

$$Z^{k+1} = \sum_i m_i Y_i \cdot V(v^{n_i}(P_i)).$$

Finally, (i) follows from properties (4) (combined with the D-property) and (5), and (ii) follows from (6). □

Exercise A.8. *Write down a complete proof for the Zero Lemma in dimension 2.*

A.3.2. *Mahler's theorem and the D-property for the Ramanujan equations*

We shall now study the foliation induced by the Ramanujan equations and prove the corresponding D-property.

We start with Mahler's theorem asserting that the j-invariant satisfies no algebraic differential equation of second order or lower (see [Mah69a]). More precisely, if we denote

$$\theta = \frac{1}{2\pi i}\frac{d}{d\tau},$$

Mahler proved that the holomorphic functions on $\mathbb{H}$

$$\tau, e^{2\pi i\tau}, j(\tau), \theta j(\tau), \theta^2 j(\tau)$$

are algebraically independent over $\mathbb{C}$.

Lemma A.3. *We have*

$$\mathbb{Q}(j, \theta j, \theta^2 j) = \mathbb{Q}(E_2, E_4, E_6).$$

Proof. Since $j \in \mathbb{Q}(E_4, E_6)$, it follows immediately from Ramanujan's equations that $\mathbb{Q}(j, \theta j, \theta^2 j) \subset \mathbb{Q}(E_2, E_4, E_6)$. Explicitly:

$$j = 1728\frac{E_4^3}{E_4^3 - E_6^2}, \quad \theta j = -1728\frac{E_4^2 E_6}{E_4^3 - E_6^2},$$

$$\theta^2 j = 288\frac{-E_2 E_4^2 E_6 + 4E_4 E_6^2 + 3E_4^4}{E_4^3 - E_6^2}.$$

The above formulas can be inverted. Recall that $\Delta = \frac{1}{1728}(E_4^3 - E_6^2)$ and that $\theta \log \Delta = E_2$ (this follows from Ramanujan's equations). Writing $j = E_4^3/\Delta$ and $j - 1728 = E_6^2/\Delta$, and using the Ramanujan equations, we get

$$\theta \log j = -\frac{E_6}{E_4}, \ \theta \log(j - 1728) = -\frac{E_4^2}{E_6}$$

so that

$$E_4 = \theta \log j \cdot \theta \log(j - 1728), \ E_6 = -(\theta \log j)^2 \cdot \theta \log(j - 1728) \in \mathbb{Q}(j, \theta j).$$

Finally,

$$E_2 = \theta \log \Delta = 3\theta \log E_4 - \theta \log j \in \mathbb{Q}(j, \theta j, \theta^2 j). \qquad \square$$

It follows from the above lemma that Mahler's theorem is equivalent to the following statement.

Theorem A.9. *The holomorphic functions on $\mathbb{H}$*

$$\tau, e^{2\pi i \tau}, E_2(\tau), E_4(\tau), E_6(\tau)$$

are algebraically independent over $\mathbb{C}$.

Our proof is different from Mahler's, but it is still fairly elementary. It relies on the following simple geometric considerations (see Remark A.9 below for the original motivation). Let T be the open affine subscheme of $\mathbb{A}^3_{\mathbb{C}} = \operatorname{Spec} \mathbb{C}[t_1, t_2, t_3]$ where $t_2^3 - t_3^2 \neq 0$ and consider the surjective map

$$\pi : \mathsf{T} \longrightarrow \mathbb{A}^1_{\mathbb{C}}, \qquad (t_1, t_2, t_3) \longmapsto 1728 \frac{t_2^3}{t_2^3 - t_3^2}.$$

If G denotes the subgroup scheme of $\mathrm{SL}(2, \mathbb{C})$ of upper triangular matrices, so that

$$\mathsf{G}(\mathbb{C}) = \left\{ \begin{pmatrix} x^{-1} & y \\ 0 & x \end{pmatrix} \in \mathrm{SL}(2, \mathbb{C}) \middle| \, x \in \mathbb{C}^\times, y \in \mathbb{C} \right\},$$

then G acts on T by

$$(t_1, t_2, t_3) \bullet \begin{pmatrix} x^{-1} & y \\ 0 & x \end{pmatrix} = (-12xy + x^2 t_1, x^4 t_2, x^6 t_3).$$

This action clearly preserves the fibres of π. In fact, T is "almost" a G-torsor over $\mathbb{A}^1_{\mathbb{C}}$. A general context for the group G and its action are presented in §§3.3 and 3.4.

Lemma A.4. *For every $z \in \mathbb{C}$ and $t \in \mathsf{T}(\mathbb{C})$ such that $\pi(t) = z$, the morphism*

$$\mathsf{G} \longrightarrow \pi^{-1}(z), \qquad \mathsf{g} \longmapsto t \bullet \mathsf{g}$$

is finite and surjective.

Proof. Exercise. □

We are now ready for our proof.

Proof. Set $X = \mathbb{A}^2_{\mathbb{C}} \times \mathsf{T}$, and denote $p = \pi \circ \mathrm{pr}_2 : X \longrightarrow \mathbb{A}^1_{\mathbb{C}}$. We must prove that the image of the holomorphic map

$$\varphi : \mathbb{H} \longrightarrow X(\mathbb{C}), \quad \tau \longmapsto (\tau, e^{2\pi i\tau}, E_2(\tau), E_4(\tau), E_6(\tau))$$

is Zariski-dense in X. Note that φ is well defined since Ramanujan's Δ function

$$\Delta(\tau) = \frac{E_4(\tau)^3 - E_6(\tau)^2}{1728}$$

never vanishes on $\mathbb{H}$; moreover, $(p \circ \varphi)(\tau) = j(\tau)$ for every $\tau \in \mathbb{H}$.

Since p and j are surjective, it suffices to prove that the image of φ is Zariski-dense in every fibre of p. It follows from Lemma A.4 that, for every $\tau \in \mathbb{H}$, the map

$$f_\tau : \mathbb{A}^2_{\mathbb{C}} \times \mathsf{G} \longrightarrow p^{-1}(j(\tau)), \qquad (a, b, \mathsf{g}) \longmapsto (a, b, (E_2(\tau), E_4(\tau), E_6(\tau)) \bullet \mathsf{g})$$

is finite and surjective, so that $\varphi(\mathbb{H}) \cap p^{-1}(j(\tau))$ is Zariski-dense in $p^{-1}(j(\tau))$ if and only if $f_\tau^{-1}(\varphi(\mathbb{H}) \cap p^{-1}(j(\tau)))$ is Zariski-dense in $\mathbb{A}^2_{\mathbb{C}} \times \mathsf{G}$.

Using the (quasi)modularity of E_2, E_4, and E_6, one easily verifies that $f_\tau^{-1}(\varphi(\mathbb{H}) \cap p^{-1}(j(\tau)))$ contains the set

$$S_\tau = \left\{ (\gamma \cdot \tau, e^{2\pi i \gamma \cdot \tau}, \mathsf{g}_{\gamma,\tau}) \in \mathbb{C}^2 \times \mathsf{G}(\mathbb{C}) \,\middle|\, \gamma = \begin{pmatrix} a & b \\ c & d \end{pmatrix} \in \mathrm{SL}(2, \mathbb{Z}),\ c\tau + d \neq 0 \right\},$$

where

$$\mathsf{g}_{\gamma,\tau} = \begin{pmatrix} (c\tau + d)^{-1} & -c/2\pi i \\ 0 & c\tau + d \end{pmatrix} \in \mathsf{G}(\mathbb{C}),$$

so that it suffices to prove that S_τ is Zariski-dense in $\mathbb{A}^2_{\mathbb{C}} \times \mathsf{G}$ for any $\tau \in \mathbb{H}$.

As a first reduction, observe that it suffices to prove that the set

$$\{(e^{2\pi i\gamma\cdot\tau}, \mathsf{g}_{\gamma,\tau}) \in \mathbb{C} \times \mathsf{G}(\mathbb{C}) \mid \gamma \in \mathrm{SL}(2,\mathbb{Z}),\, c\tau + d \neq 0\}$$

is Zariski-dense in $\mathbb{A}^1_{\mathbb{C}} \times \mathsf{G}$. Indeed, if we denote ${}_n\gamma = \begin{pmatrix} 1 & n \\ 0 & 1 \end{pmatrix} \cdot \gamma$ for every $\gamma \in \mathrm{SL}(2,\mathbb{Z})$ and $n \in \mathbb{Z}$, then

$$(\gamma_n \cdot \tau, e^{2\pi i_n\gamma\cdot\tau}, \mathsf{g}_{n\gamma,\tau}) = (\gamma\cdot\tau + n, e^{2\pi i\gamma\cdot\tau}, \mathsf{g}_{\gamma,\tau})$$

and our claim follows from the Zariski-density of $\mathbb{Z} \subset \mathbb{C}$ in $\mathbb{A}^1_{\mathbb{C}}$.

We now perform a second reduction: it suffices to prove that the set

$$\{(e^{2\pi i\frac{a}{c}}, c) \in \mathbb{C}^2 \mid (a,c) \in \mathbb{Z}^2,\, \gcd(a,c) = 1\}$$

is Zariski-dense in $\mathbb{A}^2_{\mathbb{C}}$. Indeed, let $P \in \mathbb{C}[t,x,y] \setminus \{0\}$ be such that

$$P(e^{2\pi i\gamma\tau}, \mathsf{g}_{\gamma,\tau}) = P(e^{2\pi i\gamma\tau}, c\tau + d, -c/2\pi i) = 0$$

for every $\gamma \in \mathrm{SL}(2,\mathbb{Z})$. Writing $P = \sum_{j=0}^N P_j(t,y)x^j$, with $P_N \neq 0$, we obtain that

$$\sum_{i=0}^N P_j(e^{2\pi i\gamma_n\cdot\tau}, -c/2\pi i)(c\tau + d + cn)^j = 0$$

for every $\gamma \in \mathrm{SL}(2,\mathbb{Z})$ and $n \in \mathbb{Z}$, where $\gamma_n = \gamma \cdot \begin{pmatrix} 1 & n \\ 0 & 1 \end{pmatrix}$. We can assume $c \neq 0$. Multiplying the above equation by $(c\tau + d + cn)^{-N}$ and letting $n \longrightarrow +\infty$, we obtain

$$P_N(e^{2\pi i\frac{a}{c}}, -c/2\pi i) = 0$$

for every $\gamma \in \mathrm{SL}(2,\mathbb{Z})$, and our claim follows.

Finally, suppose that there exists a polynomial $P(x,y) = \sum_{j=0}^N P_j(y)x^j \in \mathbb{C}[x,y]$, such that $P(e^{2\pi i\frac{a}{c}}, c) = 0$ for every $(a,c) \in \mathbb{Z}^2$ with $\gcd(a,c) = 1$. Taking, for instance, c to be a prime $p \geq N+1$, we see that $P(x,p)$ is a polynomial of degree N having at least $N+1$ roots: $1, e^{2\pi i\frac{1}{p}}, \ldots, e^{2\pi i\frac{p-1}{p}}$, so that $P(x,p) = 0$. As there are infinitely many prime numbers greater than $N+1$, we conclude that $P = 0$. $\square$

Remark A.9. The affine space T can be identified to the moduli space of isomorphism classes of complex elliptic curves E endowed with a basis $b = (\omega,\eta)$ of $H^1_{\mathrm{dR}}(E)$ such that $\omega \in H^0(E, \Omega^1_{X/\mathbb{C}})$ and $\langle\omega,\eta\rangle = 1$

(where $\langle\ ,\ \rangle$ denotes the cup product on algebraic de Rham cohomology) in a way that the holomorphic map $\tau \longmapsto (E_2(\tau), E_4(\tau), E_6(\tau))$ corresponds to

$$\varphi : \mathbb{H} \longrightarrow \mathsf{T}(\mathbb{C}), \qquad \tau \longmapsto [(\mathbb{C}/\mathbb{Z} + \tau\mathbb{Z}, (\omega_\tau, \eta_\tau)],$$

where $\omega_\tau = 2\pi i dz$ and $\eta_\tau = \frac{1}{2\pi i}\wp_\tau(z)dz - \frac{E_2(\tau)}{12}2\pi i dz$ (cf. Chapter 9 and [Fon21, §8]). Under such moduli-theoretic interpretation, the group scheme G acts by right multiplication on the basis b seen as a row vector. Note that T admits a natural map to the moduli stack of complex elliptic curves $\mathscr{M}_{1,1}$ and that T is a *bona fide* G-torsor over $\mathscr{M}_{1,1}$. In the above proof, we replaced $\mathscr{M}_{1,1}$ by its coarse moduli scheme $\mathbb{A}^1_{\mathbb{C}}$, at the cost of weakening this "torsor property" (cf. Lemma A.4).

Corollary A.3. *Every leaf of the holomorphic foliation on* $\mathsf{T}(\mathbb{C})$ *induced by the vector field*

$$v = \frac{(x_1^2 - x_2)}{12}\frac{\partial}{\partial x_1} + \frac{(x_1x_2 - x_3)}{3}\frac{\partial}{\partial x_2} + \frac{x_1x_3 - x_2^2}{2}\frac{\partial}{\partial x_3}$$

is Zariski-dense in T.

Proof. Let $(c, d) \in \mathbb{C}^2 \setminus \{0\}$ and define

$$\varphi_{c,d}(\tau) = \left((c\tau + d)^2 E_2(\tau) + \frac{12c}{2\pi i}(c\tau + d), (c\tau + d)^4 E_4(\tau), (c\tau + d)^6 E_6(\tau)\right).$$

One may easily check that $\varphi_{c,d}$ satisfies the differential equation

$$\theta\varphi_{c,d} = (c\tau + d)^{-2} v \circ \varphi_{c,d}$$

so that its image is a leaf of the foliation defined by v.

By Theorem A.9, the image of each $\varphi_{c,d}$ is Zariski-dense in T. Thus, to finish our proof it suffices to show that any point of T lies in the image of $\varphi_{c,d}$ for some $(c, d) \in \mathbb{C}^2 \setminus \{0\}$. Let $t \in \mathsf{T}(\mathbb{C})$ and choose $\tau \in \mathbb{H}$ such that $\pi(t) = j(\tau)$, so that t and $(E_2(\tau), E_4(\tau), E_6(\tau))$ lie in the same π-fiber. By Lemma A.4, there exists $\mathsf{g} \in \mathsf{G}(\mathbb{C})$ such that

$$t = (E_2(\tau), E_4(\tau), E_6(\tau)) \bullet g.$$

To conclude, we simply remark that any element of G is of the form

$$\mathsf{g} = \begin{pmatrix} (c\tau + d)^{-1} & -c/2\pi i \\ 0 & c\tau + d \end{pmatrix} \in \mathsf{G}(\mathbb{C})$$

for some $(c,d) \in \mathbb{C}^2 \setminus \{0\}$, so that

$$t = (E_2(\tau), E_4(\tau), E_6(\tau)) \bullet \mathrm{g} = \varphi_{c,d}(\tau). \qquad \square$$

Finally, for the next theorem, we consider E_2, E_4, E_6 as functions of $q \in D = \{z \in \mathbb{C} \mid |z| < 1\}$ under the change of variables $q = e^{2\pi i \tau}$. Note that $\theta = q\frac{d}{dq}$.

Theorem A.10. *Consider the vector field w on $\mathbb{A}^4_{\mathbb{C}}$ given by*

$$w = x_0\frac{\partial}{\partial x_0} + \frac{(x_1^2 - x_2)}{12}\frac{\partial}{\partial x_1} + \frac{(x_1x_2 - x_3)}{3}\frac{\partial}{\partial x_2} + \frac{x_1x_3 - x_2^2}{2}\frac{\partial}{\partial x_3}$$

and consider the holomorphic curve

$$\varphi : D \longrightarrow \mathbb{A}^4(\mathbb{C}), \qquad q \longmapsto (q, E_2(q), E_4(q), E_6(q))$$

satisfying the differential equation

$$\theta\varphi = w \circ \varphi.$$

Then φ satisfies Nesterenko's D-property.

Proof. Since we already know from Mahler's theorem that the image of φ is Zariski-dense in $\mathbb{A}^4_{\mathbb{C}}$, it is sufficient to prove that there's only a finite number of maximal w-invariant subvarieties containing $\varphi(0) = (0,1,1,1)$. Actually, we shall prove that every w-invariant subvariety containing p is either $V(x_0)$ or it is contained in $V(x_2^3 - x_3^2)$.

Consider the projection

$$\pi : \mathbb{A}^4_{\mathbb{C}} \longrightarrow \mathbb{A}^3_{\mathbb{C}}, \qquad (x_0, x_1, x_2, x_3) \longmapsto (x_1, x_2, x_3).$$

Let $Y \subset \mathbb{A}^4_{\mathbb{C}}$ be a w-invariant subvariety containing p. Then $\pi(Y) \neq \emptyset$ is v-invariant, where v is the 'Ramanujan vector field' of Corollary A.3. It follows from this corollary that either $\pi(Y) \subset V(x_2^3 - x_3^2)$ in $\mathbb{A}^3_{\mathbb{C}}$, or $\pi(Y) = \mathbb{A}^3_{\mathbb{C}}$. In the first case, we have that $Y \subset V(x_2^3 - x_3^2)$ in $\mathbb{A}^4_{\mathbb{C}}$. To conclude, we only need to prove that $\pi(Y) = \mathbb{A}^3_{\mathbb{C}}$ implies that $Y = V(x_0)$.

Now, if $\pi(Y) = \mathbb{A}^3_{\mathbb{C}}$, then Y has dimension at least 3, so that $Y = V(P)$ for a unique irreducible polynomial

$$P = f_n x_0^n + \cdots + f_1 x_0 + f_0$$

with $f_i \in \mathbb{C}[x_1, x_2, x_3]$, and f_n monic. Since Y is w-invariant, there exists $Q \in \mathbb{C}[x_0, x_1, x_2, x_3]$ such that $w(P) = QP$. By considering the degree in x_0, we see that $Q = g \in \mathbb{C}[x_0, x_1, x_3]$, and we get the equations

$$v(f_j) = (g - j)f_j$$

for every $j = 0, \ldots, n$. Again using Corollary A.3, we conclude (exercise) that $f_1 = 1$ and $f_j = 0$ for every $j \neq 1$, i.e., $P = x_0$. □

By combining the above theorem with the Zero Lemma (Theorem A.8), we obtain the following corollary.

Corollary A.4. *There exists a constant $C > 0$ such that, for every polynomial $P \in \mathbb{C}[x_0, x_1, x_2, x_3] \setminus \{0\}$, we have*

$$\mathrm{ord}_{q=0} P(q, E_2(q), E_4(q), E_6(q)) \leq C(\deg P)^4.$$

A.3.3. *Sketch of Nesterenko's proof*

Nesterenko's proof is rather long and intricate, but its general structure is not difficult to understand. In what follows, we outline the main steps of Nesterenko's approach.

We see E_2, E_4, E_6 as holomorphic functions on the unit disk through the change of variables $q = e^{2\pi i\tau}$. To avoid any confusion, we adopt the following convention: q will denote a generic variable in D, and $z \in D$ a point, so that $E_{2k}(q)$ is a function and $E_{2k}(z)$ is a complex number.

Let $z \in D \setminus \{0\}$. We want to prove that

$$\mathrm{trdeg}_{\mathbb{Q}} \mathbb{Q}(z, E_2(z), E_4(z), E_6(z)) \geq 3$$

and the main idea is to apply Philippon's algebraic independence criterium as stated in Theorem A.3 above. This means that we have to construct a sequence of polynomials $Q_d \in \mathbb{Z}[x_0, \ldots, x_3]$, $d \gg 0$, satisfying

(1) $\deg Q_d = O(d \log d)$,
(2) $\log \|Q_d\|_\infty = O(d(\log d)^2)$, and
(3) $-ad^4 \leq \log |Q_d(z, E_2(z), E_4(z), E_6(z)| \leq -bd^4$

for some constants $a, b > 0$ (not depending on d). The first step in obtaining the sequence $(Q_d)_{d \gg 0}$ is the construction of the so-called *auxiliary polynomials*.

Lemma A.5. *There is a constant $c > 0$ such that, for every integer $d \gg 0$ there exists a polynomial $P_d \in \mathbb{Z}[x_0, \ldots, x_3]$ satisfying*

(1) $\deg P_d = d$,
(2) $\log \|P_d\|_\infty = O(d \log d)$, *and*
(3) $\mathrm{ord}_{q=0} P_d(q, E_2(q), E_4(q), E_6(q)) \geq cd^4$.

Proof (Sketch of the proof). For every multi-index $J = (j_0, \ldots, j_3)$ with $|J| = j_0 + \cdots + j_3 \leq d$, write

$$q^{j_0} E_2(q)^{j_1} E_4(q)^{j_2} E_6(q)^{j_3} = \sum_{i \geq 0} t_{i,J} q^i \in \mathbb{Z}[\![q]\!].$$

Note that $t_{i,J}$ are rational integers because the Taylor coefficients of E_{2k} are. Let

$$P = \sum_{|J| \leq d} v_J x^J \in \mathbb{Z}[x_0, \ldots, x_3]$$

be a polynomial with 'unknown coefficients' v_J, so that

$$P(q, E_2(q), E_4(q), E_6(q)) = \sum_{i \geq 0} \left(\sum_{|J| \leq d} t_{i,J} v_J \right) q^i \in \mathbb{Z}[\![q]\!].$$

Let $r = \lfloor \frac{1}{4!} d^4 \rfloor$. To find P such that $\mathrm{ord}_{q=0} P(q, E_2(q), E_4(q), E_6(q)) \geq r$ is equivalent to solving the following r linear equations over $\mathbb{Z}$ with $s = \binom{d+4}{4}$ variables:

$$\sum_{|J| \leq d} t_{i,J} v_J = 0, \qquad i = 0, \ldots, r-1.$$

To get a bound on the size of a solution (v_J), we apply the ubiquitous Siegel's lemma.

Lemma A.6 (Siegel; cf. [Wal74, Lemme 1.3.1] or [MR14, Lemma 6.1]). *Let $s > r > 0$ be integers and $T \in M_{r \times s}(\mathbb{Z})$ be such that $\|T\|_\infty \leq b$. Then, there exists $v \in \mathbb{Z}^s \setminus \{0\}$ with $\|v\|_\infty \leq 2(2sb)^{r/(s-r)}$ such that $Tv = 0$.*

To finish, we just need to prove that $\max_{i \leq r-1, |J| \leq d} |a_{i,J}| = O(d^d)$. This follows from the fact that the Taylor coefficients of E_{2k}, which are given up to a constant by the arithmetical function $\sigma_{2k-1}(m) = \sum_{d|m} m^{2k-1}$ have polynomial growth; namely, we have the trivial bound $\sigma_{2k-1}(m) \leq m^{2k}$. □

Exercise A.10. *Complete the above proof.*

For the next step, let us denote $f_d(q) = P_d(q, E_2(q), E_4(q), E_6(q))$ and $m_d = \mathrm{ord}_{q=0} f_d$.

Lemma A.7. *There exist $\alpha > \beta > 0$ and, for $d \gg 0$, a sequence $k_d = O(d \log m_d)$ satisfying*

$$-\alpha m_d \leq \log \left| f_d^{(k_d)}(z) \right| \leq -\beta m_d.$$

This is the most technical, and most analytical, part of the proof. The main point in obtaining the bound $-\alpha m_d \leq \log \left| f_d^{(k_d)}(z) \right|$ can explained through the following intuitive argument. If all the Taylor coefficients of f_d at $q = z$ up to a sufficiently large order are too small, then its first non-zero Taylor coefficient at $q = 0$ will have absolute value < 1, thereby *contradicting its integrality* (cf. Remark A.2). Of course, the difficulty here consists in precisely quantifying "too small" and "sufficiently large order". Once this is established, the other bound $\log \left| f_d^{(k_d)}(z) \right| \leq -\beta m_d$ is a mere consequence of the Cauchy inequalities. We refer to [NP01, Chapter 3, Lemma 3.3] for details.

Finally, we use the existence of the Ramanujan equations. Consider the derivation

$$w = x_0 \frac{\partial}{\partial x_0} + \frac{(x_1^2 - x_2)}{12} \frac{\partial}{\partial x_1} + \frac{(x_1 x_2 - x_3)}{3} \frac{\partial}{\partial x_2} + \frac{x_1 x_3 - x_2^2}{2} \frac{\partial}{\partial x_3}$$

and define, for every $k \geq 1$,

$$w^{[k]} = 12^k w \circ (v - 1) \circ \cdots \circ (w - (k-1)).$$

We set

$$Q_d = w^{[k_d]} P_d \in \mathbb{Z}[x_0, \ldots, x_3].$$

Since $k_d = O(d \log m_d)$, it is clear that $\deg Q_d = O(d \log m_d)$, and that $\log \|Q_d\|_\infty = O(d(\log d)(\log m_d))$. Moreover, the identity of derivations

$$q^k \frac{d^k}{dq^k} = \theta \circ (\theta - 1) \circ \cdots \circ (\theta - (k-1)),$$

where $\theta = q\frac{d}{dq}$, implies that

$$(12q)^{k_d} f_d^{(k_d)}(q) = Q_d(q, E_2(q), E_4(q), E_6(q)).$$

Using Lemma A.7, we immediately deduce the bound

$$-\alpha m_d - \gamma k_d \leq \log |Q_d(z, E_2(z), E_4(z), E_6(z))| \leq -\beta m_d - \gamma k_d$$

where $\gamma = -\log |12z| > 0$.

To conclude, we observe that $k_d = O(d \log m_d)$, that $d^4 = O(m_d)$ by Lemma A.5, and that $m_d = O(d^4)$ by Corollary A.4.

Remark A.11. The use of "auxiliary polynomials" is an essential step in Nesterenko's proof. Similar ideas occur in most results in Diophantine approximation or transcendental number theory. The reader may consult [Mas16] for a thorough exposition of the role of auxiliary polynomials in number theory.

A.4. Periods

Roughly speaking, a period (or an arithmetic period) is an integral that appears in algebraic geometry over $\mathbb{Q}$. The set of all periods forms a countable subset of $\mathbb{C}$ — actually, a subring — containing all the algebraic numbers, but "most" of the periods are transcendental numbers. It is the subtle connection between these numbers and algebraic geometry that allows us to make sense of their structure and their symmetries.

Most of the general theory is still largely conjectural, but the omnipresence of periods in number theory (regulators, special values of L-functions, etc.) and high-energy physics (Feynman amplitudes) makes the study of periods one of the most attractive subjects in current mathematics.

A.4.1. *Definition*

There are several equivalent definitions; we start with the more elementary ones given in [KZ01].

Definition A.4. A real number ϖ is a *period* if it can be written as an absolutely convergent integral

$$\varpi = \int_\sigma \frac{P(t_1, \ldots, t_n)}{Q(t_1, \ldots, t_n)} dt_1 \cdots dt_n$$

where $P, Q \in \mathbb{Q}[t_1, \dots, t_n]$, $Q \neq 0$, and $\sigma \subset \mathbb{R}^n$ is a domain given by polynomial inequalities with rational coefficients. A complex number is a period if its real and imaginary parts are periods.

For instance,

$$\pi = \int_{t_1^2+t_2^2 \leq 1} dt_1 dt_2 \text{ and } \log 2 = \int_{1 \leq t \leq 2} \frac{dt}{t}$$

are periods. A less trivial example is

$$\zeta(3) = \sum_{n=1}^{\infty} \frac{1}{n^3} = \int_{0<t_1<t_2<t_3<1} \frac{dt_1 dt_2 dt_3}{(1-t_1)t_2 t_3}.$$

Exercise A.12. *The algebraic number $\sqrt{2}$ is a period, since it can be written as*

$$\sqrt{2} = \int_{t^2 \leq 2} \frac{dt}{2}.$$

Similarly, show that any algebraic number is a period.

One can show that we can replace rational numbers by algebraic numbers, and rational functions by algebraic functions (with algebraic coefficients) in the definition of periods.

More generally, let $\overline{\mathbb{Q}} \subset \mathbb{C}$ be the algebraic closure of $\mathbb{Q}$ in $\mathbb{C}$, and consider a tuple (X, D, ω, σ), where X is a quasi-projective variety over $\overline{\mathbb{Q}}$, Y is a closed subvariety of X, $\omega \in \Gamma(X, \Omega^n_{X/\overline{\mathbb{Q}}})$ is a closed algebraic n-form on X vanishing on Y, and $\sigma \subset X(\mathbb{C})$ is a singular (topological) n-chain with boundary $\partial\sigma \subset Y(\mathbb{C})$.

Proposition A.11. *For every tuple (X, Y, ω, σ) as above, the number*

$$\int_\sigma \omega$$

is a period. Conversely, every period is of this form.

Proof. See [HM17, 12.2]. □

Yet another way of defining periods, which makes explicit their "motivic" nature, is as follows. For a pair (X, Y) as above, we can consider its *relative algebraic de Rham cohomology* (see [HM17, Chapter 3])

$$H^n_{\mathrm{dR}}(X,Y),$$

which is a finite-dimensional $\overline{\mathbb{Q}}$-vector space, and the singular cohomology of $X(\mathbb{C})$ relative to the closed subspace $Y(\mathbb{C}) \subset X(\mathbb{C})$ with $\mathbb{Q}$-coefficients, also called *relative Betti cohomology*,

$$H^n_B(X,Y)$$

which is a finite-dimensional $\mathbb{Q}$-vector space. Then, a theorem of Grothendieck says that, after base change to $\mathbb{C}$, there is a canonical comparison isomorphism between these two cohomology groups:

$$\mathrm{comp} : H^n_{\mathrm{dR}}(X,Y) \otimes_{\overline{\mathbb{Q}}} \mathbb{C} \xrightarrow{\sim} H^n_B(X,Y) \otimes_{\mathbb{Q}} \mathbb{C}.$$

Now, a period is simply a number that appears as an entry of the matrix of comp with respect to some $\overline{\mathbb{Q}}$-basis of $H^n_{\mathrm{dR}}(X,Y)$ and some $\mathbb{Q}$-basis of $H^n_B(X,Y)$.

Note that Y can be empty, in which case we simply denote $H^n_{\mathrm{dR}}(X)$ and $H^n_B(X)$. For the proof of the equivalences between all of these different definitions of periods, we refer to [HM17, §12.2].

A.4.2. *Elliptic periods and values of quasi-modular forms*

Let E be an elliptic curve over $\overline{\mathbb{Q}}$ given by a Weierstrass equation

$$E : y^2 = 4x^3 - ux - v \qquad (u, v \in \overline{\mathbb{Q}},\ u^3 - 27v^2 \neq 0)$$

and consider the following algebraic differential forms on E:

$$\omega = \frac{dx}{y}, \quad \eta = x\frac{dx}{y}.$$

Note that ω is regular on E (form of the first kind), while η has a pole at infinity (of order 2) with vanishing residue (form of the second kind). We can see (ω, η) as a basis of the algebraic de Rham cohomology $H^1_{\mathrm{dR}}(E)$ under its classical identification with the space of forms of second kind modulo exact forms (see [HM17, Chapter 14]).

Let (γ_1, γ_2) be any oriented $\mathbb{Z}$-basis of the first homology group $H_1(E(\mathbb{C}); \mathbb{Z})$. Here, "oriented" means that the intersection product

$\gamma_1 \cdot \gamma_2 = 1$. We can then consider the four complex numbers

$$\omega_1 = \int_{\gamma_1} \omega, \quad \omega_2 = \int_{\gamma_2} \omega, \quad \eta_1 = \int_{\gamma_1} \omega, \quad \eta_2 = \int_{\gamma_2} \eta.$$

Classically, ω_1 and ω_2 are known as "periods" of E, while η_1 and η_2 are called "quasi-periods". Under the above modern definition, they are all periods.

Example A.6. Consider the elliptic curve $E : y^2 = 4x^3 - 4x$ and let $\gamma_1 \in H_1(E(\mathbb{C}); \mathbb{Z})$ be the class of the connected component of $E(\mathbb{R})$ containing the 2-torsion point $(1, 0)$.

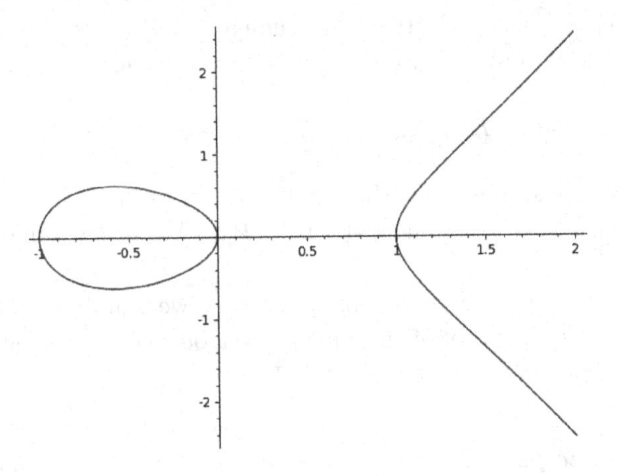

Then

$$\omega_1 = \int_{\gamma_1} \omega = 2\int_1^{\infty} \frac{dx}{\sqrt{4x^3 - 4x}} \overset{x=\frac{1}{t^2}}{=} 2\int_0^1 \frac{dt}{\sqrt{1-t^4}} = \frac{\Gamma(1/4)^2}{2\sqrt{2\pi}},$$

where to compute the last integral we used Euler's formula for the Beta function

$$B(a, b) := \int_0^1 t^{a-1}(1-t)^{b-1} dt = \frac{\Gamma(a)\Gamma(b)}{\Gamma(a+b)} \qquad (\mathrm{Re}(a), \mathrm{Re}(b) > 0) \quad \text{(A.4.1)}$$

the fact that $\Gamma(1/2) = \sqrt{\pi}$, and Euler's reflection formula:

$$\Gamma(1/4)\Gamma(3/4) = \frac{\pi}{\sin(\pi/4)}.$$

Let us remark ω_1 computed above is a very classical object in the theory of elliptic integrals: it is half of the length of Bernoulli's lemniscate $(x^2+y^2)^2 = x^2 - y^2$.

One of the fundamental problems in the theory of periods is to understand all the algebraic relations such numbers can satisfy. For instance, in the case of elliptic curves, we have Legendre's relation:

$$\omega_1\eta_2 - \omega_2\eta_1 = 2\pi i,$$

which holds for every E. Are there any other relations? In particular, are these numbers transcendental? Are some of them algebraically independent?

Conjecture A.1 (Grothendieck's period conjecture for elliptic curves). *With the above notation,*

$$\operatorname{trdeg}_{\mathbb{Q}}\mathbb{Q}(\omega_1, \omega_2, \eta_1, \eta_2) = \begin{cases} 2 & \text{if } E \text{ has complex multiplication} \\ 4 & \text{otherwise.} \end{cases}$$

Recall that E has *complex multiplication* if its endomorphism algebra $\operatorname{End}(E)$ strictly contains $\mathbb{Z}$. The geometric idea is that such extra endomorphisms of E correspond to algebraic cycles (correspondences) on $E \times E$ which force algebraic relations between periods.

Exercise A.13. *An endomorphism $\varphi : E \longrightarrow E$ defined over $\overline{\mathbb{Q}}$ induces an additive map*

$$\varphi_{B,*} : H_1(E(\mathbb{C}); \mathbb{Z}) \longrightarrow H_1(E(\mathbb{C}); \mathbb{Z})$$

preserving the intersection product, and a $\overline{\mathbb{Q}}$-linear map

$$\varphi^*_{\mathrm{dR}} : H^1_{\mathrm{dR}}(E) \longrightarrow H^1_{\mathrm{dR}}(E)$$

preserving the subspace of differentials of first kind $H^0(E, \Omega^1_{E/\overline{\mathbb{Q}}})$ and the de Rham cup product. Show that this induces an identity of the form

$$\begin{pmatrix} a & b \\ c & d \end{pmatrix} \begin{pmatrix} \omega_1 & \eta_1 \\ \omega_2 & \eta_2 \end{pmatrix} = \begin{pmatrix} \omega_1 & \eta_1 \\ \omega_2 & \eta_2 \end{pmatrix} \begin{pmatrix} r & s \\ 0 & r^{-1} \end{pmatrix}$$

with $a, b, c, d \in \mathbb{Z}$, $ad - bc = 1$, and $r, s \in \overline{\mathbb{Q}}$. Conclude that if E has complex multiplication, then τ is quadratic imaginary, $\overline{\mathbb{Q}}(\omega_1, \omega_2, \eta_1, \eta_2) = \overline{\mathbb{Q}}(\omega_1, \eta_1)$, and $\overline{\mathbb{Q}}(\omega_1, \eta_1)$ is algebraic over $\overline{\mathbb{Q}}(2\pi i, \omega_1)$.

Loosely speaking, Grothendieck conjectured that algebraic cycles in powers of some algebraic variety X are the *only way* of producing algebraic relations between periods of X (see [And04, 23.4.1] for a more precise statement). For elliptic curves, Schneider proved that each one of ω_1, ω_2, η_1

and η_2 are transcendental numbers, and Chudnovsky proved the uniform bound:

$$\mathrm{trdeg}_{\mathbb{Q}}\mathbb{Q}(\omega_1, \omega_2, \eta_1, \eta_2) \geq 2$$

for any elliptic curve, therefore establishing Grothendieck's period conjecture for complex multiplication elliptic curves (see [Wal06] and references therein).

Proposition A.12. *With notation as above, if* $\tau = \omega_2/\omega_1$, *then* $\mathrm{Im}(\tau) > 0$ *and*

$$E_2(\tau) = 12\left(\frac{\omega_1}{2\pi i}\right)\left(\frac{\eta_1}{2\pi i}\right), \; E_4(\tau) = 12u\left(\frac{\omega_1}{2\pi i}\right)^4, \; E_6(\tau) = -216v\left(\frac{\omega_1}{2\pi i}\right)^6.$$

Proof. The proof is based on the classical theory of elliptic and modular functions. Let $\Lambda = \mathbb{Z}\omega_1 + \mathbb{Z}\omega_2 \subset \mathbb{C}$. Then Λ is a lattice and it follows from Weierstrass' uniformization theorem that $u = g_2(\Lambda)$ and $v = g_3(\Lambda)$. On the other hand, if $\Lambda_\tau := \omega_1^{-1}\Lambda = \mathbb{Z} + \mathbb{Z}\tau$, then

$$g_2(\Lambda_\tau) = \frac{(2\pi i)^4}{12}E_4(\tau), \; g_3(\Lambda_\tau) = -\frac{(2\pi i)^6}{216}E_6(\tau).$$

By homogeneity, we get

$$g_2(\Lambda_\tau) = g_2(\omega_1^{-1}\Lambda) = \omega_1^4 g_2(\Lambda) = \omega_1^4 u$$

and similarly for g_3. This proves that

$$E_4(\tau) = 12u\left(\frac{\omega_1}{2\pi i}\right)^4, \; E_6(\tau) = -216v\left(\frac{\omega_1}{2\pi i}\right)^6.$$

For $E_2(\tau)$, we use the formula (see [Ser78, Eq. (46), p. 96])

$$E_2(\tau) = -\frac{12}{(2\pi i)^2}\sum_n {\sum_m}' \frac{1}{(m+n\tau)^2}$$

where $\sum'$ means that $(m,n) \neq (0,0)$.[3] Now, by using the following expression for the Weierstrass zeta function ζ_{Λ_τ} (a primitive of $-\wp_{\Lambda_\tau}$)

$$\zeta_{\Lambda_\tau}(z) = \frac{1}{z} = \sum_n {\sum_m}' \left(\frac{1}{z-m-n\tau} + \frac{1}{m+n\tau} + \frac{z}{(m+n\tau)^2}\right),$$

[3] Beware that the above sum does not converge absolutely, so that the order of the summation is important!

we obtain

$$\sum_n \sum_m{}' \frac{1}{(m+n\tau)^2} = -\int_0^1 \wp_{\Lambda_\tau}(z)dz = -\omega_1\eta_1,$$

where the last equality follows from the identity

$$\omega_1^{-2}\wp_{\Lambda_\tau}(z) = \wp_\Lambda(\omega_1 z).$$

We conclude that

$$E_2(\tau) = 12\left(\frac{\omega_1}{2\pi i}\right)\left(\frac{\eta_1}{2\pi i}\right). \qquad \square$$

It follows from the above formulas and from Nesterenko's theorem that, for every elliptic curve over $\overline{\mathbb{Q}}$,

$$\mathrm{trdeg}_{\mathbb{Q}}\mathbb{Q}\left(e^{2\pi i\frac{\omega_2}{\omega_1}}, \frac{\omega_1}{2\pi i}, \frac{\eta_1}{2\pi i}\right) = 3$$

this improves the theorem of Chudnovsky, and thus also proves Grothendieck's Period conjecture for complex multiplication elliptic curves (cf. Exercise A.13).

A.4.3. *Open problems*

Currently, the study of periods is an active and rapidly developing domain of number theory. Among the important recent achievements in the field there is Brown's theorem on multiple zeta values (see [FG] for a thorough introduction).

There remains countless open questions concerning periods. Besides Kontsevich's and Zagier's introduction [KZ01], Waldschmidt's survey [Wal06] contains a good summary of what is known and what is not regarding transcendence. Here, we focus only on values of (quasi)modular forms.

We have seen that Nesterenko's theorem proves, in particular, Grothendieck's period conjecture for complex multiplication elliptic curves. A careful analysis of the period conjecture for certain 1-motives attached to elliptic curves (see [Ber02]) suggests the following statement.

Conjecture A.2. *For any $\tau \in \mathbb{H}$, we have*

$$\mathrm{trdeg}_{\mathbb{Q}}\mathbb{Q}(2\pi i, \tau, e^{2\pi i\tau}, E_2(\tau), E_4(\tau), E_6(\tau)) \geq \begin{cases} 3 & \text{if } \tau \text{ is quadratic imaginary} \\ 5 & \text{otherwise} \end{cases}$$

Moreover, we have equality if $j(\tau) \in \overline{\mathbb{Q}}$.

This conjectural statement essentially contains everything that is known or that there is to know concerning algebraic independence of values taken by quasi-modular forms at a same $\tau \in \mathbb{H}$. For instance, both Nesterenko's theorem and Schneider's theorem on the j-function follow from the above conjecture.

Exercise A.14. *Check that Conjecture A.2 implies Nesterenko's theorem and Schneider's theorem.*

Remark A.15. A natural question is to ask what happens for elliptic modular forms of higher levels. From the point of view of transcendence, we do not get anything new. The values of higher level quasi-modular forms (or modular functions) are algebraic on values of level 1 quasi-modular forms.

A proof of Conjecture A.2 seems out of reach, but we can also turn our attention to higher dimensional notions of modular forms, such as Siegel or Hilbert modular forms. Understanding their values amounts to understanding periods of *abelian varieties*, a generalization of elliptic curves. This also includes periods of higher genus curves, since the cohomology in degree 1 (de Rham or Betti) of a curve is canonically isomorphic to the cohomology of its Jacobian.

We next exhibit some explicit examples of higher genera abelian periods.

Example A.7. Let us consider the hyperelliptic curve C over $\overline{\mathbb{Q}}$ whose affine part is given by the equation $y^2 = 1 - x^5$. The chart at ∞ is given by the equation $s^2 = t^6 - t$, where $(x, y) = (1/t, s/t^3)$. We now show how to compute the periods of C.

For $k = 1, 2, 3, 4$, define the differential forms

$$\omega_k := x^{k-1}\frac{dx}{y}.$$

One can check that ω_1 and ω_2 are of first kind (i.e., everywhere regular), whereas ω_3 and ω_4 are of the second kind (i.e., all the residues vanish). Each of these forms define an element of $H^1_{\mathrm{dR}}(C)$, and since they have distinct orders at ∞, they must be linearly independent. As $\dim H^1_{\mathrm{dR}}(C) = 2 \times \mathrm{genus}(C) = 4$, they form a basis of $H^1_{\mathrm{dR}}(C)$.

Now, consider the path

$$\epsilon : [0, 1] \longrightarrow C(\mathbb{C}), \qquad u \longmapsto (u, \sqrt{1 - u^5}).$$

Using the automorphisms $\tau : (x, y) \longmapsto (x, -y)$ and $\sigma : (x, y) \longmapsto (\zeta x, y)$ of C, where ζ denotes a primitive fifth root of unity, we may define a loop γ at $p = (0, 1) \in C(\mathbb{C})$ by

$$c := \epsilon \cdot (\tau \circ \epsilon)^{-1} \cdot (\sigma \circ \tau \circ \epsilon) \cdot (\sigma \circ \epsilon)^{-1},$$

where $\cdot$ denotes path composition and $^{-1}$ the operation on paths that reverses direction. We compute (cf. formula (A.4.1)):

$$\int_\epsilon \omega_k = \frac{1}{5} B\left(\frac{k}{5}, \frac{1}{2}\right).$$

As $\tau^* \omega_k = -\omega_k$ and $\sigma^* \omega_k = \zeta^k \omega_k$, we conclude that

$$\int_\gamma \omega_k = \int_\epsilon \omega_k - \int_\epsilon \tau^* \omega_k + \int_\epsilon (\sigma \circ \tau)^* \omega_k - \int_\epsilon \sigma^* \omega_k = \frac{2}{5}(1 - \zeta^5) B\left(\frac{k}{5}, \frac{1}{2}\right).$$

Exercise A.16. *For $l = 1, 2, 3, 4$, let $\gamma_l := \sigma_*^l \gamma \in H_1(C(\mathbb{C}), \mathbb{Q}) = H_B^1(C)^\vee$. Show that $(\gamma_1, \ldots, \gamma_4)$ forms a basis of $H_1(C(\mathbb{C}), \mathbb{Q})$, and that*

$$\int_{\gamma_l} \omega_k = \frac{2}{5} \zeta^{k(l-1)} (1 - \zeta^k) B\left(\frac{k}{5}, \frac{1}{2}\right).$$

Note that these can be expressed in terms of the Γ function via Euler's formula (A.4.1).

Here already not much is known. For instance, the period conjecture applied to the above example predicts that π, $\Gamma(1/5)$, and $\Gamma(2/5)$ should be algebraically independent over $\mathbb{Q}$, i.e.,

$$\operatorname{trdeg}_{\mathbb{Q}}(\pi, \Gamma(1/5), \Gamma(2/5)) \stackrel{?}{=} 3. \qquad \text{(A.4.2)}$$

Currently, only the weaker

$$\operatorname{trdeg}_{\mathbb{Q}}(\pi, \Gamma(1/5), \Gamma(2/5)) \geq 2$$

is proved (see [Vas96, Gri02]).

Remark A.17. If we restrict our attention to *linear relations*, instead of arbitrary algebraic relations, then a general result is known. A theorem of Wüstholz (based on his analytic subgroup theorem) shows that every $\overline{\mathbb{Q}}$-linear relation between periods of an abelian variety must come from an endomorphism of the abelian variety. This has been recently generalized by Huber–Wüstholz [HW] to 1-motives.

We have seen that interpreting periods of elliptic curves as values of modular forms can be useful, via Nesterenko's theorem, to understanding their algebraic independence properties. We can ask if a similar approach can be generalized to higher dimensions. As it was already remarked above, values of Hilbert or Siegel modular forms can be expressed in terms of periods of abelian varieties; our question then boils down to asking if Nesterenko's methods can be generalized to such modular forms of several variables.

Given the prominent role played by the Ramanujan equations in Nesterenko's method, one is naturally lead to the study of the differential equations in these higher dimensional contexts. This is indeed the point of view adopted by Pellarin [Pel05] for Hilbert modular forms. The case of Siegel modular forms was studied by Zudilin [Zud00], via explicit equations involving theta functions, and by Bertrand–Zudilin [BZ03] via derivatives of modular functions. Recently, a geometric approach to such problems has been proposed in [Fon21].

Recall that the Ramanujan equations can be interpreted as a vector field on some moduli space of elliptic curves with additional structure, and that this admits a natural generalization to moduli spaces of abelian varieties (see Chapter 11). In [Fon21], we also develop a similar theory for the Hilbert moduli problem.

For instance, consider the real quadratic field $\mathbb{Q}(\sqrt{5})$. By considering "principally polarized abelian surfaces with real multiplication by $\mathbb{Q}(\sqrt{5})$", we obtain a smooth quasi-affine variety T over $\mathbb{Q}$ of dimension 6 endowed with commuting algebraic vector fields v_1, v_2 (generalizing the Ramanujan vector fields), and a canonical analytic map

$$\varphi : \mathbb{H}^2 \longrightarrow \mathsf{T}(\mathbb{C})$$

satisfying the differential equations

$$\theta_j \varphi = v_j \circ \varphi \quad (j = 1, 2)$$

where

$$\theta_1 = \frac{1}{2\pi i}\left(\sqrt{5}^{-1}\frac{\partial}{\partial \tau_1} - \sqrt{5}^{-1}\frac{\partial}{\partial \tau_1}\right) \quad \text{and}$$
$$\theta_2 = \frac{1}{2\pi i}\left(\frac{1+\sqrt{5}}{2}^{-1}\frac{\partial}{\partial \tau_1} + \frac{1-\sqrt{5}}{2}^{-1}\frac{\partial}{\partial \tau_2}\right).$$

This differential equation satisfies many of the remarkable properties the usual Ramanujan equations satisfy. For example, φ can be shown to have integral coefficients in an appropriate q-expansion. Moreover, every leaf of the holomorphic foliation on $\mathsf{T}(\mathbb{C})$ defined by v_1 and v_2 are Zariski-dense in $B_{\mathbb{C}}$ (cf. Corollary A.3 above).

One can moreover relate the values of φ with periods of abelian surfaces with real multiplication by $\mathbb{Q}(\sqrt{5})$. This allows us to reformulate the period conjecture in terms of bounds on the transcendence degree of the fields generated by values of φ, in the same spirit of Nesterenko's theorem. As it is shown [Fon21], a successful adaptation of Nesterenko's method to this higher dimensional setting would yield in particular the conjectural statement (A.4.2).

Remark A.18. Although we do not make explicit mention to modular forms, it can be proved that φ is indeed related to modular forms and their derivatives as in the Bertrand–Zudilin approach. We refer to [Fon21, §15] for a precise statement.

There are many technical difficulties in obtaining sufficiently strong algebraic independence statements for the above higher dimensional generalizations of (E_2, E_4, E_6), the main obstacle being the presence of positive-dimensional "special subvarieties" in moduli spaces of abelian varieties.

Let us also remark that Nesterenko's method relies on a very restrictive growth condition satisfied by the Eisenstein series, namely the polynomial growth of their Fourier coefficients (see Appendix A.3.3. above). It turns out that this condition can be replaced by a much more flexible, geometric, notion of growth based on Nevanlinna theory (see [Fon19]), which is suitable to generalization.

Here, a curious problem emerges. It is shown in [Fon19] that to any collection of holomorphic functions $f_1, \ldots, f_n$ on the complex unity disk D satisfying an algebraic differential equation (with the D-property), an integrality property, and a mild growth condition, Nesterenko's method applies to give

$$\operatorname{trdeg}_{\mathbb{Q}}\mathbb{Q}(f_1(z), \ldots, f_n(z)) \geq n - 1$$

for every $z \in D \setminus \{0\}$. This comprises Nesterenko's result if we take $(f_1, f_2, f_3, f_4) = (q, E_2(q), E_4(q), E_6(q))$, but in principle it could have other applications. It turns out that no other essentially different example (i.e., not related to elliptic modular forms) is currently known. This is surprising, given the rather general shape of the hypotheses.

It could be, however, that quasi-modular forms are essentially the only functions on the disk satisfying the above conditions. If one could prove this fact, this would yield a rather exotic characterization of quasi-modular forms, making no explicit reference to modularity.

References

[Ali13a] Murad Alim. Lectures on mirror symmetry and topological string theory. *Open Problems Surveys Contemp. Math.*, 6:1, 2013.

[Ali13b] Murad Alim. Lectures on mirror symmetry and topological string theory. *Surv. Mod. Math., Int. Press.* 6:1–44, 2013.

[Ali17] Murad Alim. Algebraic structure of $t\,t*$ equations for Calabi–Yau sigma models. *Commun. Math. Phys.*, 353(3):963–1009, 2017.

[AMSY16] M. Alim, H. Movasati, E. Scheidegger, and S.-T. Yau. Gauss–Manin connection in disguise: Calabi–Yau threefolds. *Comm. Math. Phys.*, 334(3):889–914, 2016.

[And04] Yves André. *Une introduction aux motifs. Motifs purs, motifs mixtes, périodes*, volume 17. Paris: Société Mathématique de France (SMF), 2004.

[And17] Yves André. Groupes de Galois motiviques et périodes. In *Séminaire Bourbaki. Volume 2015/2016. Exposés 1104–1119. Avec table par noms d'auteurs de 1948/49 à 2015/16*, pp. 1–26, ex. Paris: Société Mathématique de France (SMF), 2017.

[Aro50] S. Aronhold. Zur Theorie der homogenen Functionen dritten Grades von drei Variablen. *J. Reine Angew. Math.*, 39:140–159, 1850.

[AvEvSZ10] Gert Almkvist, Christian van Enckevort, Duco van Straten, and Wadim Zudilin. Tables of Calabi–Yau equations. Preprint, arXiv:math/0507430, 2010.

[Bak75] Alan Baker. *Transcendental Number Theory.* London: Cambridge University Press. 1975.

[BCOV93] M. Bershadsky, S. Cecotti, H. Ooguri, and C. Vafa. Holomorphic anomalies in topological field theories. *Nucl. Phys. B*, 405(2–3):279–304, 1993.

[BCOV94] M. Bershadsky, S. Cecotti, H. Ooguri, and C. Vafa. Kodaira–Spencer theory of gravity and exact results for quantum string amplitudes. *Comm. Math. Phys.*, 165(2):311–427, 1994.

[BDGP96] Katia Barré-Sirieix, Guy Diaz, François Gramain, and Georges Philibert. Une preuve de la conjecture de Mahler-Manin. *Invent. Math.*, 124(1–3):1–9, 1996.

[Ber02] Cristiana Bertolin. Périodes de 1-motifs et transcendance. *J. Number Theory*, 97(2):204–221, 2002.

[Ber12] Daniel Bertrand. Le théorème de Siegel–Shidlovsky revisité. In *Number Theory, Analysis and Geometry. In Memory of Serge Lang*, pp. 51–67. Berlin: Springer, 2012.

[BH89] F. Beukers and G. Heckman. Monodromy for the hypergeometric function ${}_nF_{n-1}$. *Invent. Math.*, 95(2):325–354, 1989.

[Bin14] G. Binyamini. Multiplicity estimates, analytic cycles, and newton polytopes. Preprint https://arxiv.org/abs/1407.1183, 2014.

[BJB66] Walter L. Baily Jr. and Armand Borel. Compactification of arithmetic quotients of bounded symmetric domains. *Ann. of Math. (2)*, 84:442–528, 1966.

[Bor91] Armand Borel. *Linear Algebraic Groups*, volume 126 of Graduate Texts in Mathematics. Springer-Verlag, New York, second edition, 1991.

[Bor97] Ciprian Borcea. $K3$ surfaces with involution and mirror pairs of Calabi–Yau manifolds. In *Mirror Symmetry, II*, volume 1 of *AMS/IP Stud. Adv. Math.*, pp. 717–743. Amer. Math. Soc., Providence, RI, 1997.

[Bor01] Armand Borel. *Essays in the History of Lie Groups and Algebraic Groups*, volume 21 of History of Mathematics. American Mathematical Society, Providence, RI; London Mathematical Society, Cambridge, 2001.

[Bos01] J.-B. Bost. Algebraic leaves of algebraic foliations over number fields. *Publ. Math. Inst. Hautes Études Sci.*, 93:161–221, 2001.

[Bv95] Victor V. Batyrev and Duco van Straten. Generalized hypergeometric functions and rational curves on Calabi–Yau complete intersections in toric varieties. *Commun. Math. Phys.*, 168(3):493–533, 1995.

[BZ01] D. Bertrand and W. Zudilin. Derivatives of Siegel modular forms and exponential functions. *Izv. Math.*, 65(4):659–671, 2001.

[BZ03] D. Bertrand and W. Zudilin. On the transcendence degree of the differential field generated by Siegel modular forms. *J. Reine Angew. Math.*, 554:47–68, 2003.

[CD12] Adrian Clingher and Charles F. Doran. Lattice polarized K3 surfaces and Siegel modular forms. *Adv. Math.*, 231(1):172–212, 2012.

[CDF+93] Anna T. Ceresole, R. D'Auria, Sergio Ferrara, Wolfgang Lerche, J. Louis, and Tullio Eugenio Regge. Picard–Fuchs equations, special geometry and target space duality. In *Essays on Mirror Manifolds*, pp. 1–89, 1993.

[CDF+97] A. Ceresole, R. D'Auria, S. Ferrara, W. Lerche, J. Louis, and T. Regge. Picard–Fuchs equations, special geometry and target space duality. In *Mirror Symmetry, II*, volume 1 of *AMS/IP Stud. Adv. Math.*, pp. 281–353. *Amer. Math. Soc.*, Providence, RI, 1997.

[CDK95] Eduardo H. Cattani, Pierre Deligne, and Aroldo G. Kaplan. On the locus of Hodge classes. *J. Amer. Math. Soc.*, 8(2):483–506, 1995.

[CdlO91] Philip Candelas and Xenia de la Ossa. Moduli Space of Calabi–Yau Manifolds. *Nucl. Phys.*, B355:455–481, 1991.

[CDLOGP91a] Philip Candelas, Xenia C. De La Ossa, Paul S. Green, and Linda Parkes. A pair of Calabi–Yau manifolds as an exactly soluble superconformal theory. *Nucl. Phys.*, B359:21–74, 1991.

[CdlOGP91b] Philip Candelas, Xenia C. de la Ossa, Paul S. Green, and Linda Parkes. A pair of Calabi–Yau manifolds as an exactly soluble superconformal theory. *Nucl. Phys. B*, 359(1):21–74, 1991.

[CDLW09] Adrian Clingher, Charles F. Doran, Jacob Lewis, and Ursula Whitcher. Normal forms, $K3$ surface moduli, and modular parametrizations. In *Groups and Symmetries*, volume 47 of *CRM Proc. Lecture Notes*, pp. 81–98. *Amer. Math. Soc.*, Providence, RI, 2009.

[CG80] James A. Carlson and Phillip A. Griffiths. Infinitesimal variations of Hodge structure and the global Torelli problem. *Journees de geometrie algebrique, Angers/France* 1979:51–76, 1980.

[CGGH83] James Carlson, Mark Green, Phillip Griffiths, and Joe Harris. Infinitesimal variations of Hodge structure. I, II, III. *Compositio Math.*, 50(2–3):109–205, 1983.

[CK78] Eduardo H. Cattani and Aroldo G. Kaplan. Horizontal SL_2-orbits in flag domains. *Math. Ann.*, 235(1):17–35, 1978.

[CKS86] Eduardo Cattani, Aroldo Kaplan, and Wilfried Schmid. Degeneration of Hodge structures. *Ann. of Math. (2)*, 123(3):457–535, 1986.

[CL85] C. Camacho and A. Lins Neto. *Geometric Theory of Foliations.* Transl. from the Portuguese by Sue E. Goodman. Boston–Basel–Stuttgart: Birkhäuser. V, 205 p. DM 98.00 (1985), 1985.

[CL16] D. Cerveau and A. Lins Neto. Codimension two holomorphic foliations. Preprint, May 2016.

[Cla] Pete L. Clark. *Commutative Algebra.* http://math.uga.edu/~pete/integral.pdf.

[CMN+18] J. A. Cruz Morales, H. Movasati, Y. Nikdelan, R. Roychowdhury, and M. A. C. Torres. Manifold ways to Darboux–Halphen System. *SIGMA, 14, 003*, September 2018.

[CS87] César Camacho and Paulo Sad. *Pontos singulares de equações diferenciais analíticas.* 16So Colóquio Brasileiro de Matemática. [16th Brazilian Mathematics Colloquium]. Instituto de Matemática Pura e Aplicada (IMPA), Rio de Janeiro, 1987.

[CS11] François Charles and Christian Schnell. Notes on absolute hodge classes. Preprint arXiv:1101.3647 [math.AG], 2011.

[CTC01] Y. K. Leong and C. T. Chong. An interview with Jean-Pierre Serre. In *Mathematical Conversations*. Springer, New York, NY, 2001.

[Cv17] S. Cynk and D. van Straten. Picard–Fuchs operators for octic arrangements I (The case of orphans). Preprint, September 2017.

[Dar78a] G. Darboux. Mémoire sur les équations différentielles algébriques du premier ordre et du premier degré. *Darboux Bull. (2)*, 2:60–96, 123–144, 151–200, 1878.

[Dar78b] G. Darboux. Sur la théorie des coordonnées curvilignes et les systèmes orthogonaux. *Ann Ecole Norm. Sup.*, 7:101–150, 1878.

[Del68] P. Deligne. Théorème de Lefschetz et critères de dégénérescence de suites spectrales. *Publ. Math. Inst. Hautes Études Sci.*, (35):259–278, 1968.

[Del69] Pierre Deligne. Travaux de griffiths. In *Séminaire Bourbaki, 23ème année (1969/70), Exp. No. 376.* 1969.

[Del71a] Pierre Deligne. Théorie de Hodge. II. *Publ. Math. Inst. Hautes Études Sci.*, 40:5–57, 1971.

[Del71b] Pierre Deligne. Travaux de Shimura. In *Séminaire Bourbaki, 23ème année (1970/71), Exp. No. 389*, pp. 123–165. *Lecture Notes in Mathematics*, volume 244. Springer, Berlin, 1971.

[Del74] Pierre Deligne. Théorie de Hodge. III. *Publ. Math. Inst. Hautes Études Sci.*, 44:5–77, 1974.

[Del79] Pierre Deligne. Variétés de Shimura: interprétation modulaire, et techniques de construction de modèles canoniques. In *Automorphic forms, representations and L-functions (Proc. Sympos. Pure Math., Oregon State Univ., Corvallis, Ore., 1977), Part 2*, Proc. Sympos. Pure Math., XXXIII, pp. 247–289. *Amer. Math. Soc.*, Providence, RI, 1979.

[Del81] Pierre Deligne. Cristaux ordinaires et coordonnées canoniques (avec la collaboration de L. Illusie). Surfaces algébriques, Sémin. de géométrie algébrique, Orsay 1976–78, *Lect. Notes Math.* 868:80–137, 1981.

[DG70] Michel Demazure and Alexander Grothendieck, editors. *Schémas en groupes. II: Groupes de type multiplicatif, et structure des schémas en groupes généraux. Exposés VIII à XVIII. Séminaire de Géométrie Algébrique du Bois Marie 1962/64 (SGA 3) dirigé par Michel Demazure et Alexander Grothendieck. Revised reprint*, volume 152. Springer, Cham, 1970.

[DGMS13] C. Doran, T. Gannon, H. Movasati, and K. Shokri. Automorphic forms for triangle groups. *Communications in Number Theory and Physics*, 7(4):689–737, 2013.

[Die72] J. Dieudonné. The historical development of algebraic geometry. *Amer. Math. Monthly*, 79:827–866, 1972.

[DL90] Olivier Debarre and Yves Laszlo. Le lieu de Noether–Lefschetz pour les variétés abéliennes. *C. R. Acad. Sci. Paris Sér. I Math.*, 311(6):337–340, 1990.

[dM77] Airton S. de Medeiros. Structural stability of integrable differential forms. Geom. Topol., III. Lat. Am. Sch. Math., Proc., Rio de Janeiro 1976, Lect. Notes Math. 597, 395–428, 1977.

[dM00] Airton S. de Medeiros. Singular foliations and differential p-forms. *Ann. Fac. Sci. Toulouse, Math. (6)*, 9(3):451–466, 2000.

[DMOS82] Pierre Deligne, James S. Milne, Arthur Ogus, and Kuang-Yen Shih. *Hodge cycles, motives, and Shimura varieties*, volume 900 of *Lecture Notes in Mathematics*. Springer-Verlag, Berlin, 1982. Philosophical Studies Series in Philosophy, 20.

[DMWH16] Ch. Doran, H. Movasati, U. Whitcher, and A. Harder. Humbert surfaces and the moduli of lattice polarized K3 surfaces. *Proceedings of Symposia in Pure Mathematics, String-Math Conference Proceeding, 2014*, 93, 2016.

[Dol96] I.V. Dolgachev. Mirror symmetry for lattice polarized $K3$ surfaces. *J. Math. Sci.*, New York, 81(3):2599–2630, 1996.

[EH00] David Eisenbud and Joe Harris. *The geometry of schemes*, volume 197 of *Graduate Texts in Mathematics*. Springer-Verlag, New York, 2000.

[EK93] Jens Erler and Albrecht Klemm. Comment on the generation number in orbifold compactifications. *Comm. Math. Phys.* 153(3):579–604, 1993.

[FC90] Gerd Faltings and Ching-Li Chai. *Degeneration of Abelian Varieties.* Berlin: Springer-Verlag, 1990.

[FG] J. Fresán and J. I. Burgos Gil. Multiple zeta values: from numbers to motives. *Clay Mathematics Proceedings, to appear.*

[FGK05] Stéphane Fischler, Éric Gaudron, and Samy Khémira, editors. *Formes modulaires et transcendance. Colloque jeunes.*, volume 12. Paris: Société Mathématique de France, 2005.

[Fon19] Tiago J. Fonseca. Algebraic independence for values of integral curves. *Algebra Number Theory*, 13(3):643–694, 2019.

[Fon21] Tiago J. Fonseca. Higher Ramanujan equations and periods of abelian varieties. *To appear in Memoirs of the AMS*, 2021.

[Fre83] Eberhard Freitag. *Siegelsche Modulfunktionen*, volume 254 of *Grundlehren der Mathematischen Wissenschaften.* Springer-Verlag, Berlin, 1983.

[Gas10] C. Gasbarri. Analytic subvarieties with many rational points. *Math. Ann.*, 346(1):199–243, 2010.

[Gas13] C. Gasbarri. Horizontal sections of connections on curves and transcendence. *Acta Arith.*, 158(2):99–128, 2013.

[Ghy00] É. Ghys. À propos d'un théorème de J.-P. Jouanolou concernant les feuilles fermées des feuilletages holomorphes. *Rend. Circ. Mat. Palermo (2)*, 49(1):175–180, 2000.

[GMP95] Brian R. Greene, David R. Morrison, and M. Ronen Plesser. Mirror manifolds in higher dimension. *Comm. Math. Phys.*, 173(3):559–597, 1995.

[GMZ12] V. González-Aguilera, J. M. Munoz-Porras, and A. G. Zamora. On the irreducible components of the singular locus of A_g. II. *Proc. Am. Math. Soc.*, 140(2):479–492, 2012.

[God91] Claude Godbillon. *Feuilletages*, volume 98 of *Progress in Mathematics*. Birkhäuser Verlag, Basel, 1991. Études géométriques. [Geometric studies], With a preface by G. Reeb.

[GP07] Gert-Martin Greuel and Gerhard Pfister. *A Singular introduction to commutative algebra. With contributions by Olaf Bachmann, Christoph Lossen and Hans Schönemann. 2nd extended edition* Berlin: Springer, 2nd extended edition, 2007.

[Gri69] Phillip A. Griffiths. On the periods of certain rational integrals. I, II. *Ann. of Math. (2) 90 (1969), 460–495; ibid. (2)*, 90:496–541, 1969.

[Gri70] Phillip A. Griffiths. Periods of integrals on algebraic manifolds: Summary of main results and discussion of open problems. *Bull. Amer. Math. Soc.*, 76:228–296, 1970.

[Gri02] Pierre Grinspan. Measures of simultaneous approximation for quasi-periods of Abelian varieties. *J. Number Theory*, 94(1):136–176, 2002.

[Gro66] Alexander Grothendieck. On the de Rham cohomology of algebraic varieties. *Publ. Math. Inst. Hautes Études Sci.*, 29:95–103, 1966.

[GS69] P. Griffiths and W. Schmid. Locally homogeneous complex manifolds. *Acta Math.*, 123:253–302, 1969.

[Gui07] A. Guillot. Sur les équations d'Halphen et les actions de $\mathrm{SL}_2(\mathbf{C})$. *Publ. Math. Inst. Hautes Études Sci.*, 105:221–294, 2007.

[GvG10] Alice Garbagnati and Bert van Geemen. The Picard–Fuchs equation of a family of Calabi–Yau threefolds without maximal unipotent monodromy. *Int. Math. Res. Not. IMRN*, 16:3134–3143, 2010.

[Hal81a] G. H. Halphen. Sur certains systéme d'équations differetielles. *C. R. Acad. Sci. Paris*, 92:1404–1407, 1881.

[Hal81b] G. H. Halphen. Sur des fonctions qui proviennent de l'équation de gauss. *C. R. Acad. Sci. Paris*, 92:856–859, 1881.

[Hal81c] G. H. Halphen. Sur une systéme d'équations differetielles. *C. R. Acad. Sci. Paris*, 92:1101–1103, 1881.

[Hal86] G. H. Halphen. *Traité des fonctions elliptiques et de leurs applications*, volume 1. Gauthier-Villars, Paris, 1886.

[Ham17] Mark J. D. Hamilton. *Mathematical Gauge Theory. With Applications to the Standard Model of Particle Physics.* Cham: Springer, 2017.

[Har75] Robin Hartshorne. On the De Rham cohomology of algebraic varieties. *Inst. Hautes Étud. Sci. Publ. Math.*, 45:5–99, 1975.

[Har77] Robin Hartshorne. *Algebraic Geometry.* Springer-Verlag, New York, 1977. Graduate Texts in Mathematics, No. 52.

[Har85] Joe Harris. An introduction to infinitesimal variations of Hodge structures. In *Workshop Bonn 1984 (Bonn, 1984)*, volume 1111 of *Lecture Notes in Mathematics*, pp. 51–58. Springer, Berlin, 1985.

[Has00] Brendan Hassett. Special cubic fourfolds. *Compos. Math.*, 120(1):1–23, 2000.

[Hel01] Sigurdur Helgason. *Differential Geometry, Lie Groups, and Symmetric Spaces*, volume 34 of Graduate Studies in Mathematics. American Mathematical Society, Providence, RI, 2001. Corrected reprint of the 1978 original.

[Her] M. Herblot. Algebraic points on meromorphic curves. Preprint https://arxiv.org/abs/1204.633, 2012.

[HLTY18] Shinobu Hosono, Bong H. Lian, Hiromichi Takagi, and Shing-Tung Yau. K3 surfaces from configurations of six lines in $\mathbb{P}^2$ and mirror symmetry I. Preprint arXiv:1810.00606, 2018.

[HM17] Annette Huber and Stefan Müller-Stach. *Periods and Nori motives*, volume 65. Cham: Springer, 2017.

[HMY17] B. Haghighat, H. Movasati, and S.-T. Yau. Calabi–Yau modular forms in limit: Elliptic fibrations. *Communications in Number Theory and Physics*, 11:879–912, 2017.

[Hum93] G. Humbert. Théorie générale des surfaces hyperelliptiques. *J. Math. (4)*, 9:29–170, 1893.

[HW] A. Huber and G. Wüstholz. Transcendence and linear relations of 1-periods. Preprint arXiv:1805.10104, 2018.

[Igu62] Jun-ichi Igusa. On Siegel modular forms of genus two. *Amer. J. Math.*, 84:175–200, 1962.

[IKSY91] Katsunori Iwasaki, Hironobu Kimura, Shun Shimomura, and Masaaki Yoshida. *From Gauss to Painlevé.* Aspects of Mathematics, E16. Friedr. Vieweg & Sohn, Braunschweig, 1991. A modern theory of special functions.

[IM10] Donatella Iacono and Marco Manetti. An algebraic proof of Bogomolov–Tian–Todorov theorem. In *Deformation Spaces. Perspectives on algebro-geometric moduli. Including papers from the workshops held at the Max-Planck-Institut für Mathematik, Bonn, Germany, July 2007 and August 2008*, pp. 113–133. Wiesbaden: Vieweg+Teubner, 2010.

[Jor80] C. Jordan. Memoire sur l'equivalence des formes. *J. Ecole Polytechnique*, 48:112–150, 1880.

[Jou79] J. P. Jouanolou. *Équations de Pfaff algébriques*, volume 708 of Lecture Notes in Mathematics. Springer, Berlin, 1979.

[Kat73] N. M. Katz. p-adic properties of modular schemes and modular forms. In *Modular functions of one variable, III (Proc. Internat. Summer School, Univ. Antwerp, Antwerp, 1972)*, pp. 69–190. Volume 350 of Lecture Notes in Mathematics. Springer, Berlin, 1973.

[Kat76] N. M. Katz. p-adic interpolation of real analytic Eisenstein series. *Ann. of Math.* (2), 104(3):459–571, 1976.

[Ker14] Matt Kerr. Shimura varieties: a Hodge-theoretic perspective. In *Hodge theory. Based on lectures delivered at the summer school on Hodge theory and related topics, Trieste, Italy, June 14–July 2, 2010*, pp. 531–575. Princeton, NJ: Princeton University Press, 2014.

[Kli90] Helmut Klingen. *Introductory Lectures on Siegel Modular Forms*, volume 20 of Cambridge Studies in Advanced Mathematics. Cambridge University Press, Cambridge, 1990.

[KO68] Nicholas M. Katz and Tadao Oda. On the differentiation of de Rham cohomology classes with respect to parameters. *J. Math. Kyoto Univ.*, 8:199–213, 1968.

[KS58] Kunihiko Kodaira and D.C. Spencer. On deformations of complex analytic structures. I. *Ann. Math.* (2), 67:328–401, 1958.

[KZ95] Masanobu Kaneko and Don Zagier. A generalized Jacobi theta function and quasimodular forms. In *The Moduli Space of Curves (Texel Island, 1994)*, volume 129 of *Progr. Math.*, pp. 165–172. Birkhäuser Boston, Boston, MA, 1995.

[KZ01] Maxim Kontsevich and Don Zagier. Periods. In *Mathematics unlimited—2001 and beyond*, pp. 771–808. Springer, Berlin, 2001.

[LB92] Herbert Lange and Christina Birkenhake. *Complex abelian varieties*, volume 302 of Grundlehren der Mathematischen Wissenschaften. Springer-Verlag, Berlin, 1992.

[Lio51] J. Liouville. Sur des classes très-étendues de quantités dont la valeur n'est ni algébrique, ni même reductible à des irrationnelles algébriques. *J. Math. Pures Appl.*, 16:133–142, 1851.

[LM87] Paulette Libermann and Charles-Michel Marle. Symplectic geometry and analytical mechanics. Transl. from the French by Bertram Eugene Schwarzbach. Volume 35 of Mathematics and its Applications, Dordrecht etc.: D. Reidel Publishing Company, a member of the Kluwer Academic Publishers Group. XVI, 89.00; 73.50, 1987.

[LN99] Alcides Lins Neto. A note on projective Levi flats and minimal sets of algebraic foliations. *Ann. Inst. Fourier (Grenoble)*, 49(4):1369–1385, 1999.

[LNS] A. Lins Neto and B. Scárdua. *Introdução á Teoria das Folhações Algébricas Complexas.* Available online at IMPA's website.

[Lor06] Frank Loray. Pseudo-groupe d'une singularité de feuilletage holomorphe en dimension deux. Preprint http://hal.archives-ouvertes.fr/ccsd-00016434, 2006.

[LPT11] F. Loray, J. V. Pereira, and F. Touzet. Singular foliations with trivial canonical class. *To appear in Inv. Math.*, July 2011.

[LPT13] Frank Loray, Jorge Vitório Pereira, and Frédéric Touzet. Foliations with trivial canonical bundle on Fano 3-folds. *Math. Nachr.*, 286(8–9):921–940, 2013.

[Maa71] Hans Maass. *Siegel's Modular Forms and Dirichlet Series.* Springer-Verlag, Berlin, 1971. Volume 216 of Lecture Notes in Mathematics.

[Mah69a] K. Mahler. On algebraic differential equations satisfied by automorphic functions. *J. Aust. Math. Soc.*, 10:445–450, 1969.

[Mah69b] K. Mahler. Remarks on a paper by W. Schwarz. *J. Number Theory*, 1:512–521, 1969.

[Mas16] David Masser. *Auxiliary Polynomials in Number Theory*, volume 207. Cambridge: Cambridge University Press, 2016.

[MFK94] D. Mumford, J. Fogarty, and F. Kirwan. *Geometric invariant theory*, volume 34 of Ergebnisse der Mathematik und ihrer Grenzgebiete (2). Springer-Verlag, Berlin, third edition, 1994.

[Mila] James Milne. *Commutative Algebra.* www.jmilne.org.

[Milb] James Milne. *Introduction to Shimura Varieties.* Lecture notes available at www.jmilne.org.

[Mil86] James Milne. Abelian varieties. In *Arithmetic geometry*, pp. 79–101. Springer-Verlag, New York, 1986. Papers from the conference held at the University of Connecticut, Storrs, Connecticut, July 30–August 10, 1984.

[MM64] Hideyuki Matsumura and Paul Monsky. On the automorphisms of hypersurfaces. *J. Math. Kyoto Univ.*, 3:347–361, 1963/1964.

[MN18] H. Movasati and Y. Nikdelan. Gauss–Manin Connection in Disguise: Dwork Family. *To appear in J. Diff. Geom.*, March 2018.

[Mov08a] H. Movasati. Moduli of polarized Hodge structures. *Bull. Braz. Math. Soc. (N.S.)*, 39(1):81–107, 2008.

[Mov08b] H. Movasati. On differential modular forms and some analytic relations between Eisenstein series. *Ramanujan J.*, 17(1):53–76, 2008.

[Mov08c] H. Movasati. On elliptic modular foliations. *Indag. Math. (N.S.)*, 19(2):263–286, 2008.

[Mov11a] H. Movasati. Eisenstein type series for Calabi–Yau varieties. *Nucl. Phys. B*, 847:460–484, 2011.

[Mov11b] H. Movasati. *Multiple Integrals and Modular Differential Equations.* 28th Brazilian Mathematics Colloquium. Instituto de Matemática Pura e Aplicada, IMPA, 2011.

[Mov12a] H. Movasati. On Ramanujan relations between Eisenstein series. *Manuscripta Math.*, 139(3–4):495–514, 2012.

[Mov12b] H. Movasati. Quasi-modular forms attached to elliptic curves, I. *Ann. Math. Blaise Pascal*, 19(2):307–377, 2012.

[Mov13] H. Movasati. Quasi-modular forms attached to Hodge structures. *Fields Commun. Ser., 67*, 2013.

[Mov15a] H. Movasati. Modular-type functions attached to mirror quintic Calabi–Yau varieties. *Math. Zeit.*, 281, Issue 3, pp. 907–929(3):907–929, 2015.

[Mov15b] H. Movasati. Quasi-modular forms attached to elliptic curves: Hecke operators. *J. Number Theory*, 157:424–441, 2015.

[Mov17a] H. Movasati. *A Course in Complex Geometry and Holomorphic Foliations.* Available at author's webpage. 2017.

[Mov17b] H. Movasati. Gauss–Manin connection in disguise: Calabi–Yau modular forms. *Surveys of Modern Mathematics*, Int. Press, Boston, 2017.

[Mov17c] H. Movasati. Gauss–Manin connection in disguise: Noether–Lefschetz and Hodge loci. *Asian J. Math.*, 21(3):463–482, 2017.

[Mov18] H. Movasati. On elliptic modular foliations ii. Preprint, 2018.

[Mov19] H. Movasati. *A Course in Hodge Theory: with Emphasis on Multiple Integrals.* Available at author's webpage. 2019.

[Mov20a] H. Movasati. *An Advanced Course in Hodge Theory.* Available at author's webpage. 2020.

[Mov20b] H. Movasati. Gauss–Manin connection in disguise: Genus two curves. 21, 2020.

[MR05] François Martin and Emmanuel Royer. Formes modulaires et périodes. In *Formes modulaires et transcendance*, volume 12 of Sémin. Congr., pp. 1–117. Soc. Math. France, Paris, 2005.

[MR06] H. Movasati and S. Reiter. Hypergeometric series and Hodge cycles of four dimensional cubic hypersurfaces. *Int. J. Number Theory*, 2(6), 2006.

[MR14] M. Ram Murty and Purusottam Rath. *Transcendental Numbers.* New York, NY: Springer, 2014.

[MS14] H. Movasati and K. M. Shokri. Automorphic forms for triangle groups: Integrality properties. *J. Number Theory*, 145:67–78, 2014.

[Muk03] Shigeru Mukai. *An introduction to invariants and moduli*, volume 81 of Cambridge Studies in Advanced Mathematics. Cambridge University Press, Cambridge, 2003. Translated from the 1998 and 2000 Japanese editions by W. M. Oxbury.

[Mum66] D. Mumford. On the equations defining Abelian varieties. I-III. *Invent. Math.*, 1:287–354, 1966.

[Mum77] David Mumford. Stability of projective varieties. *Enseignement Math. (2)*, 23(1–2):39–110, 1977.

[Mum08] David Mumford. *Abelian Varieties. With Appendices by C. P. Ramanujam and Yuri Manin. Corrected Reprint of the 2nd ed. 1974*. New Delhi: Hindustan Book Agency/distrib. by American Mathematical Society (AMS); Bombay: Tata Institute of Fundamental Research, corrected reprint of the 2nd ed. 1974 edition, 2008.

[Mur05] Fiona Murnaghan. Linear algebraic groups. In *Harmonic analysis, the trace formula, and Shimura varieties*, volume 4 of Clay Math. Proc., pp. 379–391. Amer. Math. Soc., Providence, RI, 2005.

[Nes96] Yu. V. Nesterenko. Modular functions and transcendence questions. *Sb. Math.*, 187(9):1319–1348, 1996.

[Net07] Lins Neto. *Espaços de Folheações Algébricas de Codimenso Um.* 26th Brazilian Mathematics Colloquium. 2007.

[New78] P.E. Newstead. Lectures on introduction to moduli problems and orbit spaces. Published for the Tata Institute of Fundamental Research, Bombay. Tata Institute of Fundamental Research Lectures on Mathematics and Physics. Mathematics. 51. Berlin–Heidelberg–New York: Springer-Verlag, 1978.

[Nik15] Younes Nikdelan. Darboux–Halphen–Ramanujan vector field on a moduli of Calabi–Yau manifolds. *Qual. Theory Dyn. Syst.*, 14(1):71–100, 2015.

[NP01] Y.V. Nesterenko and P. Philippon. *Introduction to algebraic independence theory*, volume 1752 of Lecture Notes in Mathematics. Springer-Verlag, Berlin, 2001. With contributions from F. Amoroso, D. Bertrand, W. D. Brownawell, G. Diaz, M. Laurent, Yuri V. Nesterenko, K. Nishioka, Patrice Philippon, G. Rémond, D. Roy and M. Waldschmidt, Edited by Nesterenko and Philippon.

[OR16] John J. O'Connor and Edmund F. Robertson. *MacTutor History of Mathematics archive.* http://www-groups.dcs.st-and.ac.uk/~history/, 2016.

[Pel05] Federico Pellarin. Introduction aux formes modulaires de Hilbert et à leurs propriétés différentielles. In *Formes modulaires et transcendance. Colloque jeunes*, pp. 215–269. Paris: Société Mathématique de France, 2005.

[Phi86] Patrice Philippon. Critères pour l'indépendance algébrique. *Publ. Math. Inst. Hautes Étud. Sci.*, 64:5–52, 1986.

[Poi87] H. Poincaré. Sur les résidus des intégrales doubles. *Acta Math.*, 9(1):321–380, 1887.

[Ram16] S. Ramanujan. On certain arithmetical functions. *Trans. Cambridge Philos. Soc.*, 22:159–184, 1916.

[Res70a] H.L. Resnikoff. A differential equation for an Eisenstein series of genus two. *Proc. Natl. Acad. Sci. USA*, 65:495–496, 1970.

[Res70b] H.L. Resnikoff. Generating some rings of modular forms in several complex variables by one element. *Rice Univ. Stud.*, 56(2):115–126, 1970.

[Roh09] Jan Christian Rohde. *Cyclic coverings, Calabi–Yau manifolds and complex multiplication*, volume 1975 of Lecture Notes in Mathematics. Springer-Verlag, Berlin, 2009.

[RS11] Martin Raussen and Christian Skau. Interview with Abel Laureate John Tate. *Notices Am. Math. Soc.*, 58(3):444–452, 2011.

[RS14] Martin Raussen and Christian Skau. Interview with Pierre Deligne. *Notices Amer. Math. Soc.*, 61(2):177–185, 2014.

[Sai01] Kyoji Saito. Primitive automorphic forms. In *Mathematics unlimited—2001 and Beyond*, pp. 1003–1018. Springer, Berlin, 2001.

[Sas74] T. Sasai. Monodromy representations of homology of certain elliptic surfaces. *J. Math. Soc. Japan*, 26:296–305, 1974.

[Sch57] Theodor Schneider. *Einführung in die transzendenten Zahlen*, volume 81. Springer, Berlin, 1957.

[Sch73] Wilfried Schmid. Variation of Hodge structure: the singularities of the period mapping. *Invent. Math.*, 22:211–319, 1973.

[Sch88] Chad Schoen. On fiber products of rational elliptic surfaces with section. *Math. Z.*, 197(2):177–199, 1988.

[Ser56] Jean-Pierre Serre. Géométrie algébrique et géométrie analytique. *Ann. Inst. Fourier, Grenoble*, 6:1–42, 1955–1956.

[Ser78] Jean-Pierre Serre. A course in arithmetic. Translation of "Cours d'arithmetique". 2nd corr. print. Graduate Texts in Mathematics, Vol. 7. New York, Heidelberg, Berlin: Springer-Verlag, 1978.

[Ser97] Jean-Pierre Serre. *Lectures on the Mordell–Weil theorem. Transl. and ed. by Martin Brown from notes by Michel Waldschmidt.* Wiesbaden: Vieweg, 3rd edition, 1997.

[Shu18] Alok Shukla. Codimensions of the spaces of cusp forms for Siegel congruence subgroups in degree two. *Pac. J. Math.*, 293(1):207–244, 2018.

[Sil94] Joseph H. Silverman. *Advanced topics in the arithmetic of elliptic curves*, volume 151 of Graduate Texts in Mathematics. Springer-Verlag, New York, 1994.

[Siu83] Y.-T. Siu. Every K3 surface is Kähler. *Invent. Math.*, 73:139–150, 1983.

[Spr98] T. A. Springer. *Linear Algebraic Groups*, volume 9 of Progress in Mathematics. Birkhäuser Boston Inc., Boston, MA, second edition, 1998.

[Str90] Andrew Strominger. Special geometry. *Commun. Math. Phys.*, 133:163–180, 1990.

[SXZ13] Mao Sheng, Jinxing Xu, and Kang Zuo. Maximal families of Calabi–Yau manifolds with minimal length Yukawa coupling. *Commun. Math. Stat.*, 1(1):73–92, 2013.

[SZ10] Mao Sheng and Kang Zuo. Polarized variation of Hodge structures of Calabi–Yau type and characteristic subvarieties over bounded symmetric domains. *Math. Ann.*, 348(1):211–236, 2010.

[Tia87] Gang Tian. Smoothness of the universal deformation space of compact Calabi–Yau manifolds and its Peterson–Weil metric. Mathematical aspects of string theory, Proc. Conf., San Diego/Calif. 1986, Adv. Ser. Math. Phys. 1, 629–646, 1987.

[Tod03] Andrey Todorov. Local and global theory of the moduli of polarized Calabi–Yau manifolds. In *Proceedings of the International Conference on Algebraic Geometry and Singularities (Spanish) (Sevilla, 2001)*, volume 19, pp. 687–730, 2003.

[Vas96] K. G. Vasil'ev. On the algebraic independence of periods of abelian integrals. *Mat. Zametki*, 60(5):681–691, 799, 1996.

[Vie95] Eckart Viehweg. *Quasi-projective moduli for polarized manifolds*, volume 30 of Ergebnisse der Mathematik und ihrer Grenzgebiete (3) [Results in Mathematics and Related Areas (3)]. Springer-Verlag, Berlin, 1995.

[Voi93] Claire Voisin. Miroirs et involutions sur les surfaces $K3$. *Astérisque*, (218):273–323, 1993. *Journées de Géométrie Algébrique d'Orsay* (Orsay, 1992).

[Voi02] Claire Voisin. *Hodge Theory and Complex Algebraic Geometry. I*, volume 76 of Cambridge Studies in Advanced Mathematics. Cambridge University Press, Cambridge, 2002. Translated from the French original by Leila Schneps.

[Voi03] Claire Voisin. *Hodge Theory and Complex Algebraic Geometry. II*, volume 77 of Cambridge Studies in Advanced Mathematics. Cambridge University Press, Cambridge, 2003. Translated from the French by Leila Schneps.

[Voi07] Claire Voisin. Hodge loci and absolute Hodge classes. *Compos. Math.*, 143(4):945–958, 2007.

[Voi13] Claire Voisin. Hodge loci. In *Handbook of Moduli, Vol. III*, volume 26 of Adv. Lect. Math. (ALM), pp. 507–546. Int. Press, Somerville, MA, 2013.

[Wal74] Michel Waldschmidt. *Nombres Transcendants*, volume 402. Springer, Cham, 1974.

[Wal06] Michel Waldschmidt. Transcendence of periods: the state of the art. *Pure Appl. Math. Q.*, 2(2):435–463, 2006.

[Wei77] Abelian varieties and the hodge ring. *André Weil: Collected Papers III*, pp. 421–429, 1977.

[Wei79] André Weil. *Collected Papers III*. Springer, Berlin/Heidelberg, 1979.

[Wei87] R. Weissauer. Divisors of the Siegel Modular Variety. Number Theory, Semin. New York 1984/85, Lect. Notes Math. 1240, 304–324, 1987.

[Wei99] André Weil. *Elliptic functions according to Eisenstein and Kronecker. Reprint of the 1976 edition.* Berlin: Springer, reprint of the 1976 edition edition, 1999.

[Yau93] Shing-Tung Yau. A splitting theorem and an algebraic geometric characterization of locally Hermitian symmetric spaces. *Commun. Anal. Geom.*, 1(3):473–486, 1993.

[Yu13] J.-D. Yu. On ordinary crystals with logarithmic poles. Preprint, October 2013.

[Yui78] N. Yui. Explicit form of the modular equation. *J. Reine Angew. Math.*, 299/300:185–200, 1978.

[Zud00] V. V. Zudilin. Thetanulls and differential equations. *Sb. Math.*, 191(12):1827–1871, 2000.

Index

H

I

K

L

Y

Z

Printed in the United States
by Baker & Taylor Publisher Services